Inference and Prediction in Large Dimensions

Inference and Prediction in Large Dimensions

Denis Bosq, Delphine Blanke
Université Pierre et Marie Curie - Paris 6
Paris, France

John Wiley & Sons, Ltd

DUNOD

This work is in the Wiley-Dunod Series co-published between Dunod and John Wiley & Sons, Ltd.

This Work is in the Wiley-Dunod Series co-published between Dunod and John Wiley & Sons, Ltd.

Telephone (+44) 1243 779777

Email (for orders and customer service enquiries): cs-books@wiley.co.uk
Visit our Home Page on www.wileyeurope.com or www.wiley.com

Other Wiley Editorial Offices

John Wiley & Sons Inc., 111 River Street, Hoboken, NJ 07030, USA
Jossey-Bass, 989 Market Street, San Francisco, CA 94103-1741, USA
Wiley-VCH Verlag GmbH, Boschstr. 12, D-69469 Weinheim, Germany
John Wiley & Sons Australia Ltd, 42 McDougall Street, Milton, Queensland 4064, Australia
John Wiley & Sons (Asia) Pte Ltd, 2 Clementi Loop #02-01, Jin Xing Distripark, Singapore 129809
John Wiley & Sons Canada Ltd, 6045 Freemont Blvd, Mississauga, ONT, L5R 4J3

Wiley also publishes its books in a variety of electronic formats. Some content that appears in print may not be available in electronic books.

Anniversary Logo Design: Richard J. Pacifico

A catalogue record for this book is available from the British Library

ISBN-13 978-0470-01761-6

Typeset in10/12pt Times by Thomson Digital Noida
Printed and bound in Great Britain by TJ International Ltd, Padstow, Cornwall
This book is printed on acid-free paper responsibly manufactured from sustainable forestry in which at least two trees are planted for each one used for paper production.

Contents

List of abbreviations

⋆···⋆	see the Appendix
■	end of a proof
◇	end of an example, a remark
a.e.	almost everywhere
cadlag	continuous on the right and limit on the left
iff	if and only if
i.i.d. r.v.	independent and identically distributed random variables
A(i)–(iii)	conditions A(i) to A(iii)
A(i),(iii)	conditions A(i) and A(iii)
AR	AutoRegressive process
ARH (ARB)	AutoRegressive Hilbertian (Banach) process
BUE (BUP)	Best Unbiased Estimator (Predictor)
EDE	Empirical Density Estimator
EUE (EUP)	Efficient Unbiased Estimator (Predictor)
FLP	Functional Linear Process
GSM	Geometrically Strongly Mixing process
LCS	Linearly Closed Spacc
LIL	Law of the Iterated Logarithm
LPH(WS)	Linear Process in H (in the Wide Sense)
MA	Moving Average process
MAH	Moving Average Hilbertian process
MLE	Maximum Likelihood Estimator
MWS	Markovian process in the Wide Sense
OU	Ornstein–Uhlenbeck process
QPE	Quadratic Prediction Error
WSH	Weakly Stationary Hilbertian process
$\mathbb{N}$, $\mathbb{Z}$, $\mathbb{Q}$, $\mathbb{R}$, $\mathbb{C}$	sets of natural numbers, integers, rational numbers, real numbers, complex numbers.
$[a, b]$, $]a, b[$	closed interval, open interval
$\mathring{A}$, $\overline{A}$, A^c	interior, closure and complement of A
$\sigma(\cdots)$	σ-algebra generated by $(\cdots)$
$\mathbb{1}_A$	indicator of A: $\mathbb{1}_A(x) = 1$, $x \in A$; $\mathbb{1}_A(x) = 0$, $x \in A^c$

$\ln x$	logarithm of x
$[x]$	integer part (floor) of x
x_+, x_-	$x_+ = \max(0, x)$, $x_- = \max(0, -x)$
$f \otimes g$	for all (x,y), $(f \otimes g)(x, y) = f(x)g(y)$
$u_n \sim v_n$	$u_n/v_n \to 1$
$u_n \simeq v_n$	there exist constants c_1 and c_2 such that, for large enough n $0 < c_1 v_n < u_n < c_2 v_n$
$u_n = o(v_n)$	$u_n/v_n \to 0$
$u_n = \mathcal{O}(v_n)$	there is a $c > 0$ such that $u_n \leq c v_n$, $n \geq 1$
$X \perp\!\!\!\perp Y$	the random variables X and Y are stochastically independent
$X \overset{d}{=} Y$	the random variables X and Y have the same distribution
$\xrightarrow{d}$, $\xrightarrow{w}$	convergence in distribution, weak convergence
$\xrightarrow{p}$, $\xrightarrow{a.s.}$	convergence in probability, almost sure convergence

Introduction

The purpose of the present work is to investigate inference and statistical prediction when the data and (or) the unknown parameter have large or infinite dimension.

The *data* in question may be curves, possibly interpolated from discrete observations, or sequences associated with a large number of items.

The *parameters* are functions (distribution function, density, regression, spectral density, ...) or operators (covariance or correlation operators, ...).

Grenander's well-known book 'Abstract Inference' was devoted to such topics. However our aim is rather different: whereas Grenander studies a maximum likelihood-type method (the 'sieves'), we must often use nonparametric instruments.

More precisely we focus on adaptive projection and kernels methods and, since prediction is our main goal, we study these methods for correlated data.

Statistical prediction theory

In the first part we provide some elements of statistical prediction theory.

A priori a prediction problem is similar to an estimation problem since it deals with 'approximation' of a nonobserved random variable by using an observed one. Despite this analogy, prediction theory is rarely developed from a statistical point of view.

In Chapter 1 we study properties of the prediction model: sufficient statistics for prediction, unbiased predictors, Rao–Blackwell Theorem and optimality, Cramér-Rao type inequalities, efficiency and extended exponential model, and extensions to function spaces. Various applications to prediction of discrete or continuous time processes are given.

Chapter 2 deals with asymptotics. We present the Blackwell algorithm for prediction of 0-1 sequences; and results concerning pointwise convergence and limit in distribution for predictors associated with estimators of the unknown parameter. In addition we briefly present prediction for small and large time lags.

Inference and Prediction in Large Dimensions D. Bosq and D. Blanke

Inference by projection

The second part considers statistical models dependent on an infinite-dimensional functional parameter, say φ. The main assumption is that φ is the limit of a sequence (φ_k) of finite-dimensional parameters that are the expectation of known functions of data. This is the case for density, spectral density and covariance operators, among others.

For this type of parameter one may construct genuine estimators using empirical means, the main problem being to determine a suitable truncation index k_n which depends on the size n of the sample.

In a Hilbertian context these estimators may be interpreted as 'projection estimators'. Now, the choice of k_n depends on the norm of $||\varphi - \varphi_k||$, $k \geq 1$. Since this norm is, in general, unknown, we study the behaviour of an adaptive version of the projection estimator. It can be shown that this estimator reaches a parametric rate on a subclass dense in the parameter space and the optimal rate, up to a logarithm, elsewhere (Chapter 3).

Chapter 4 is devoted to tests of fit based on projection estimators. If the functional parameter is density, these tests appear as extensions of the $\mathcal{X}^2$-test, the latter being associated with a histogram. Various properties of these functional tests are considered: asymptotic level and power under local hypotheses, asymptotic efficiency with respect to the Neyman–Pearson test, and Bahadur efficiency.

Finally, Chapter 5 deals with a class of nonparametric predictors based on regression projection estimators. The general framework of the study allows application of the results to approximation of the conditional distribution of the future, given the past. As an application we construct prediction intervals.

Inference by kernels

Part three deals with the popular kernel method. This consists of regularization of the empirical measure by convolution.

Chapter 6 presents the method in a discrete time context. Since our final goal is prediction we concentrate on the case where data are correlated. Optimal rates of convergence for density and regression estimators are given, with application to prediction of Markov and more general processes.

Continuous time is considered in Chapter 7. On one hand, one may obtain optimal (minimax) rates similar to those of the discrete case; on the other hand, irregularity of sample paths provides additional information which leads to 'superoptimal' rates. This phenomenon occurs if, for example, one observes a stationary diffusion process. In a multidimensional context the situation is rather intricate and various rates may appear, in particular for multidimensional Gaussian or diffusion processes. Finally, if regularity of sample paths is unknown, it is possible to construct adaptive estimators that reach optimal rates, up to a logarithm.

Now, in practice, observation of a continuous time process with irregular sample path is somewhat difficult. Thus, in general, the Statistician works with sampled data. Chapter 8 examines that situation. We especially investigate high rate sampling. We show that, for density estimation, there exists an optimal sampling rate which depends on regularity and dimension of sample paths. Various simulations confirm the theoretical results. Analogous results are obtained for regression estimation.

Local time

In this part, we consider the special case where the empirical measure of a continuous time process has a density with respect to Lebesgue measure. This 'empirical density' (or 'local time') does exist in many cases, in particular for real stationary diffusion processes. It defines an unbiased density estimator, which reaches the superoptimal rate and has various nice asymptotic properties. Moreover, the projection and kernel density estimators are approximations of local time.

Linear processes in high dimensions

Let $\xi = (\xi_t,\ t \in \mathbb{R})$ be a real continuous time process. One may cut ξ into pieces by setting $X_n = (\xi_{n\delta+t},\ 0 \leq t \leq \delta),\ n \in \mathbb{Z}\ (\delta > 0)$. So ξ may be interpreted as a sequence of random variables with values in some function space. This representation is a motivation for studying discrete time processes in such a space.

In Chapter 10 we consider linear processes in Hilbert spaces. We use a definition based on Wold decomposition which is more general than the classical one. Then, having in mind statistical applications, we focus on autoregressive and moving average processes in Hilbert spaces. Some extensions to Banach space are also presented.

Finally, Chapter 11 deals with inference for functional linear processes. Estimation of the mean is studied in detail. To estimate the covariance operator of an autoregressive process we use the fact that the empirical covariance operator is the empirical mean of as $\mathcal{S}$-valued autoregressive process, where $\mathcal{S}$ denotes the space of Hilbert–Schmidt operators. Empirical cross-covariance operators enjoy similar properties. The above considerations lead to estimation of the autocorrelation operator, allowing us to construct a suitable statistical predictor.

General comments

Prerequisites for this book are knowledge of probability based on measure theory, a basic knowledge of stochastic processes, and classical theory of statistics. Some reminders appear in the appendix.

A large number of results presented in this book are new or very recent. Some of them are improvements of results that appear in Bosq (1998, 2000, 2005b).

For the references we have had to make a difficult choice among thousands of papers devoted to the various topics developed in this book.

Acknowledgments

Part of the material presented in this book is based on our own published work. We are greatly indebted to our collaborators and colleagues from whom we learned a great deal. We also thank J. Pavlevski, S. Barclay, W. Hunter, S. Kiddle, A. Bourguignon and E. d'Engenières for their kind encouragements and support, as well as S. Lightfoot for the preparation of the book. Finally, we especially acknowledge C. Lendrem for her very careful rereading.

Part I
Statistical Prediction Theory

1

Statistical prediction

1.1 Filtering

Filtering is searching information provided by observed events on nonobserved events. These events are assumed to be associated with random experiments.

To describe such a problem, one may define a *probability space* $(\Omega, \mathcal{A}, P)$ and two sub-σ-algebras of $\mathcal{A}$, say $\mathcal{B}$ and $\mathcal{C}$, respectively observed and non-observed. '$\mathcal{B}$-observed' means that, for all $B \in \mathcal{B}$, the observer knows whether B occurs or not. Information provided by $\mathcal{B}$ on $\mathcal{C}$ may be quantified by $P^{\mathcal{B}}$, the *conditional probability* with respect to $\mathcal{B}$. In general the statistician does not know $P^{\mathcal{B}}$.

In the following, we are mainly interested in *prediction* (or *forecasting*). This signifies that $\mathcal{B}$ is associated with the past and $\mathcal{C}$ with the future. In practice one tries to predict a $\mathcal{C}$-measurable *random variable* Y from a $\mathcal{B}$-measurable random variable X.

If X is partly controllable, one can replace prediction by *foresight*, which consists in preparing for the future by constructing *scenarios* and then by selecting the most favourable option. It is then possible to make a *plan* for the future. We refer to Kerstin (2003) for a discussion about these notions.

Predictions can also sometimes modify the future. For example the publication of economic forecasts may change the behaviour of the economic agents involved and, therefore, influences future data. This phenomenon is discussed in Armstrong (2001), among others.

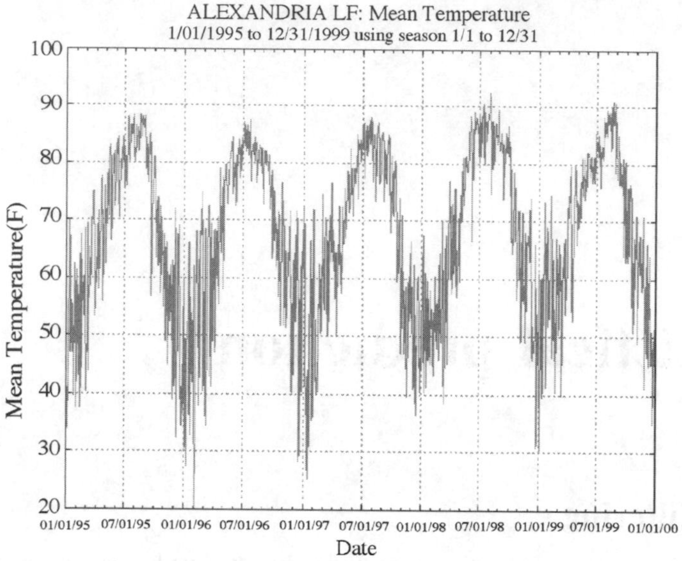

Figure 1.1 Example of a time series.

In this book we don't deal with this kind of problem, but we focus on *statistical prediction*, that is, prediction of future from a *time series* (i.e. a family of data indexed by time, see figure 1.1).[1]

1.2 Some examples

We now state some prediction problems. The following examples show that various situations can arise.

Example 1.1 *(The Blackwell's problem)*
The purpose is to predict occurrence or nonoccurrence of an event associated with an experiment when n repetitions of this experiment are available. Let $(x_n, n \geq 1)$ be the sequence of results ($x_n = 1$ if the nth event occurs, $= 0$ if not) and note by $p_n = p_n(x_1, \ldots, x_n)$ any predictor of x_{n+1}. The problem is to choose a 'good' p_n. Note that the sequence (x_n) may be deterministic or random. ◇

Example 1.2 *(Forecasting a discrete time process)*
Let $(X_n, n \geq 1)$ be a real square integrable *random process*. One observes $X = (X_1, \ldots, X_n)$ and intends to predict $Y = X_{n+h}$ $(h \geq 1)$; h is called the *forecasting horizon.*

[1]Provided by the NOAA-CIRES Climate Diagnostics Center: http://www.cdc.noaa.gov

This scheme corresponds to many practical situations: prediction of temperature, unemployment rate, foreign exchange rate,... ◇

Example 1.3 *(Forecasting a continuous time process)*
Let $(X_t, t \in \mathbb{R}_+)$ be a real square integrable ⋆continuous time random process⋆. One wants to predict $Y = X_{t+h}$ $(h > 0)$ from the observed piece of sample path $X = (X_t, 0 \leq t \leq T)$ or $X = (X_{t_1}, \ldots, X_{t_n})$.

The difference with the above discrete time scheme is the possibility of considering small horizon forecasting (i.e. $h \to 0(+)$).

Classical applications are: prediction of electricity consumption, evolution of market prices during a day's trading, prediction of the ⋆counting process⋆ associated with a ⋆point process⋆, ... ◇

Example 1.4 *(Predicting curves)*
If the data are *curves* one may interpret them as realizations of random variables with values in a suitable function space. Example 1.3 can be rewritten in this new framework.

This kind of model appears to be useful when one wishes to predict the future evolution of a quantity during a full time interval. For example, electricity consumption for a whole day or variations of an electrocardiogram during one minute. ◇

Example 1.5 *(Prediction of functionals)*
In the previous examples one may predict functionals linked with the future of the observed process. For instance:

- The *conditional density* f_Y^X or the *conditional distribution* P_Y^X of Y given X.
- The *next crossing* at a given level, that is

$$Y_x = \min\{\tau : \tau > n, X_\tau = x\},$$

and, more generally, the *next visit* to a given Borel set,

$$Y_B = \min\{\tau : \tau > n, X_\tau \in B\}.$$

Notice that the forecasting horizon is not defined for these visit questions.
- Finally it is interesting to construct *prediction intervals* of the form $P(Y \in [a(X), b(X)])$. ◇

We now specify the prediction model, beginning with the real case.

1.3 The prediction model

Let $(\Omega, \mathcal{A}, P_\theta, \theta \in \Theta)$ be a *statistical model*, where $\mathcal{A}$ is the σ-algebra of events on Ω and (P_θ) a family of probability measures on $\mathcal{A}$ indexed by the unknown parameter θ. $\mathcal{B} = \sigma(X)$ is the observed *σ-algebra* and $\mathcal{C}$ the nonobserved one. X takes its values in some *measurable space* $(E_0, \mathcal{B}_0)$.

A priori one intends to predict a $\mathcal{C}$-measurable real random variable Y. Actually, for reasons that are specified below, we extend the problem by considering prediction of $g(X, Y, \theta) \in \bigcap_{\theta\in\Theta} L^2(P_\theta)$, where g is known and Y nonobserved.

If g only depends on Y we will say that one deals with *pure prediction*, if it depends only on θ it is an *estimation problem*; finally, if g is only a function of X the topic is *approximation* (at least if g is difficult to compute!). The other cases are *mixed*. So, prediction theory appears as an extension of estimation theory.

Now a *statistical predictor* of $g(X, Y, \theta)$ is a known real measurable function of X, say $p(X)$. Note that the *conditional expectation* $\mathrm{E}^X_{\theta,g}$ is not in general a statistical predictor. In the following we assume that $p(X) \in \bigcap_{\theta\in\Theta} L^2(P_\theta)$, unless otherwise stated.

In order to evaluate the accuracy of p one may use the *quadratic prediction error* (QPE) defined as

$$R_\theta(p, g) = \mathrm{E}_\theta(p(X) - g(X, Y, \theta))^2, \theta \in \Theta.$$

This *risk function* induces the *preference relation*

$$p_1 \prec p_2 \Leftrightarrow R_\theta(p_1, g) \le R_\theta(p_2, g), \theta \in \Theta. \tag{1.1}$$

If (1.1) is satisfied we will say that 'the predictor p_1 is preferable to the predictor p_2 for predicting g', and write $p_1 \prec p_2\ (g)$.

The QPE is popular and easy to handle, it has withstood the critics because it is difficult to find a good substitute. However, some other preference relations will be considered in Section 1.7 of this chapter.

Now, let $\mathcal{P}_G$ be the class of statistical predictors p such that $p(X) \in G = \bigcap_{\theta\in\Theta} G_\theta$ where G_θ is some closed linear space of $L^2(P_\theta)$, with *orthogonal projector* Π^θ, $\theta \in \Theta$. The following lemma shows that a prediction problem is, in general, mixed.

Lemma 1.1 *(Decomposition of the QPE)*
If $p \in \mathcal{P}_G$, its QPE has decomposition

$$\mathrm{E}_\theta(p - g)^2 = \mathrm{E}_\theta(p - \Pi^\theta g)^2 + \mathrm{E}_\theta(\Pi^\theta g - g)^2, \theta \in \Theta. \tag{1.2}$$

Hence, $p_1 \prec p_2$ for predicting g if and only if $p_1 \prec p_2$ for predicting $\Pi^\theta g$.

PROOF:
Decomposition (1.2) is a straightforward application of the Pythagoras theorem.

Therefore

$$\mathrm{E}_\theta(p_1 - g)^2 \leq \mathrm{E}_\theta(p_2 - g)^2 \Leftrightarrow \mathrm{E}_\theta(p_1 - \Pi^\theta g)^2 \leq \mathrm{E}_\theta(p_2 - \Pi^\theta g)^2, \theta \in \Theta.$$

■

Lemma 1.1 is simple but crucial: one must focus on the *statistical prediction error* $\mathrm{E}_\theta(p - \Pi^\theta g)^2$, since the *probabilistic prediction error* $\mathrm{E}_\theta(\Pi^\theta g - g)^2$ is not controllable by the statistician.

Thus, predicting $g(X, Y, \theta)$ or predicting $\Pi^\theta g(X, Y, \theta)$ is the same activity. In particular, to predict Y is equivalent to predicting $\Pi^\theta Y$; this shows that a nondegenerated prediction problem is mixed.

Example 1.6
Suppose that $X_1, \ldots, X_n$ are observed temperatures. One wants to know if X_{n+1} is going to exceed some threshold θ; then $g = \mathbb{1}_{X_{n+1} > \theta}$ and the problem is mixed. ◇

Let us now give some classical examples of spaces G.

Example 1.7
If $G = \bigcap_{\theta \in \Theta} L^2(\Omega, \mathcal{B}, P_\theta)$, then Π^θ is the *conditional expectation* given $\mathcal{B}$. ◇

Example 1.8
Suppose that $X = (X_1, \ldots, X_n)$ where $X_i \in L^2(P_\theta)$, $1 \leq i \leq n$ and $G = \mathrm{sp}\{1, X_1, \ldots, X_n\}$ does not depend on θ. Then $\Pi^\theta g$ is the *affine regression* of g on X. ◇

Example 1.9
If $X = (X_1, \ldots, X_n)$ with $X_i \in L^{2k}(P_\theta)$, $1 \leq i \leq n$, $(k \geq 1)$ and $G = \mathrm{sp}\{1, X_i, \ldots, X_i^k, 1 \leq i \leq n\}$ does not depend on θ. Then $\Pi^\theta g$ is a *polynomial regression*:

$$\Pi^\theta g(X_1, \ldots, X_n) = a_0 + \sum_{i=1}^{n} \sum_{j=1}^{k} b_{ji} X_i^j,$$

where a_0 and (b_{ji}) only depend on θ. ◇

1.4 P-sufficient statistics

As in estimation theory, sufficient statistics play a significant role in prediction. The definition is more restrictive.

Definition 1.1
A statistic $S(X)$ is said to be sufficient for predicting $g(X,Y,\theta)$ *(or* P-sufficient) *if*

(i) *$S(X)$ is sufficient in the usual sense, i.e. the conditional distribution $P_X^{S(X)}$* of *X with respect to $S(X)$ does not depend on θ.*

(ii) *For all θ, X* and *$g(X,Y,\theta)$ are *conditionally independent* given $S(X)$.*

Condition (ii) means that

$$P_\theta^{S(X)}(X \in B, g \in C) = P_\theta^{S(X)}(X \in B)P_\theta^{S(X)}(g \in C),$$

$\theta \in \Theta, B \in \mathcal{B}_{\mathbb{R}}, C \in \mathcal{B}_{\mathbb{R}}$.

Note that, if $g(X,Y,\theta) = g(S(X),\theta)$, (ii) is automatically satisfied. Moreover one may show (see Ash and Gardner 1975, p. 188) that (ii) is equivalent to
(ii)$'P_{\theta,g}^X = P_{\theta,g}^{S(X)}, \theta \in \Theta$.

We now give a statement which connects sufficiency and P-sufficiency.

Theorem 1.1
*Suppose that (X,Y) has a strictly positive density $f_\theta(x,y)$ with respect to a *σ-finite* measure $\mu \otimes \nu$. Then*

(a) *If $S(X)$ is P-sufficient for predicting Y, $(S(X),Y)$ is sufficient in the statistical model associated with (X,Y).*

(b) *Conversely if $f_\theta(x,y) = h_\theta(S(x),y)c(x)d(y)$, then $S(X)$ is P-sufficient for predicting Y.*

PROOF:

(a) Consider the decomposition

$$f_\theta(x,y) = f_\theta(x)f_\theta(y|x),$$

where $f_\theta(\cdot)$ is the density of X and $f_\theta(\cdot|\cdot)$ the conditional density of Y given $X = \cdot$. If $S(X)$ is P-sufficient, the factorization theorem (see Lehmann and Casella 1998, p. 35) yields $f_\theta = \varphi_\theta(S(x))\psi(x)$, where ψ does not depend on θ.

Now (ii)$'$ entails $f_\theta(y|x) = \gamma_\theta(y|S(x))$ where $\gamma_\theta(\cdot|\cdot)$ is the conditional density of Y given $S(X) = \cdot$.

Finally $f_\theta(x,y) = \varphi_\theta(S(x))\gamma_\theta(y|S(x))\psi(x)$ and the factorization theorem gives sufficiency of $(S(X),Y)$ in the model associated with (X,Y).

(b) Conversely, the relation

$$f_\theta(x,y) = h_\theta(S(x),y)c(x)d(y)$$

may be replaced by $f_\theta(x, y) = h_\theta(S(x), y)$ by substituting μ and ν for $c \cdot \mu$ and $d \cdot \nu$ respectively.

This implies

$$f_\theta(x) = \int h_\theta(S(x), y)\mathrm{d}\nu(y) := H_\theta(S(x)),$$

thus, $S(X)$ is sufficient, and since

$$f_\theta(y|x) = \frac{h_\theta(S(x), y)}{H_\theta(S(x))},$$

(ii)′ holds, hence $S(X)$ is P-sufficient. ■

We now give some examples and counterexamples concerning P-sufficiency.

Example 1.10

Let $X_1, \ldots, X_{n+1}$ be independent random variables with common density $\theta e^{-\theta x} \mathbb{1}_{x>0}$ ($\theta > 0$). Set $X = (X_1, \ldots, X_n)$ and $Y = X_{n+1} - X_n$. Then $\sum_{i=1}^n X_i$ is sufficient for X but $(\sum_{i=1}^n X_i, Y)$ is not sufficient for (X, Y).

This shows that, in Theorem 1.1 (a), P-sufficiency of $S(X)$ cannot be replaced by sufficiency. ◇

Example 1.11

Let $X = (X_1, X_2)$ and Y be such that the density of (X, Y) is

$$f_\theta(x_1, x_2, y) = \theta^{3/2} e^{-\theta(x_1+y)} e^{-\frac{\pi^2}{4} x_2^2 y} \mathbb{1}_{x_1>0,\, x_2>0,\, y>0}, \ (\theta > 0).$$

Then (X_1, Y) is sufficient in the model (X, Y) but X_1 is not sufficient in the model (X_1, X_2).

This shows that, in Theorem 1.1 (b), the result is not valid if the special form $f_\theta(x, y) = h_\theta(S(x), y)c(x)d(y)$ does not hold. ◇

Example 1.12 *(discrete time Gaussian process)*

Let $(X_i, i \geq 1)$ be a ⋆Gaussian process⋆, where $X_i \stackrel{d}{=} \mathcal{N}(\theta, 1)$, $i \geq 1$ ($\theta \in \mathbb{R}$). Suppose that the covariance matrix C_n of $X = (X_1, \ldots, X_n)$ is known and invertible for each n, and set

$$C_{n+1} = \begin{bmatrix} C_n & \gamma_{n+1} \\ \gamma_{n+1} & 1 \end{bmatrix}.$$

Then, in order to predict $Y = X_{n+1}$, the statistic

$$(C_n^{-1} u_n X, C_n^{-1} \gamma_{n+1} X),$$

where $u_n = (1, \ldots, 1)'$, is P-sufficient. The statistic $C_n^{-1} u_n X$ is sufficient but not P-sufficient. ◇

Example 1.13 *(Poisson process)*
Let $N = (N_t, t \geq 0)$ be a homogeneous *Poisson process* with intensity λ, and observed over $[0, T]$. Then N_T is sufficient. Now, since N is a *Markov process* the σ-algebras $\sigma(N_s, s \leq T)$ and $\sigma(N_s, s > T)$ are independent given N_T. Hence N_T is P-sufficient for predicting N_{T+h} $(h > 0)$. ◇

Example 1.14 *(Ornstein–Uhlenbeck process)*
Consider an *Ornstein–Uhlenbeck process* (OU) defined as

$$X_t = \int_{-\infty}^{t} \mathrm{e}^{-\theta(t-s)} \mathrm{d}W(s), t \in \mathbb{R}, (\theta > 0),$$

where $W = (W_t, t \in \mathbb{R})$ is a *standard bilateral Wiener process* and the integral is taken in *Ito* sense. (X_t) is a zero-mean stationary Gaussian Markov process. Here the observed variable is $X_{(T)} = (X_t, 0 \leq t \leq T)$.

The likelihood of $X_{(T)}$ with respect to the distribution of $W_{(T)} = (W_t, 0 \leq t \leq T)$ in the space $\mathcal{C}([0, T])$ of continuous real functions defined on $[0, T]$ with uniform norm is

$$L(X_{(T)}, \theta) = \exp\left(-\frac{\theta}{2}(X_T^2 - X_0^2 - T) - \frac{\theta^2}{2}\int_0^T X_t^2 \mathrm{d}t\right), \tag{1.3}$$

(See Liptser and Shiraev 2001). Then, the factorization theorem yields sufficiency of the statistics $(X_0^2, X_T^2, \int_0^T X_t^2 \mathrm{d}t)$. Consequently $Z_T = (X_0, X_T, \int_0^T X_t^2 \mathrm{d}t)$ is also sufficient.

Now, since (X_t) is Markovian, we have

$$\sigma(X_{(T)}) \perp\!\!\!\perp \sigma(X_{T+h}) | \sigma(X_T),$$

then

$$P_{\theta, X_{T+h}}^{Z_T} = P_{\theta, X_{T+h}}^{X_T} = P_{\theta, X_{T+h}}^{X_{(T)}}$$

hence (ii)′ holds and Z_T is P-sufficient for predicting X_{T+h}. ◇

The next statement shows that one may use P-sufficient statistics to improve a predictor. It is a Rao–Blackwell type theorem (See Lehmann 1991, p. 47).

Theorem 1.2 *(Rao–Blackwell theorem)*
Let $S(X)$ be a P-sufficient statistic for $g(X, Y, \theta)$ and $p(X)$ a statistical predictor of g. Then $\mathrm{E}^{S(X)}p(X)$ is preferable to $p(X)$ for predicting g.

PROOF:
$P_{\theta,X}^{S(X)}$ does not depend on θ, thus $q(X) = \mathrm{E}^{S(X)}p(X)$ is again a statistical predictor. Furthermore

$$\mathrm{E}_\theta(p-g)^2 = \mathrm{E}_\theta(p-q)^2 + \mathrm{E}_\theta(q-g)^2 + 2\mathrm{E}_\theta((p-q)(q-g)). \qquad (1.4)$$

Now, by definition of conditional expectation $\mathrm{E}_\theta((p-q)q) = 0$ and, from condition (ii) in Definition 1.1,

$$\mathrm{E}_\theta[(p-q)g] = \mathrm{E}_\theta[\mathrm{E}_\theta^s(p-q)g] = \mathrm{E}_\theta^s(p-q)\mathrm{E}_\theta^s(g) = 0$$

since $\mathrm{E}_\theta^s(p) = q$; thus (1.4) gives $\mathrm{E}_\theta(p-g)^2 \geq \mathrm{E}_\theta(q-g)^2$. ■

Note that, if (ii) does not hold, the result remains valid provided $\mathrm{E}_\theta[(p-q)g] = 0$, $\theta \in \Theta$.

Finally it is noteworthy that, if $S(X)$ is P-sufficient to predict g it is also P-sufficient to predict $\mathrm{E}_\theta^{S(X)}(g)(= \mathrm{E}_\theta^X(g))$; actually this conditional expectation is a function of $S(X)$, hence $X \perp\!\!\!\perp \mathrm{E}_\theta^{S(X)}(g)|S(X)$.

1.5 Optimal predictors

A statistical predictor p_O is said to be *optimal* in the family $\mathcal{P}$ of predictors of g if

$$p_O \prec p, p \in \mathcal{P}$$

that is

$$\mathrm{E}_\theta(p_O - g)^2 \leq \mathrm{E}_\theta(p-g)^2; \quad \theta \in \Theta, p \in \mathcal{P}.$$

Notice that, in the family of all square integrable predictors, such a predictor does not exist as soon as $\mathrm{E}_\theta^X(g)$ is not constant in θ. Indeed, $p_1(X) = \mathrm{E}_{\theta_1}^X(g)$ is optimal at θ_1 when $p_2(X) = \mathrm{E}_{\theta_2}^X(g)$ is optimal at θ_2 which is impossible if $\mathrm{E}_{\theta_i}(\mathrm{E}_{\theta_1}^X(g) - \mathrm{E}_{\theta_2}^X(g))^2 \neq 0$; $i = 1, 2$.

Thus, in order ro construct an optimal predictor, it is necessary to restrict $\mathcal{P}$. For example one may consider only *unbiased* predictors.

Definition 1.2
A predictor $p(X)$ of $g(X, Y, \theta)$ is said to be unbiased if

$$\mathrm{E}_\theta(p(X)) = \mathrm{E}_\theta(g(X, Y, \theta)), \theta \in \Theta. \qquad (1.5)$$

Condition (1.5) means that p is an unbiased estimator of the parameter $\mathrm{E}_\theta(g)$.

Example 1.15 *(Autoregressive process of order 1)*
Consider an autoregressive process of order 1 (AR(1)) defined as

$$X_n = \sum_{j=0}^{\infty} \theta^j \varepsilon_{n-j}, n \in \mathbb{Z} \tag{1.6}$$

where $(\varepsilon_n, n \in \mathbb{Z})$ is *strong white noise* (i.e. a sequence of i.i.d. random variables such that $0 < \mathrm{E}\varepsilon_n^2 = \sigma^2 < \infty$ and $\mathrm{E}\varepsilon_n = 0$) and $0 < |\theta| < 1$. The series converges in mean square and almost surely.

From (1.6) it follows that

$$X_n = \theta X_{n-1} + \varepsilon_n, n \in \mathbb{Z},$$

then

$$\mathrm{E}_{\theta,\sigma^2}^{\sigma(X_i, i \leq n)}(X_{n+1}) = \theta X_n.$$

For convenience we suppose that $\sigma^2 = 1 - \theta^2$, hence $\mathrm{E}_{\theta,\sigma^2}(X_n^2) = 1$. Now a natural unbiased estimator of θ, based on $X_1, \ldots, X_n$ (n > 1) is

$$\hat{\theta}_n = \frac{1}{n-1} \sum_{i=1}^{n-1} X_i X_{i+1},$$

hence a predictor of X_{n+1} is defined as

$$\hat{X}_{n+1} = \hat{\theta}_n X_n.$$

Now we have

$$\mathrm{E}_\theta(\hat{X}_{n+1}) = \theta^2 \frac{1 - \theta^{n-1}}{(1-\theta)(n-1)} \mathrm{E}_\theta(X_0^3),$$

thus $\hat{X}_{n+1}$ is unbiased if and only if (iff)

$$\mathrm{E}_\theta(X_0^3) = 0, 0 < |\theta| < 1.$$ ◇

We now give an extension of the classical Lehmann–Scheffé theorem (see Lehmann 1991, p. 88).

First recall that a statistic S is said to be *complete* if

$$\mathrm{E}_\theta(U) = 0, \theta \in \Theta \text{ and } U = \varphi(S) \Rightarrow U = 0, P_\theta \text{ a.s. for all } \theta.$$

Theorem 1.3 *(Lehmann–Scheffé theorem)*
If S is a complete P-sufficient statistic for g and p is an unbiased predictor of g, then $\mathrm{E}^S(p)$ is the unique optimal unbiased predictor of g (P_θ a.s. for all θ).

PROOF:
From Theorem 1.2 any optimal unbiased predictor of g is a function of S. Thus $\mathrm{E}^S(p)$ is a candidate since $\mathrm{E}_\theta[\mathrm{E}^S p] = \mathrm{E}_\theta(p) = g$. But completeness of S entails uniqueness of an unbiased predictor of g as a function of S, hence $\mathrm{E}^S(p)$ is optimal. ■

Example 1.13 *(continued)*
N_T is a complete P-sufficient statistic for N_{T+h}, the consequently unbiased predictor

$$p(N_T) = \frac{T+h}{T} N_T$$

is optimal for predicting N_{T+h}. Its quadratic error is

$$\mathrm{E}_\lambda(p(N_T) - N_{T+h})^2 = \lambda h\left(1 + \frac{h}{T}\right).$$

It is also optimal unbiased for predicting

$$\mathrm{E}_\lambda^{N_T}(N_{T+h}) = \lambda h + N_T$$

with quadratic error

$$\mathrm{E}_\lambda(p(N_T) - \mathrm{E}_\lambda^{N_T}(N_{T+h}))^2 = \frac{\lambda h^2}{T}. \qquad \diamond$$

The following statement gives a condition for optimality of an unbiased predictor.

Theorem 1.4
Set $\mathcal{U} = \{U(X) : \mathrm{E}_\theta U^2(X) < \infty, \mathrm{E}_\theta U(X) = 0; \theta \in \Theta\}$. *Then an unbiased predictor* p *of* g *is optimal iff*

$$\mathrm{E}_\theta[(p-g)U] = 0; U \in \mathcal{U}, \theta \in \Theta. \tag{1.7}$$

PROOF:
If p is optimal, we set

$$q = p + \alpha U, U \in \mathcal{U}, \alpha \in \mathbb{R},$$

then

$$\mathrm{E}_\theta(p + \alpha U - g)^2 \geq \mathrm{E}_\theta(p-g)^2,$$

therefore

$$\alpha^2 \mathrm{E}_\theta U^2 + 2\alpha \mathrm{E}_\theta((p-g)U) \geq 0, U \in \mathcal{U}, \alpha \in \mathbb{R}$$

which is possible only if (1.7) holds.

Conversely, if p satisfies (1.7) and p' denotes another unbiased predictor, then $p' - p \in \mathcal{U}$, therefore

$$\mathrm{E}_\theta[(p-g)(p'-p)] = 0,$$

which implies

$$\begin{aligned}\mathrm{E}_\theta[(p'-g)^2 - (p-g)^2] &= \mathrm{E}_\theta(p'^2 - p^2) - 2\mathrm{E}_\theta[(p'-p)g] \\ &= \mathrm{E}_\theta(p'^2 - p^2) - 2\mathrm{E}_\theta[(p'-p)p] \\ &= \mathrm{E}_\theta(p'-p)^2 \geq 0,\end{aligned}$$

thus p is preferable to p'. ■

Note that such a predictor is unique. Actually, if p' is another optimal unbiased predictor, (1.7) yields

$$\mathrm{E}_\theta((p'-p)U) = 0, U \in \mathcal{U}, \theta \in \Theta$$

which shows that $p' - p$ is an optimal unbiased predictor of 0. But 0 is an optimal unbiased predictor of 0, with quadratic error 0, thus $\mathrm{E}_\theta(p'-p)^2 = 0$, $\theta \in \Theta$ hence $p' = p$, P_θ a.s. for all θ.

Now, since an unbiased predictor of g is an unbiased estimator of $\mathrm{E}_\theta g$, it is natural to ask whether the best unbiased estimator (BUE) of $\mathrm{E}_\theta g$ and the best unbiased predictor (BUP) of g coincide or not.

The next theorem gives an answer to this question.

Theorem 1.5
The BUE of $\mathrm{E}_\theta g$ *and the BUP of* g *coincide iff*

$$\mathrm{E}_\theta(gU) = 0, U \in \mathcal{U}, \theta \in \Theta. \tag{1.8}$$

PROOF:
First suppose that (1.8) holds. If p is the BUE of $\mathrm{E}_\theta g$, it is also the BUP of $\mathrm{E}_\theta g$, then Theorem 1.4 implies that, for all $U \in \mathcal{U}$ and all $\theta \in \Theta$, we have

$$\mathrm{E}_\theta((p - \mathrm{E}_\theta g)U) = 0, \tag{1.9}$$

therefore $\mathrm{E}_\theta(pU) = 0$, and from (1.8) it follows that

$$\mathrm{E}_\theta[(p-g)U] = 0$$

then (1.7) holds and p is the BUP of g.

Conversely a BUP of g satisfies (1.7), and, if it coincides with the BUE of $E_\theta g$, (1.9) holds. Finally, (1.7) and (1.9) give (1.8). ■

Note that, if $E_\theta g = 0, \theta \in \Theta$, the BUE of $E_\theta g$ is $p = 0$. Thus 0 is the BUP of g if and only if (1.8) holds. For example, if g is zero-mean and independent of X, 0 is the BUP of g and its quadratic error is $E_\theta g^2$.

Example 1.13 *(continued)*
If $X = N_T$ is observed we have $\mathcal{U} = \{0\}$, hence $p(N_T) = N_T(T+h)/T$ is the BUP of N_{T+h} and the BUE of $E_\lambda(N_{T+h}) = \lambda(T+h)$. ◇

We now indicate a method that allows us to construct a BUP in some special cases.

Theorem 1.6
If g is such that

$$E_\theta^X(g) = \phi(X) + \psi(\theta), \theta \in \Theta$$

where $\phi \in L^2(P_{\theta,X})$ and ψ are known, and if $s(X)$ is the BUE of $\psi(\theta)$; then

$$p(X) = \phi(X) + s(X)$$

is the BUP of g.

PROOF:
First note that $E_\theta(p - E_\theta g)^2 = E_\theta(s - \psi(\theta))^2$. Now, let q be another unbiased predictor of g, then $q - \phi$ is an unbiased estimator of $\psi(\theta)$, hence

$$E_\theta(q - \phi - \psi(\theta))^2 \geq E_\theta(s - \psi(\theta))^2$$

that is

$$E_\theta(q - E_\theta^X(g))^2 \geq E_\theta(p - E_\theta^X(g))^2$$

thus, $p \prec q$, for predicting $E_\theta^X(g)$ and, from Lemma 1.1, it follows that $p \prec q$ for predicting g. ■

Example 1.10 *(continued)*
Here we have $E_\theta^X(Y) = \theta - X_n$ and, by the Lehmann–Scheffé theorem, $\bar{X}_n$ is the BUE of θ. Thus Theorem 1.6 shows that $p(X) = \bar{X}_n - X_n$ is the BUP of Y. ◇

Example 1.16 *(Semi-martingales)*
Let $(X_t, t \in \mathbb{R}_+)$ be a real square integrable process and $m(\theta, t)$ a deterministic function, such that

$$Y_t = X_t + m(\theta, t), t \in \mathbb{R}_+ (\theta \in \Theta \subset \mathbb{R})$$

is a *martingale* with respect to $\mathcal{F}_t = \sigma(X_s, 0 \le s \le t)$, $t \in \mathbb{R}_+$, i.e.

$$\mathrm{E}_\theta^{\mathcal{F}_s}(Y_t) = Y_s, 0 \le s \le t, \theta \in \Theta.$$

In order to predict X_{T+h} $(h > 0)$ given the data $X_{(T)} = (X_t, 0 \le t \le T)$ we write

$$\begin{aligned}\mathrm{E}_\theta^{X_{(T)}}(X_{T+h}) &= \mathrm{E}_\theta^{\mathcal{F}_T}(Y_{T+h} - m(\theta, T+h)) \\ &= X_T + [m(\theta, T) - m(\theta, T+h)],\end{aligned}$$

and it is possible to apply Theorem 1.6 if $\psi(\theta) = m(\theta, T) - m(\theta, T+h)$ possesses a BUE.

In particular if (X_t) has *independent increments* then $(X_t - \mathrm{E}_\theta(X_t))$ becomes a martingale with $m(\theta, t) = -\mathrm{E}_\theta(X_t)$.

A typical example is again the Poisson process: (N_t) has independent increments, then $(N_t - \lambda t)$ is a martingale and $\mathrm{E}_\lambda^{N_T} = N_T + \lambda h$; applying Theorem 1.6 one again obtains that

$$N_T + \frac{N_T}{T}h = \frac{T+h}{T}N_T$$

is the BUP of N_{T+h}. ◇

The next lemma shows that existence of unbiased predictors does not imply existence of a BUP.

Lemma 1.2
Let $(X_t, t \in I)(I = \mathbb{Z}$ or $\mathbb{R})$ be a square integrable, zero-mean real Markov process with

$$\mathrm{E}_\theta^{\mathcal{F}_T}(X_{T+h}) = \varphi_T(\theta, X_T), \theta \in \Theta,$$

where $\mathcal{F}_T = \sigma(X_s, s \le T)$.

Suppose that

(i) $\mathrm{E}_{\theta'}[\varphi_T(\theta, X_T)] = 0, \theta \in \Theta, \theta' \in \Theta$,

*(ii) there exist θ_1 and θ_2 in Θ such that P_{θ_1, X_T} and P_{θ_2, X_T} are *equivalent* and*

$$P_{\theta_1}[\varphi_T(\theta_1, X_T) \ne \varphi_T(\theta_2, X_T)] > 0.$$

Then, the class of unbiased predictors of X_{T+h}, given $X = (X_t, 0 \le t \le T)$ does not contain a BUP.

PROOF:
Consider the statistical predictor $p_1(X) = \varphi_T(\theta_1, X_T)$; from (i) it follows that it is unbiased, and a BUP p_O must satisfy

$$p_O(X) = p_1(X)\ P_{\theta_1} \quad \text{a.s.} \tag{1.10}$$

Similarly, if $p_2(X) = \varphi_T(\theta_2, X_T)$, we have

$$p_O(X) = p_2(X)\ P_{\theta_2} \quad \text{a.s.} \tag{1.11}$$

and (ii) shows that (1.10) and (1.11) are incompatible. ■

Example 1.14 *(continued)*
For the OU process, $X_T \stackrel{d}{=} \mathcal{N}(0, 1/2\theta)$ and $\mathrm{E}_\theta^{\mathcal{F}_T}(X_{T+h}) = \mathrm{e}^{-\theta h} X_T$, then (i) and (ii) hold and there is no BUP. ◇

Example 1.15 *(continued)*
If (ε_n) is Gaussian, $X_n \stackrel{d}{=} \mathcal{N}(0, 1)$, and, since $\mathrm{E}_\theta^{\mathcal{F}_T}(X_{T+1}) = \theta X_T$, no BUP may exist. In particular $\hat{X}_{T+1}$ is not BUP. ◇

1.6 Efficient predictors

Under regularity conditions it is easy to obtain a Cramér–Rao type inequality for unbiased predictors. More precisely we consider the following assumptions:

Assumptions 1.1 (A1.1)
$\Theta \subset \mathbb{R}$ is an open set, the model associated with X is dominated by a σ-finite measure μ, the density $f(x, \theta)$ of X is such that $\{x : f(x, \theta) > 0\}$ does not depend on θ, $\partial f(x, \theta)/\partial\theta$ does exist. Finally the Fisher information

$$I_X(\theta) = \mathrm{E}_\theta \left(\frac{\partial}{\partial\theta} \ln f(X, \theta) \right)^2$$

satisfies $0 < I_X(\theta) < \infty$, $\theta \in \Theta$.

Theorem 1.7 *(Cramér–Rao inequality)*
If A1.1 holds, p is an unbiased predictor, and the equality

$$\int p(x) f(x, \theta) \mathrm{d}\mu(x) = \mathrm{E}_\theta(g(X, Y, \theta))$$

can be differentiated under the integral sign, then

$$\mathrm{E}_\theta(p - g)^2 \geq \mathrm{E}_\theta(g - \mathrm{E}_\theta^X g)^2 + \frac{[\gamma'(\theta) - \mathrm{E}_\theta(\mathrm{E}_\theta^X(g) \frac{\partial}{\partial\theta} \ln f(X, \theta))]^2}{I_X(\theta)}, \tag{1.12}$$

where $\gamma(\theta) = \mathrm{E}_\theta g(X, Y, \theta)$.

PROOF:
Clearly it suffices to show that $E_\theta(p - E_\theta^X g)^2$ is greater than or equal to the second term in the right hand side of (1.12).

Now the *Cauchy–Schwarz* inequality yields

$$\mathrm{Cov}\left(p - E_\theta^X g, \frac{\partial}{\partial\theta}\ln f\right) \leq [E_\theta(p - E_\theta^X g)^2]^{1/2}[I_X(\theta)]^{1/2}$$

moreover

$$\mathrm{Cov}\left(p - E_\theta^X g, \frac{\partial}{\partial\theta}\ln f\right) = E_\theta\left(p\frac{\partial}{\partial\theta}\ln f\right) - E_\theta\left(E_\theta^X(g)\frac{\partial}{\partial\theta}\ln f\right)$$
$$= \gamma'(\theta) - E_\theta\left(E_\theta^X(g)\frac{\partial}{\partial\theta}\ln f\right),$$

hence (1.12). ■

The next statement gives an inequality very similar to the classical Cramér–Rao inequality.

Corollary 1.1
If, in addition, the equality

$$\gamma(\theta) = \int E_\theta^{X=x}(g) f(x,\theta)\mathrm{d}\mu(x) \tag{1.13}$$

is differentiable under the integral sign, then

$$E_\theta(p(X) - E_\theta^X(g))^2 \geq \frac{\left[E_\theta\left(\frac{\partial E_\theta^X(g)}{\partial\theta}\right)\right]^2}{I_X(\theta)}, \theta \in \Theta. \tag{1.14}$$

PROOF:
Differentiating (1.13) one obtains

$$\gamma'(\theta) = \int \frac{\partial}{\partial\theta}(E_\theta^{X=x}(g))f(x,\theta)\mathrm{d}\mu(x) + \int E_\theta^{X=x}(g)\frac{\partial \ln f(x,\theta)}{\partial\theta}f(x,\theta)\mathrm{d}\mu(x)$$

hence (1.14) from (1.12) where g is replaced by $E_\theta^X g$. ■

Of course, (1.12) and (1.14) reduce to the classical Cramér–Rao inequality (see Lehmann 1991, p. 120) if g only depends on θ. So we will say that p is *efficient* if (1.12) is an equality. Note that p is efficient for predicting g if and only if it is efficient for predicting $E_\theta^X(g)$.

We now give a result similar to Theorem 1.6.

Theorem 1.8
Under A1.1 and conditions in Theorem 1.6, if s(X) is an efficient unbiased estimator (EUE) of $\psi(\theta)$, then $p(X) = \phi(X) + s(X)$ is an efficient unbiased predictor (EUP) of g.

PROOF:
Efficiency of $s(X)$ means that ψ is differentiable and

$$\mathrm{E}_\theta(s(x) - \psi(\theta))^2 = \frac{[\psi'(\theta)]^2}{I_X(\theta)}, \tag{1.15}$$

now we have

$$\mathrm{E}_\theta(p(X)) = \mathrm{E}_\theta(\mathrm{E}_\theta^X g) = \gamma(\theta) \tag{1.16}$$

and

$$\mathrm{E}_\theta(p(X) - \mathrm{E}_\theta^X(g))^2 = \mathrm{E}_\theta(s(X) - \psi(\theta))^2.$$

Noting that

$$\begin{aligned}\frac{\partial}{\partial\theta}\mathrm{E}_\theta(\phi(X)) &= \frac{\partial}{\partial\theta}\int \phi(x)f(x,\theta)\mathrm{d}\mu(x)\\ &= \mathrm{E}_\theta\left[\phi(X)\frac{\partial \ln f(X,\theta)}{\partial\theta}\right]\end{aligned}$$

and

$$\mathrm{E}_\theta\left(\psi(\theta)\frac{\partial \ln f(X,\theta)}{\partial\theta}\right) = 0$$

we obtain

$$\gamma'(\theta) - \mathrm{E}_\theta\left(\mathrm{E}_\theta^X(g)\frac{\partial \ln f(X,\theta)}{\partial\theta}\right) = \psi'(\theta)$$

then, the lower bound in (1.12) is $[\psi'(\theta)]^2/I_X(\theta)$ and efficiency of $p(X)$ follows from (1.15) and (1.16). ■

Example 1.10 *(continued)*
It is easy to verify that $\bar{X}_n - X_n$ is an efficient predictor of $X_{n+1} - X_n$. ◇

Example 1.17 *(Signal with Gaussian noise)*
Consider the process

$$X_t = \theta \int_0^t f(s)\mathrm{d}s + W_t, t \geq 0$$

where $\theta \in \mathbb{R}$, f is a locally square integrable function such that $\int_0^t f^2(s)\mathrm{d}s > 0$ for $t > 0$ and (W_t) is a standard Wiener process.

One can show (see Ibragimov and Hasminskii (1981)) that the maximum likelihood estimator of θ based on $X = (X_t, 0 \leq t \leq T)$ is

$$\hat{\theta}_T = \frac{\int_0^T f(t)\mathrm{d}X_t}{\int_0^T f^2(t)\mathrm{d}t} = \theta + \frac{\int_0^T f(t)\mathrm{d}W_t}{\int_0^T f^2(t)\mathrm{d}t}.$$

$\hat{\theta}_T$ is an EUE with variance $\left(\int_0^T f^2(t)\mathrm{d}t\right)^{-1}$.

Now, for $h > 0$, we have

$$\mathrm{E}_\theta^X(X_{T+h}) = X_T + \theta \int_T^{T+h} f(t)\mathrm{d}t := \phi(X) + \psi(\theta).$$

Applying Theorem 1.8 we obtain the EUP

$$\hat{X}_{T+h} = X_T + \frac{\int_0^T f(t)\mathrm{d}X_t}{\int_0^T f^2(t)\mathrm{d}t} \int_T^{T+h} f(t)\mathrm{d}t$$

with quadratic error

$$\mathrm{E}_\theta(\hat{X}_{T+h} - X_{T+h})^2 = \frac{\left(\int_T^{T+h} f(t)\mathrm{d}t\right)^2}{\int_0^T f^2(t)\mathrm{d}t} + h.$$ ◇

The next example shows that A1.1 does not imply existence of an EUP.

Example 1.18
If $X = (X_1, \ldots, X_n) \stackrel{d}{=} \mathcal{N}(\theta, 1)^{\otimes n}$ $(\theta \in \mathbb{R})$ and $g = \theta X_n$, the Cramér–Rao bound is θ^2/n. Then an EUP $p(X)$ must satisfy $\mathrm{E}_0(p(X) - 0)^2 = 0$ hence $p(X) = 0$ P_θ a.s. for all θ, which is contradictory. ◇

We now study conditions for existence of an EUP, beginning with a necessary condition.

Theorem 1.9 *(Extended exponential model)*
If A1.1 holds, $\partial \ln f/\partial\theta$ and $\mathrm{E}_\theta((\partial \ln f/\partial\theta)(p - \mathrm{E}_\theta^X g))/I_X(\theta)$ are continuous in θ and if p is EUP for g with

$$\mathrm{E}_\theta(p(X) - \mathrm{E}_\theta^X g)^2 > 0, \theta \in \Theta$$

then

$$f(x, \theta) = \exp(A(\theta)p(x) - B(x, \theta) + C(x)) \tag{1.17}$$

where

$$\frac{\partial B(x, \theta)}{\partial\theta} = A'(\theta)\mathrm{E}_\theta^X(g). \tag{1.18}$$

PROOF:
If p is efficient for g, it is efficient for $\mathrm{E}_\theta^X(g)$ and the Cauchy–Schwarz inequality becomes an equality. Hence $\partial \ln f(X, \theta)/\partial\theta$ and $p(X) - \mathrm{E}_\theta^X(g)$ are collinear, and since they are not degenerate, this implies that

$$\frac{\partial \ln f(X, \theta)}{\partial\theta} = a(\theta)(p(X) - \mathrm{E}_\theta^X(g)) \tag{1.19}$$

where

$$a(\theta) = \frac{I_X(\theta)}{\mathrm{E}_\theta\left(\frac{\partial \ln f}{\partial\theta}(p - \mathrm{E}_\theta^X g)\right)}.$$

Using the continuity assumptions one may integrate (1.19) to obtain (1.17) and (1.18). ■

Note that decomposition (1.17) is not unique. Actually one may rewrite it under the form

$$f(x, \theta) = \exp(A(\theta)(p(x) + h(x)) - [B(x, \theta) + A(\theta)h(x)] + C(x))$$

but the prediction problem is then different: $p(X) + h(X)$ is EUP for $\mathrm{E}_\theta^X(g) + h(X)$.

The next statement gives a converse of Theorem 1.9 in a canonical case.

Theorem 1.10
Consider the model

$$f(x, \theta) = \exp(\theta p(x) - B(x, \theta) + C(x)), \theta \in \Theta.$$

If A1.1 holds and the equality $\int f(x,\theta)\mathrm{d}\mu(x) = 1$ is twice differentiable under the integral sign, then $p(X)$ is an EUP of $\partial B(X,\theta)/\partial\theta$.

PROOF:
By differentiating $\int f(x,\theta)\mathrm{d}\mu(x) = 1$ one obtains

$$\int [p(x) - \frac{\partial B}{\partial \theta}(x,\theta)]f(x,\theta)\mathrm{d}\mu(x) = 0,$$

therefore

$$\mathrm{E}_\theta p(X) = \mathrm{E}_\theta\left(\frac{\partial B}{\partial \theta}(X,\theta)\right).$$

Differentiating again leads to

$$\int \left[p(x) - \frac{\partial B}{\partial \theta}(x,\theta)\right]^2 f(x,\theta)\mathrm{d}\mu(x) = \int \frac{\partial^2 B}{\partial \theta^2}(x,\theta)f(x,\theta)\partial\mu(x),$$

that is

$$\mathrm{E}_\theta\left(p(X) - \frac{\partial B}{\partial \theta}(X,\theta)\right)^2 = \mathrm{E}\left(\frac{\partial^2 B(X,\theta)}{\partial \theta^2}\right),$$

it is then easy to verify that $\mathrm{E}_\theta(\partial^2 B(X,\theta)/\partial\theta^2)$ is the Cramér–Rao bound: p is EUP. ■

Example 1.13 *(continued)*
Here the likelihood takes the form

$$f_1(N_T,\lambda) = \exp(-\lambda T + N_T \ln\lambda + c(N_T)),$$

putting $\theta = \ln\lambda$ yields

$$f(N_T,\theta) = \exp(\theta N_T - T\mathrm{e}^\theta + c(N_T))$$

hence N_T is EUP for $\partial(T\mathrm{e}^\theta)/\partial\theta = \lambda T$, thus $(N_T/T)h$ is EUP for λh. From Theorem 1.8 it follows that $(N_T/T)h + N_T$ is EUP for $\lambda h + N_T = \mathrm{E}_\lambda^{N_T}(N_{T+h})$ and for N_{T+h}. ◇

Example 1.14 *(continued)*
Taking into account the form of the likelihood we set

$$p(X) = \frac{T - X_T^2 + X_0^2}{2} \quad \text{and} \quad B(X,\theta) = \frac{\theta^2}{2}\int_0^T X_t^2 \mathrm{d}t,$$

then, Theorem 1.10 gives efficiency of $p(X)$ for predicting $\theta\int_0^T X_t^2\mathrm{d}t$.

Taking $\theta' = \theta^2$ as a new parameter, one obtains

$$f_1(X, \theta') = \exp\left(-\frac{\theta'}{2}\int_0^T X_t^2 \mathrm{d}t - \sqrt{\theta'}\frac{X_T^2 - X_0^2 - T}{2}\right),$$

hence

$$-\frac{1}{2}\int_0^T X_t^2 \,\mathrm{d}t$$

is efficient for predicting

$$\frac{1}{2\sqrt{\theta'}}\frac{X_T^2 - X_0^2 - T}{2} = \frac{1}{4\theta}(X_T^2 - X_0^2 - T).$$

This means that the empirical moment of order 2,

$$\frac{1}{T}\int_0^T X_t^2 \mathrm{d}t$$

is efficient for predicting

$$\frac{1}{2\theta}\left(1 - \frac{X_T^2}{T} + \frac{X_0^2}{T}\right).$$

It can be shown that it is not efficient for estimating $\mathrm{E}_\theta(X_0^2) = 1/2\theta$. In fact, its variance is

$$\frac{1}{2\theta^3 T} + \frac{1}{4\theta^2 T^2}(1 - \mathrm{e}^{-2\theta T})$$

when the Cramér–Rao bound is $1/(2\theta^3 T)$.

Note that the above predicted variables are not natural conditional expectations. Of course the genuine problem is to predict $\mathrm{E}_\theta^{\mathcal{F}_T}(X_{T+h}) = \mathrm{e}^{-\theta h} X_T$, but we have seen in Section 1.5 that there is no BUP in that case. We will consider this question from an asymptotic point of view in the next chapter. ◇

Example 1.19 *(Noncentered Ornstein–Uhlenbeck process)*
Consider the process

$$X_t = m + \int_{-\infty}^t \mathrm{e}^{-\theta(t-s)}\mathrm{d}W(s), t \in \mathbb{R}\,(\theta > 0, m \in \mathbb{R})$$

where W is a standard bilateral Wiener process.

Suppose that θ is known and m is unknown, and consider the process

$$X_{0,t} = \int_{-\infty}^{t} \mathrm{e}^{-\theta(t-s)} \mathrm{d}W(s), t \in \mathbb{R}.$$

On the space $C[0,T]$ the likelihood of $(X_t, 0 \leq t \leq T)$ with respect to $(X_{0,t}, 0 \leq t \leq T)$ has the form

$$f(X,m) = \exp\left[-\frac{\theta m^2}{2}(2+\theta T) + \theta m\left(X_0 + X_T + \theta \int_0^T X_s \mathrm{d}s\right)\right],$$

(see Grenander 1981), therefore $\theta\left(X_0 + X_T + \theta \int_0^T X_s \mathrm{d}s\right)$ is EUE for $\theta m(2+\theta T)$ and

$$m_T = \frac{1}{(2+\theta T)}\left[X_0 + X_T + \int_0^T X_s \mathrm{d}s\right]$$

is EUE for m.

Now $\mathrm{E}_\theta^{\mathcal{F}_T}(X_{T+h}) = \mathrm{e}^{-\theta h}(X_T - m) + m = m(1-\mathrm{e}^{-\theta h}) + \mathrm{e}^{-\theta h} X_T n$, and, from Theorem 1.8 it follows that $m_T(1-\mathrm{e}^{-\theta h}) + \mathrm{e}^{-\theta h} X_T$ is EUP for $\mathrm{E}_\theta^{\mathcal{F}_T}(X_{T+h})$ and for X_{T+h}.

Finally, since θ is known, the efficient predictor is

$$\hat{X}_{T+h} = \mathrm{e}^{-\theta h} X_T + \frac{1-\mathrm{e}^{-\theta h}}{2+\theta T}\left[X_0 + X_T + \int_0^T X_s \mathrm{d}s\right].$$

◇

Example 1.13 *(continued)*
Suppose that one wants to predict $\mathbb{1}_{\{N_{T+h}=0\}}$. Then $p = (-h/T)^{N_T}$ is the unique unbiased predictor function of N_T. It is optimal but not efficient. Moreover the naive predictor $\mathbb{1}_{\{N_T=0\}}$ is not unbiased but preferable to p.

Thus an optimal predictor is not always efficient and an unbiased optimal predictor is not always a good predictor. ◇

1.7 Loss functions and empirical predictors

The quadratic prediction error is not the single interesting risk function. Other preference relations are also used in practice.

Another matter is optimality; it is sometimes convenient to use predictors that are suboptimal but easy to compute and (or) robust.

In the current section we glance at various preference relations and some empirical predictors.

1.7.1 Loss function

A *loss function* $L : \mathbb{R} \times \mathbb{R} \to \mathbb{R}$ is a positive measurable function such that $L(a,a) = 0$, $a \in \mathbb{R}$. It generates a *risk function* defined as

$$R_\theta(g,p) = \mathrm{E}_\theta[L(g(X,Y,\theta),p(X))], \theta \in \Theta$$

which measures the accuracy of p when predicting g.

The resulting preference relation is

$$p_1 \prec p_2 \Leftrightarrow R_\theta(g,p_1) \leq R_\theta(g,p_2), \theta \in \Theta.$$

The following extension of the Rao–Blackwell theorem holds.

Theorem 1.11
Let $\prec$ be a preference relation defined by a loss function $L(x,y)$ which is convex with respect to y. Then, if S is P-sufficient for g and p is an integrable predictor, we have

$$\mathrm{E}_p^S \prec p. \tag{1.20}$$

PROOF:
We only give a sketch of the proof; for details we refer to Adke and Ramanathan (1997).

First we have the following preliminary result:

Let $(\Omega, \mathcal{A}, P)$ be a Probability space; $\mathcal{F}_1$, $\mathcal{F}_2$, $\mathcal{F}_3$ three sub-σ-algebras of $\mathcal{A}$, then

$$\mathcal{F}_1 \perp\!\!\!\perp \mathcal{F}_2 | \mathcal{F}_3 \Rightarrow \sigma(\mathcal{F}_1 \cup \mathcal{F}_3) \perp\!\!\!\perp \sigma(\mathcal{F}_2 \cup \mathcal{F}_3 | \mathcal{F}_3).$$

A consequence is:

if U_1, U_2 are real random variables such that $U_1 \in L^1(\Omega, \sigma(\mathcal{F}_1 \cup \mathcal{F}_3), P)$, $U_2 \in L^0(\sigma(\mathcal{F}_2 \cup \mathcal{F}_3))$, the space of real and $\sigma(\mathcal{F}_2 \cup \mathcal{F}_3)$-measurable applications, and $U_1 U_2 \in L^1(\Omega, \mathcal{A}, P)$, then

$$\mathrm{E}^{\mathcal{F}_3}(U_1) = 0 \Rightarrow \mathrm{E}(U_1 U_2) = 0. \tag{1.21}$$

Now, by using convexity of L, one obtains

$$L(g,p) \geq L(g, \mathrm{E}^S(p)) + L'(g, \mathrm{E}^S(p))(p - \mathrm{E}^S(p)) \tag{1.22}$$

where L' is the right derivative of $L(g, \cdot)$.

Choosing $\mathcal{F}_1 = \sigma(X)$, $\mathcal{F}_2 = \sigma(g)$, $\mathcal{F}_3 = \sigma(S)$ and applying (1.21) one obtains (1.20) by taking expectations in (1.22). ■

Note that an immediate consequence of Theorem 1.11 is an extension of the Lehmann–Scheffé theorem (Theorem 1.3). We let the reader verify that various optimality results given above remain valid if the quadratic error is replaced by a convex loss function.

1.7.2 Location parameters

Suppose that L is associated with a location parameter μ defined by

$$\mathrm{E}L(Z, \mu) = \min_{a \in \mathbb{R}} \mathrm{E}L(Z, a)$$

with $P_Z \in \mathcal{P}_L$, a family of probability measures on $\mathbb{R}$. Then, since

$$\mathrm{E}_\theta[L(g, p)] = \mathrm{E}_\theta[\mathrm{E}_\theta^X L(g, p)], \tag{1.23}$$

if $P_{\theta,g}^X \in \mathcal{P}_L$, the right side of (1.23) is minimum for $p_0(X) = \mu_\theta(X)$, where $\mu_\theta(X)$ is the location parameter associated with $P_{\theta,g}^X$. If $x \mapsto \mu_\theta(x)$ is measurable, one obtains

$$\mathrm{E}_\theta L(g, p_0(X)) = \min_{p \in L^0(\mathcal{B})} \mathrm{E}_\theta L(g, p(X)), \theta \in \Theta.$$

Three particular cases are classical:

- If $L(u, v) = (v - u)^2$, $\mu_\theta(X) = \mathrm{E}_\theta^X(g)$,
- if $L(u, v) = |v - u|$, $\mu_\theta(X)$ is a *median* of $P_{\theta,g}^X$,
- if $L(u, v) = \mathbb{1}_{|v-u| \geq \varepsilon}$ ($\varepsilon > 0$), $\mu_\theta(X)$ is a *mode* of $P_{\theta,g}^X$. Note that this last loss function is not convex with respect to v.

In order to construct a statistical predictor based on these location parameters, one may use a *plug-in method*: this consists of replacing θ by an estimator $\hat{\theta}(X)$ to obtain the predictor

$$p(X) = \mu_{\hat{\theta}(X)}(X).$$

Such a predictor is not optimal in general but it may have sharp asymptotic properties if $\hat{\theta}$ is a suitable estimator of θ. We give some details in Chapter 2.

In a *nonparametric framework* the approach is somewhat different: it uses direct estimators of the conditional location parameter. For example the regression kernel estimator allows us to construct a predictor associated with conditional expectation (see

Chapter 6). For the conditional mode (respectively median) it may be estimated by taking the mode (respectively median) of a nonparametric conditional density estimator.

1.7.3 Bayesian predictors

In the Bayesian scheme one interprets θ as a random variable with a prior distribution τ.

Then if (X, Y) has density $f(x, y, \theta)$ with respect to $\mu \otimes \nu$ σ-finite, (X, Y, θ) has density $f(x, y, \theta)$ with respect to $\mu \otimes \nu \otimes \tau$. Thus, the marginal density of (X, Y) is given by

$$\varphi(x, y) = \int f(x, y, \theta) \mathrm{d}\tau(\theta).$$

Now, if L is a loss function with risk R_θ, the associated *Bayesian risk* for predicting Y is defined as

$$\begin{aligned} r(Y, p(X)) &= \int R_\theta(Y, p(X)) \mathrm{d}\tau(\theta) \\ &= \int L(y, p(x)) f(x, y, \theta) \mathrm{d}\mu(x) \mathrm{d}\nu(y) \mathrm{d}\tau(\theta) \\ &= \int L(y, p(x)) \varphi(x, y) \mathrm{d}\mu(x) \mathrm{d}\nu(y). \end{aligned}$$

As before a solution of $\min_p r(Y, p(X))$ is a location parameter associated with P_Y^X. If L is quadratic error, one obtains the *Bayesian predictor*

$$\tilde{Y} = \mathrm{E}^X(Y) = \int y \varphi(y|X) \mathrm{d}\nu(y) \tag{1.24}$$

where $\varphi(\cdot|X)$ is the marginal conditional density of Y given X.

Of course the recurrent problem is choice of the prior distribution. We refer to Lehmann and Casella (1998) for a comprehensive discussion.

Now the Bayesian approach turns out to be very useful if X does not provide enough information about Y. We give an example.

Example 1.20
Consider the model

$$Y = \theta X + \varepsilon, \ (|\theta| < 1)$$

where X and Y have distribution $\mathcal{N}(0, 1)$ and $\varepsilon \perp\!\!\!\perp X$. Then (1.24) gives

$$\tilde{Y} = \mathrm{E}(\theta) X = \int_{-1}^{1} \theta \mathrm{d}\tau(\theta) \cdot X,$$

while no reasonable nonbayesian predictor is available if θ is unknown. ◇

1.7.4 Linear predictors

Let $(X_t, t \geq 1)$ be a real square integrable stochastic process; assume that $X = (X_1, \ldots, X_n)$ is observed and $Y = X_{n+1}$ has to be predicted.

A commonly used empirical predictor is the linear predictor

$$p(X) = \sum_{i=1}^{n} a_i X_i. \tag{1.25}$$

If $\sum_{i=1}^{n} a_i = 1$ and $X_1, \ldots, X_{n+1}$ have the same expectation, $p(X)$ is unbiased. A typical predictor of this form is the *simple exponential smoothing* given by

$$p_\beta(X) = \frac{X_n + \beta X_{n-1} + \cdots + \beta^{n-1} X_1}{1 + \beta + \cdots \lambda + \beta^{n-1}},$$

with $0 \leq \beta \leq 1$.

If $\beta = 0$, it reduces to the *naive predictor*

$$p_0(X) = X_n,$$

if $\beta = 1$, one obtains the *empirical mean*

$$p_1(X) = \frac{1}{n} \sum_{i=1}^{n} X_i,$$

if $0 < \beta < 1$ and n is large enough, practitioners use the classical form

$$\tilde{p}_\beta(X) = (1 - \beta)(X_n + \beta X_{n-1} + \cdots + \beta^{n-1} X_1).$$

The naive predictor is interesting if the X_t's are highly locally correlated. In particular, if (X_t) is a martingale, we have

$$X_n = \mathrm{E}_\theta^{\sigma(X_t, t \leq n)}(X_{n+1}), \theta \in \Theta$$

and X_n is the BUP with null statistical prediction error.

In contrast, the empirical mean is BUP when the X_t's are i.i.d. and $(\sum_{i=1}^{n} X_i)/n$ is the BUE of $\mathrm{E}_\theta X_1$ (cf. Theorem 1.5).

Finally, if $0 < \beta < 1$, one has

$$\tilde{p}_\beta(X) = \mathrm{E}_\beta^{\mathcal{F}_n}(X_{n+1})$$

provided (X_t) is an ARIMA (0,1,1) process defined by

$$\begin{cases} X_t - X_{t-1} = \varepsilon_t - \beta\varepsilon_{t-1}; & t \geq 2 \\ X_1 = \varepsilon_1. \end{cases}$$

and (ε_t) is a strong white noise. In fact this property remains valid for the wider range $-1 < \beta < 1$, see Chatfield (2000).

Concerning the choice of β, a general empirical method is *validation*. Consider the empirical prediction error

$$\delta_n(\beta) = \sum_{k=k_n}^{n} |X_k - p_{\beta,k-1}(X_1, \ldots, X_{k-1})|^2$$

where $p_{\beta,k-1}$ is the simple exponential smoothing constructed with the data $X_1, \ldots, X_{k-1}$ and $k_n < n$ is large enough. Then an estimator of β is

$$\hat{\beta} = \arg\min_{0 \leq \beta \leq 1} \delta_n(\beta).$$

If the model is known to be an ARIMA (0,1,1) one may also estimate β in this framework. If in addition, (X_t) is Gaussian, the maximum likelihood estimator (MLE) of β is asymptotically efficient as $n \to \infty$, see Brockwell and Davis (1991).

Finally, concerning the general predictor defined by (1.25) one may use *linear regression* techniques for estimating $(a_1, \ldots, a_n)$. Similarly as above, specific methods are available if (X_t) is an ARIMA (p,d,q) process, see Brockwell and Davis (1991).

1.8 Multidimensional prediction

We now consider the case where θ and (or) g take their values in a multidimensional space, or, more generally, in an infinite-dimensional space (recall that X takes its value in an arbitrary measurable space $(E_0, \mathcal{B}_0)$).

For example $X = (X_1, \ldots, X_n)$ where the X_i's are $\mathbb{R}^d$-valued with common density θ, and g is the conditional density $f^X_{\theta,Y}(\cdot)$.

A general enough framework is the case where $\theta \in \Theta \subset \Theta_0$ and g is B-valued where Θ_0 and B are separable Banach spaces with respective norms $\| \cdot \|_0$ and $\| \cdot \|$. Now, assume that $\mathrm{E}_\theta \| g \|^2 < \infty$, $\theta \in \Theta$ and denote by B^* the topological dual space of B. Then a natural preference relation between predictors is defined by

$$p_1 \prec p_2 \Leftrightarrow \mathrm{E}_\theta(x^*(p_1 - g))^2 \leq \mathrm{E}_\theta(x^*(p_2 - g))^2, \theta \in \Theta, x^* \in B^*.$$

This means that p_1 is preferable to p_2 for predicting g, if and only if $x^*(p_1)$ is preferable to $x^*(p_2)$ for predicting $x^*(g)$ for all x^* in B^*, with respect to the preference relation (1.1).

Now let $\mathcal{P}_\mathcal{G}(B)$ be the class of B-valued predictors such that $x^*(p) \in \mathcal{P}_\mathcal{G}$ for each $x^* \in B^*$, where $\mathcal{P}_\mathcal{G}$ is defined in Section 1.3. Then we have the following extension of Lemma 1.1:

Lemma 1.3
If $p \in \mathcal{P}_\mathcal{G}(B)$, then

$$\mathrm{E}_\theta(x^*(p) - x^*(g))^2 = \mathrm{E}_\theta(x^*(p) - \Pi^\theta x^*(g))^2 + \mathrm{E}_\theta(\Pi^\theta x^*(g) - x^*(g))^2, \\ \theta \in \Theta, x^* \in B^*. \tag{1.26}$$

In particular, if Π^θ is conditional expectation, it follows that $p_1 \prec p_2$ for predicting g if and only if $p_1 \prec p_2$ for predicting $\mathrm{E}_\theta^X(g)$.

PROOF:
It suffices to apply Lemma 1.1 to $x^*(p)$ and $x^*(g)$ and then to use the property

$$x^*(\mathrm{E}_\theta^X(g)) = \mathrm{E}_\theta^X(x^*(g))(\text{a.s.}).$$

■

Now, if $B = H$ (a Hilbert space) we have the additional property

$$\mathrm{E}_\theta \parallel p - g \parallel^2 = \mathrm{E}_\theta \parallel p - \mathrm{E}_\theta^X(g) \parallel^2 + \mathrm{E}_\theta \parallel \mathrm{E}_\theta^X(g) - g \parallel^2, \theta \in \Theta,$$

which is obtained by applying (1.26) to $x^* = e_j, j \geq 1$ where (e_j) is an orthonormal basis of H, and by summing the obtained equalities.

In this context one may use the simpler but less precise preference relation:

$$p_1 \prec_1 p_2 \Leftrightarrow \mathrm{E}_\theta \parallel p_1 - g \parallel^2 \leq \mathrm{E}_\theta \parallel p_2 - g \parallel^2, \theta \in \Theta.$$

Clearly

$$p_1 \prec p_2 \Rightarrow p_1 \prec_1 p_2.$$

Now, the results concerning sufficient statistics remain valid in the multidimensional case. In particular, application of Theorem 1.2 to $x^*(p)$ and $x^*(g)$ shows that, if $S(X)$ is P-sufficient, one has $\mathrm{E}^S(p) < p$. A similar method allows us to extend the results concerning BUP. Details are left to the reader.

We now turn to efficiency. The next theorem gives a multidimensional Cramér–Rao inequality for predictors. We consider the following set of assumptions.

Assumptions 1.2 (A1.2)
Θ is open in Θ_0, X has a strictly positive density $f(x, \theta)$ with respect to a σ-finite measure, and

(i) $(\forall\theta \in \Theta)$, $(\exists U_\theta \in \Theta_0)$, $(\forall u \in U_\theta)$, $\exists V_{\theta,u}$ a neighbourhood of 0 in $\mathbb{R}$: $(\forall\delta \in V_{\theta,u})$, $\theta + \delta u \in \Theta$ and $\partial f(x, \theta + \delta u)/\partial\delta$ does exist. Then, one sets

$$\dot{f}_u(x,\theta) = \frac{\partial}{\partial\delta} f(x, \theta + \delta u)|_{\delta=0}.$$

(ii) The relation

$$\int f(x, \theta + \delta u)\mathrm{d}\mu(x) = 1, \quad \theta \in \Theta, u \in U_\theta, \delta \in V_{\theta,u}, \tag{1.27}$$

is differentiable with respect to δ under the integral sign.

(iii) The B-valued predictor p is such that

$$\int x^*(p(x)) f(x, \theta + \delta u)\mathrm{d}\mu(x) = x^*(\gamma(\theta + \delta u)), \tag{1.28}$$

$x^* \in B^*$, $u \in U_\theta$, $\delta \in V_{\theta,u}$, where $\gamma : \Theta_0 \mapsto B$ is linear. Moreover this equality is differentiable with respect to δ under the integral sign.

(iv) $(\forall\theta \in \Theta)$, $(\forall u \in U_\theta)$,

$$I_\theta(X,u) = \mathrm{E}_\theta\left(\frac{\dot{f}_u(X,\theta)}{f(X,\theta)}\right)^2 \in]0,\infty[.$$

Then:

Theorem 1.12
If A1.2 holds, we have the bound

$$\mathrm{E}_\theta(x^*(p-g))^2 \geq \mathrm{E}_\theta(x^*(g - \mathrm{E}_\theta^X g))^2 + \frac{x^*\left[\gamma(u) - \mathrm{E}_\theta\left(\mathrm{E}_\theta^X(g)\frac{\dot{f}_u(X,\theta)}{f(X,\theta)}\right)\right]^2}{I_\theta(X,u)},$$

$$\theta \in \Theta, u \in U_\theta, x^* \in B^*. \tag{1.29}$$

PROOF:
Differentiating (1.27) and taking $\delta = 0$, one obtains

$$\mathrm{E}_\theta\left(\frac{\dot{f}_u}{f}\right) = \int \dot{f}_u(x,\theta)\mathrm{d}\mu(x) = 0;$$

the same operations applied to (1.28), and linearity of γ give

$$\begin{aligned}\mathrm{E}_\theta\left(x^*(p)\frac{\dot{f}_u}{f}\right) &= \int x^*(p(x))\frac{\dot{f}_u(x,\theta)}{f(x,\theta)} f(x,\theta)\mathrm{d}\mu(x) \\ &= x^*(\gamma(u)).\end{aligned}$$

Now the Cauchy–Schwarz inequality entails

$$\left[\mathrm{E}_\theta\left(x^*(p-\mathrm{E}_\theta^X g)\cdot\frac{\dot{f}_u}{f}\right)\right]^2 \le \mathrm{E}_\theta(x^*(p-\mathrm{E}_\theta^X g))^2\cdot \mathrm{E}_\theta\left(\frac{\dot{f}_u}{f}\right)^2,$$

collecting the above results and using Lemma 1.3 one arrives at (1.29). ■

In a Hilbert space it is possible to obtain a global result.

Corollary 1.2
If $B = H$, a Hilbert space, then

$$\mathrm{E}_\theta \parallel p - g \parallel^2 \ \ge\ \mathrm{E}_\theta \parallel p - \mathrm{E}_\theta^X(g) \parallel^2 + \frac{\parallel \gamma(u) - \mathrm{E}_\theta\left(g\frac{\dot{f}_u}{f}\right) \parallel^2}{I_\theta(X,u)}, \tag{1.30}$$

$\theta \in \Theta, u \in U_\theta$.

PROOF:
Apply (1.29) to $x^* = e_j$, $j \ge 1$ where (e_j) is a complete orthonormal system in H, and take the sum. ■

In (1.29) and (1.30), that are slight extensions of the Grenander inequality (1981, p. 484), the choice of u is arbitrary. Of course it is natural to choose a u that maximizes the lower bound. If the lower bound is achieved p is said to be efficient, but, in general, this only happens for some specific values of (x^*, u), as shown in the next example.

Example 1.21 *(Sequence of Poisson processes)*
Let $(N_{t,j}, t \ge 0)$, $j \ge 0$ be a sequence of independent homogeneous Poisson processes with respective intensity λ_j, $j \ge 1$ such that $\sum_j \lambda_j < \infty$.

Since $\mathrm{E}N_{t,j}^2 = \lambda_j t(1 + \lambda_j t)$ it follows that $\sum_j N_{t,j}^2 < \infty$ a.s., therefore $M_t = (N_{t,j}, j \ge 0)$ defines an ℓ^2-valued random variable, where ℓ^2 is the Hilbert space $\{(x_j) \in \mathbb{R}^{\mathbb{N}}, \sum_j x_j^2 < \infty\}$ with norm $\parallel (x_j) \parallel = \left(\sum_j x_j^2\right)^{1/2}$.

One observes $M_{(T)} = (M_t, 0 \le t \le T)$ and wants to predict M_{T+h} $(h > 0)$. It is easy to see that M_T is a P-sufficient statistic, then one only considers predictors of the form $p(M_T)$.

Now let $\mathcal{N} \subset \ell^2$ the family of sequences (x_j) such that $(x_j) \in \mathbb{N}^{\mathbb{N}}$ and $x_j = 0$ for j large enough. This family is countable, hence the counting measure μ on $\mathcal{N}$, extended by $\mu(\ell^2 - \mathcal{N}) = 0$, is σ-finite.

Then, note that $\sum_j N_{T,j}^2 < \infty$ a.s. yields $N_{T,j} = 0$ almost surely for j large enough. Thus M_T is $\mathcal{N}$-valued (a.s.). This implies that M_T has a density with respect to μ and the corresponding likelihood is

$$f(M_T(\omega), (\lambda_j)) = \sum_{j=0}^{J(T,\omega)} \mathrm{e}^{-\lambda_j T} \frac{(\lambda_j T)^{N_{T,j}(\omega)}}{N_{T,j}(\omega)!} \mathrm{e}^{-\sum_{j=0}^{J(T,\omega)} \lambda_j T}, \omega \in \Omega$$

where $J(T, \omega)$ is such that $N_{T,j}(\omega) = 0$ for $j > J(T, \omega)$.

Hence the MLE of $\theta = (\lambda_j)$ is

$$\hat{\theta}_T(\omega) = \left(\frac{N_{T,j}}{T}, 0 \leq j \leq J(T, \omega) \right) = \left(\frac{N_{T,j}(\omega)}{T}, j \geq 0 \right).$$

Then, an unbiased predictor of M_{T+h} should be

$$\hat{M}_{T+h} = \left(\frac{T+h}{T} N_{T,j}, j \geq 0 \right).$$

In order to study its efficiency we consider the loglikelihood:

$$\ln f(M_T, \theta + \delta u) = \sum_{j=0}^{\infty} [-(\lambda_j + \delta u_j)T + N_{T,j} \ln((\lambda_j + \delta u_j)T) + \ln(N_{T,j}!)]$$

since if $u = (u_j) \in \ell^2$, then $\sum (\lambda_j + \delta u_j)^2 < \infty$. Therefore

$$\frac{\dot{f}_u(M_T, \theta)}{f(M_T, \theta)} = \sum_{j=0}^{\infty} u_j \left(\frac{N_{T,j}}{\lambda_j} - T \right),$$

and

$$I_\theta(X, \mu) = T \sum_{j=0}^{\infty} \frac{u_j^2}{\lambda_j},$$

which belongs to $]0, \infty[$ if $\sum_j u_j^2 > 0$ and $\sum_j u_j^2/\lambda_j < \infty$.
Now, on one hand we have

$$\mathrm{E}_{\theta+\delta u}(x^*(\hat{M}_{T+h})) = (T+h) \sum_{j=0}^{\infty} (\lambda_j + \delta u_j) x_j,$$

thus, in (1.28), $\gamma : \ell^2 \mapsto \ell^2$ may be defined by

$$\gamma(v) = (T+h)v, v \in \ell^2,$$

on the other hand

$$\mathrm{E}_\theta^X(g) = \mathrm{E}_0^{M_T}(M_{T+h}) = (N_{T,j} + \lambda_j h, j \geq 0),$$

hence

$$\mathrm{E}_\theta\left(\frac{\dot{f}_u}{f}\mathrm{E}_\theta^X(g)\right) = T(u_j)$$

and

$$x^*\left[\gamma(u) - \mathrm{E}_\theta\left(\frac{\dot{f}_u}{f}\mathrm{E}_\theta^X g\right)\right] = h\sum_j x_j u_j.$$

Finally, since

$$\mathrm{E}_\theta[x^*(p - \mathrm{E}_\theta^X g)]^2 = \frac{h^2}{T}\sum_j \lambda_j x_j^2,$$

(1.29) shows that $\hat{M}_{T+h}$ is efficient if and only if

$$\sum_j \lambda_j x_j^2 = \frac{(\sum_j x_j u_j)^2}{\sum_j \frac{u_j^2}{\lambda_j}}. \tag{1.31}$$

In particular, if $x^* = (0, \ldots, 0, x_{j_0}, 0, \ldots)$ efficiency holds, provided $0 < \sum_j u_j^2/\lambda_j < \infty$.

More generally, set $x_j = \alpha_j/\sqrt{\lambda_j}$ and $u_j = \sqrt{\lambda_j}, \beta_j, j \geq 0$. If (α_j) and (β_j) are in ℓ^2, (1.31) gives

$$(\sum \alpha_j^2)(\sum \beta_j^2) = (\sum \alpha_j \beta_j)^2,$$

thus (α_j) and (β_j) are collinear, i.e. $(\lambda_j x_j)$ and (u_j) are collinear. ◇

Notes

As far as we know a systematic exposition of the theory of statistical prediction is not available in the literature. In this Chapter we have tried to give some elements of this topic.

Presentation of the prediction model is inspired by Yatracos (1992). Lemma 1.1 belongs to folklore but it is fundamental since it shows that the statistician may only predict $\Pi^\theta g$, rather than g.

Definition of P-sufficient statistics appear in Takeuchi and Akahira (1975). Also see Bahadur (1954); Johansson (1990) and Torgersen (1977) among others.

Theorem 1.1 is simple but useful and probably new, while Theorem 1.2 is in Johansson. The study of optimal unbiased predictors comes from Yatracos (1992) and Adke and Ramanathan (1997).

Theorem 1.7 is also in Yatracos but the more compact Corollary 1.1 and results concerning efficiency seem to be new.

Theorem 1.11 is taken from Adke and Ramanathan (1997) and other results of Section 1.7 are classical.

The elements of multidimensional prediction theory stated in Section 1.8 are natural extensions of the one-dimensional theory. The bound in Theorem 1.12 is an extension of the Cramér–Rao type bound of Grenander (1981). The application to sequences of Poisson processes is new.

2

Asymptotic prediction

2.1 Introduction

In various situations no optimal predictor may exist (see Lemma 1.2). It is then rather natural to adopt an asymptotic point of view by studying consistency and limit in distribution of predictors as size of dataset tends to infinity.

The basic question is prediction of sequences of events that may be random or deterministic. The Blackwell algorithm gives an asymptotically minimax solution of this problem without any special assumption.

The natural extension is prediction of discrete or continuous time stochastic processes. In a parametric framework a classical method consists in replacing the unknown parameter with a suitable estimator; the obtained plug-in predictor has nice asymptotic properties.

The results are obtained with a fixed prediction horizon. Some indications concerning forecasting for small or large horizons are given at the end of the current chapter.

2.2 The basic problem

Let $(A_n, n \geq 1)$ be a sequence of events and $X_n = \mathbb{1}_{A_n}$, $n \geq 1$ the associated sequence of 0-1 random or deterministic variables. The question is: how to construct a general algorithm of prediction for X_{n+1}, given $X_1, \ldots, X_n$; $n \geq 1$?

A plausible predictor should be

$$\widetilde{X}_{n+1} = \mathbb{1}_{\{\overline{X}_n > \frac{1}{2}\}}, \quad n \geq 1 \tag{2.1}$$

Inference and Prediction in Large Dimensions D. Bosq and D. Blanke

where $\overline{X}_n = \left(\sum_{i=1}^n X_i\right)/n$. In order to measure quality of this algorithm, one puts $\widetilde{E}_n = \left(\sum_{i=1}^n \mathbb{1}_{\{\widetilde{X}_i = X_i\}}\right)/n$ where $\widetilde{X}_1 = 0$ (for example).

Algorithm (2.1) is asymptotically optimal if (X_n) is an i.i.d. sequence of Bernoulli random variables with $\mathrm{E}\, X_n = p (0 \leq p \leq 1)$. Actually, the strong law of large numbers allows us to establish that $\widetilde{E}_n \underset{n\to\infty}{\longrightarrow} \max(p, 1-p)$ a.s. when, if p is known, the best strategy consists in systematically predicting 1 if $p > 1/2$, 0 if $p < 1/2$, and arbitrarily 0 or 1 if $p = 1/2$. This strategy supplies the same asymptotic behaviour as $(\widetilde{X}_n)$.

However (2.1) fails completely for the cyclic sequence $X_1 = 1,\ X_2 = 0,\ X_3 = 1, \ldots$

Now the *Blackwell algorithm* is a remedy for this deficiency. In order to introduce it we use Figure 2.1

The predictor $(\widehat{X}_n)$ is defined recursively through the sequence

$$M_n = (\overline{X}_n, \overline{E}_n), \quad n \geq 1$$

where

$$\overline{E}_n = \frac{1}{n} \sum_{i=1}^n \mathbb{1}_{\{\widehat{X}_i = X_i\}}$$

and $\widehat{X}_1 = 0$.

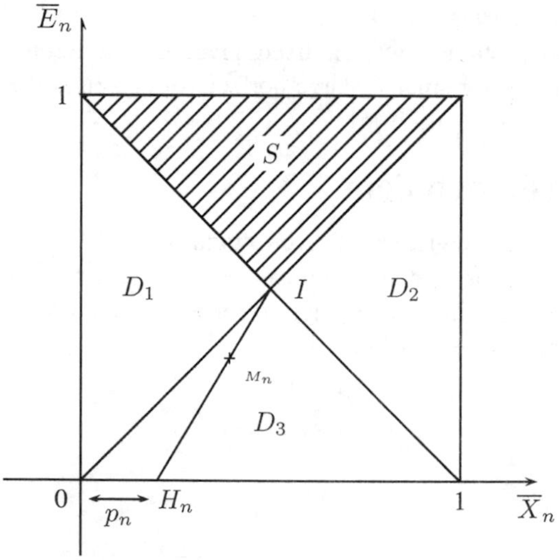

Figure 2.1 The Blackwell algorithm.

Actually we have

$$P(\widehat{X}_{n+1} = 1) = \begin{cases} 0 & \text{if} \quad M_n \in D_1, \\ 1 & \text{if} \quad M_n \in D_2, \\ p_n & \text{if} \quad M_n \in D_3, \end{cases}$$

with

$$p_n = \frac{1}{2} \frac{\overline{X}_n - \overline{E}_n}{1/2 - \overline{E}_n},$$

finally, if $M_n \in \overset{\circ}{S} \cup \{I\}$, $\widehat{X}_{n+1}$ is arbitrary.

Note that this algorithm is random, even if (X_n) is not random.

Now, let d be the Euclidian metric over $\mathbb{R}^2$, we have the following outstanding result.

Theorem 2.1

$$d(M_n, S) \underset{n\to\infty}{\longrightarrow} 0 \quad a.s. \tag{2.2}$$

PROOF: (*Outline*)
A geometric interpretation of the algorithm allows us to prove that

$$\mathrm{E}^{\mathcal{F}_n}[d(M_{n+1}, S)]^2 \leq \left(\frac{n}{n+1}\right)^2 [d(M_n, S)]^2 + \frac{1}{2(n+1)^2},$$

where $\mathcal{F}_n = \sigma(M_k, k \leq n), \quad n \geq 1$.

Thus $([d(M_n, S)]^2, n \geq 1)$ is an 'almost supermartingale', hence (2.2) from a convergence theorem in Robbins and Siegmund (1971).

For details we refer to Lerche and Sarkar (1998). ■

Theorem 2.1 has an asymptotically minimax aspect because (2.2) is asymptotically optimal for i.i.d. Bernoulli random variables and these sequence are the hardest to predict.

Now, in order to derive more precise results, one needs information about the distribution of the process $(X_n, n \geq 1)$. We study this situation in the next section.

2.3 Parametric prediction for stochastic processes

Consider a stochastic process with distribution depending on a finite-dimensional parameter θ. We study the plug-in predictor obtained by substituting an estimator for θ in the best probabilistic predictor. The results hold as well in discrete time as in continuous time.

Let $X = (X_t,\ t \in I)$, where $I = \mathbb{Z}$ or $\mathbb{R}$, be an $\mathbb{R}^d$-valued stochastic process, defined on the probability space $(\Omega, \mathcal{A}, P_\theta)$, where the unknown parameter $\theta \in \Theta$, an open set in $\mathbb{R}^{d'}$. Euclidian norm and scalar products in $\mathbb{R}^d$ or $\mathbb{R}^{d'}$ are denoted by $\|\cdot\|$ and $\langle\cdot,\cdot\rangle$ respectively.

$X_{(T)} = (X_t,\ 0 \le t \le T,\ t \in I)$ is observed, and one intends to predict $X_{T+h}\ (h > 0,\ T \in I,\ T+h \in I)$. In the following 'belonging to I' is omitted and h is fixed.

Define the σ-algebras

$$\mathcal{F}_a^b = \sigma(X_t,\ a \le t \le b), \quad -\infty \le a \le b \le +\infty,$$

with the conventions $X_{-\infty} = X_{+\infty} = 0$, and suppose that $\mathrm{E}_\theta \| X_{T+h} \|^2 < \infty$ and that the conditional expectation X^*_{T+h} with respect to $\mathcal{F}^T_{-\infty}$ has the form

$$X^*_{T+h} = \mathrm{E}_\theta^{\mathcal{F}^T_{-\infty}}(X_{T+h}) = r_{T,h}(\theta, Y_T) + \eta_{T,h}(\theta), \tag{2.3}$$

$T > 0, \theta \in \Theta$, where Y_T is a $\mathcal{F}^T_{\phi(T)}$-measurable random variable with values in some measurable space $(\mathcal{F}, \mathcal{C})$; $(\phi(T))$ is such that $0 < \phi(T) \le T$ and $\lim_{T\to\infty} \phi(T)/T = 1$, and we have

$$\lim_{T\to\infty} \mathrm{E} \| \eta_{T,h}(\theta) \|^2 = 0, \quad \theta \in \Theta;$$

finally $r_{T,h}(\cdot, Y_T)$ is assumed to be known.

Note that if $\phi(T) = T$ and $\eta_{T,h}(\theta) = 0$, (2.3) turns out to be a Markov type condition.

In order to construct a statistical predictor, one uses an estimator $\widehat{\theta}_{\psi(T)}$ of θ based on $(X_t,\ 0 \le t \le \psi(T))$ where $0 < \psi(T) < T$, $\lim_{T\to\infty} \psi(T)/T = 1$ and $\lim_{T\to\infty}(\phi(T) - \psi(T)) = +\infty$.

The plug-in predictor is then defined as

$$\widehat{X}_{T+h} = r_{T,h}(\widehat{\theta}_{\psi(T)}, Y_T).$$

Now, consider the following assumptions.

Assumptions 2.1 (A2.1)

(i) $\| r_{T,h}(\theta'', Y_T) - r_{T,h}(\theta', Y_T) \| \le Z_T \| \theta'' - \theta' \|; \theta', \theta'' \in \Theta$, where Z_T is a random variable such that, for all $\theta \in \Theta$,

$$Z_T \in \mathcal{L}^{2kq}(\Omega, \mathcal{F}^T_{\phi(T)}, P_\theta), \quad T > 0\ (k \ge 1,\ q > 1),$$
$$\overline{\lim_{T\to\infty}}\ a(T)\mathrm{E}_\theta(Z_T^{2k}) < \infty\ (a(T) > 0),$$

and

$$\overline{\lim_{T\to\infty}}\ \mathrm{E}_\theta \parallel Z_T \parallel^{2kq} < \infty.$$

(ii) X is *strongly mixing*, that is $\lim_{u\to\infty} \alpha_\theta(u) = 0$, $\theta \in \Theta$, where

$$\alpha_\theta(u) = \sup_{t\in I} \sup_{B\in\mathcal{F}_{-\infty}^{t}, C\in\mathcal{F}_{t+u}^{\infty}} |P_\theta(B\cap C) - P_\theta(B)P_\theta(C)|.$$

(iii) There exists an integer $k' \geq 1$ and $s > 1$ such that

$$\overline{\lim_{T\to\infty}}\ b^{k''}(T)\mathrm{E}_\theta \parallel \widehat{\theta}_T - \theta \parallel^{2k''} < \infty, \quad 1 \leq k'' \leq k' \ (b(T) > 0)$$

and

$$\overline{\lim_{T\to\infty}}\ \mathrm{E}_\theta \parallel \widehat{\theta}_T - \theta \parallel^{2k's} < \infty.$$

(iv) $r_{T,h}$ has partial derivatives with respect to $\theta = (\theta_1, \ldots, \theta_{d'})$ and, as $T \to \infty$,

$$\mathrm{grad}\ r_{T,h}(\theta, Y_T) \xrightarrow{\mathcal{D}} U \stackrel{d}{=} Q_\theta, \ \theta \in \Theta.$$

(v) The estimator $(\widehat{\theta}_T)$ satisfies

$$\widehat{\theta}_T \to \theta \quad \text{a.s.},$$

and

$$c(T)(\widehat{\theta}_T - \theta) \xrightarrow{\mathcal{D}} V \stackrel{d}{=} R_\theta \quad (c(T) > 0).$$

The above regularity conditions hold for various classical discrete or continuous time processes (ARMA processes,*diffusion processes*, ...) and various estimators (maximum likelihood, Bayesian estimators, ...). Examples are studied in Section 2.4.

We now study asymptotic behaviour of $\widehat{X}_{T+h}$. The first statement provides rate of convergence.

Theorem 2.2
If A2.1(i)–(iii) hold with $k' = k$ and $1/q + 1/s < 1$ and if, for all θ,

$$\overline{\lim_{T\to\infty}}\ [a(T)b(T)]^k [\alpha_\theta(\phi(T) - \psi(T))]^{1-\frac{1}{q}-\frac{1}{s}} < \infty, \tag{2.4}$$

$$\frac{b(T)}{b(\psi(T))} \xrightarrow[T\to\infty]{} 1,$$

$$\overline{\lim_{T\to\infty}}\ [a(T)b(T)]^k \mathrm{E}_\theta(\parallel \eta_{T,h}(\theta) \parallel^{2k}) < \infty \tag{2.5}$$

then

$$\overline{\lim_{T\to\infty}}\, [a(T)b(T)]^k \mathrm{E}_\theta(\| \widehat{X}_{T+h} - X_{T+h} \|^{2k}) < \infty. \tag{2.6}$$

PROOF:
First note that

$$\begin{aligned}
&[a(T)b(T)]^k \mathrm{E}_\theta \| \widehat{X}_{T+h} - X^*_{T+h} \|^{2k} \\
&\leq 2^{2k-1}[[a(T)b(T)]^k \mathrm{E}_\theta \| r_{T,h}(\widehat{\theta}_{\psi(T)}, Y_T) - r_{T,h}(\theta, Y_T) \|^{2k} \\
&+ [a(T)b(T)]^k \mathrm{E}_\theta \| \eta_{T,h}(\theta) \|^{2k}] \\
&:\leq 2^{2k-1}[\Delta_{1,T} + \Delta_{2,T}].
\end{aligned}$$

Then, from (2.5), it suffices to study the asymptotic behaviour of $\Delta_{1,T}$. Now, A 2.1(i) entails

$$\Delta_{1,T} \leq \mathrm{E}_\theta[(a^k(T)Z_T^{2k})(b^k(T) \| \widehat{\theta}_{\psi(T)} - \theta \|^{2k})]$$

and from the *Davydov inequality* it follows that

$$\Delta_{1,T} \leq \mathrm{E}_\theta(a^k(T)Z_T^{2k}) \cdot \mathrm{E}_\theta(b^k(T) \| \widehat{\theta}_{\psi(T)} - \theta \|^{2k}) + \delta_T$$

where

$$\begin{aligned}
\delta_T = 2p[2\alpha_\theta(\phi(T) - \psi(T))]^{1/p}\, a^k(T)(\mathrm{E}_\theta |Z_T|^{2kq})^{1/q} \\
\times\, b^k(T)(\mathrm{E}_\theta \| \widehat{\theta}_{\psi(T)} - \theta \|^{2kr})^{1/r}
\end{aligned}$$

with $1/p = 1 - 1/q - 1/r$.

Then, A2.1(i)–(iii) and (2.4) yield (2.6). ■

Note that, if $I = \mathbb{Z}$, and the model contains i.i.d. sequences, we have, for these sequences,

$$X^*_{T+h} = \mathrm{E}_\theta(X_{T+h}) := r(\theta).$$

Consequently, if $b(T)^{-1}$ is the minimax quadratic rate for estimating θ and $r(\theta)$, it is also the minimax quadratic rate for predicting X_{T+h}.

We now turn to limit in distribution. For convenience we suppose that $d = 1$. Then we have

Theorem 2.3
If A2.1(ii),(iv)–(v) hold, $\frac{c(T)}{c(\psi(T))} \to 1$ *and* $c(T) \cdot \eta_{T,h}(\theta) \xrightarrow{p} 0$, *then*

$$(c(T)(\widehat{X}_{T+h} - X^*_{T+h})) \xrightarrow{\mathcal{D}} \langle U, V \rangle$$

where $(U, V) \stackrel{d}{=} Q_\theta \otimes R_\theta$.

PROOF:
From A2.1(iv) it follows that

$$r_{T,h}(\widehat{\theta}_T, Y_T) - r_{T,h}(\theta, Y_T) = \sum_{i=1}^{d'} (\widehat{\theta}_{\psi(T),i} - \theta_i)\left(\frac{\partial r_{T,h}}{\partial \theta_i}(\theta, Y_T) + \delta_{T,i}\right)$$

where $\delta_T = (\delta_{T,1}, \ldots, \delta_{T,d'}) \to \vec{0} = (0, \ldots, 0)$ as $\widehat{\theta}_{\psi(T)} \to \theta$, hence $\delta_T \to \vec{0}$ a.s.

Now let us set

$$U_T = c(T)(\widehat{\theta}_{\psi(T)} - \theta),$$

and

$$V_T = \mathrm{grad}_\theta r_{T,h}(\widehat{\theta}_{\psi(T)}, Y_T).$$

The strong mixing condition A2.1(ii) entails

$$\sup_{A \in \mathcal{B}_{\mathbb{R}^d}, B \in \mathcal{B}_{\mathbb{R}^d}} |P_\theta(U_T \in A, V_T \in B) - P_\theta(U_T \in A) P_\theta(V_T \in B)|$$
$$\leq \alpha_\theta(\varphi(T) - \psi(T)) \to 0;$$

then, by using classical properties of *weak convergence* (see Billingsley 1968) one deduces that

$$\mathcal{L}(\langle U_T, V_T \rangle) \xrightarrow{w} \mathcal{L}(\langle U, V \rangle)$$

and since

$$\mathcal{L}(\langle U_T, \delta_T \rangle) \xrightarrow{w} \delta_{(0)},$$

and $c(T)\eta_{T,h}(\theta) \xrightarrow{p} 0$, the desired result follows. ■

2.4 Predicting some common processes

We now apply the previous asymptotic results to some common situations.

Example 1.15 *(continued)*
Let $(X_t, t \in \mathbb{Z})$ be an AR(1) process such that $\mathrm{Var}_\theta X_0 = 1$ and $\mathrm{E}_\theta |X_0|^m < \infty$ where $m > 4$. If ε_0 has density with respect to Lebesgue measure, such a process is geometrically strongly mixing (GSM), that is

$$\alpha(u) \leq [a\mathrm{e}^{-bu}], \quad u \geq 0 \ (a > 0, b > 0)$$

(see Rosenblatt 2000). Here we have $X_{T+1}^* = \theta X_T := r(\theta, X_T)$. Now, the unbiased estimator

$$\widehat{\theta}_T = \frac{1}{T}\sum_{t=1}^{T} X_{t-1}X_t$$

produces the statistical predictor $\widehat{X}_{T+h} = \widehat{\theta}_{\psi(T)} \cdot X_T$ with $\psi(T)$ given by $\psi(T) = [T - \ln T \cdot \ln\ln T]$.

The GSM condition and Davydov inequality imply

$$|\mathrm{Cov}(X_{t-1}X_t, X_{t+u}X_{t+1+u})| = \mathcal{O}(\mathrm{e}^{-cu}), \ (c > 0),$$

then a straightforward computation gives

$$T \cdot \mathrm{E}_\theta(\widehat{\theta}_T - \theta)^2 \to \gamma_\theta = \sum_{-\infty}^{\infty} \mathrm{Cov}(X_0X_1, X_{1+u}X_{2+u})$$

and

$$\mathrm{E}_\theta|\widehat{\theta}_T - \theta|^m = \mathcal{O}(T^{-\frac{m}{2}}).$$

Thus, Theorem 2.2 entails

$$\mathrm{E}_\theta(\widehat{X}_{T+h} - X_{T+h}^*)^2 = \mathcal{O}\left(\frac{1}{T}\right).$$

Concerning limit in distribution one may use the central limit theorem for strongly mixing processes, see Rio (2000), to obtain

$$\sqrt{T}\,(\widehat{\theta}_T - \theta) \xrightarrow{\mathcal{D}} N \stackrel{d}{=} \mathcal{N}(0, \gamma_\theta),$$

and, since X_0 has a fixed distribution Q_θ, Theorem 2.3 gives

$$\sqrt{T}\,(\widehat{X}_{T+h} - X_{T+h}^*) \xrightarrow{\mathcal{D}} U \cdot V$$

where

$$(U, V) \stackrel{d}{=} Q_\theta \otimes \mathcal{N}(0, \gamma_\theta). \qquad \diamond$$

Example 2.1 *(ARMA(p,q) process)*
Consider the process

$$X_t - \varphi_1 X_{t-1} + \cdots + \varphi_p X_{t-p} = \varepsilon_t - \gamma_1\varepsilon_{t-1} - \cdots - \gamma_q\varepsilon_{t-q}, \quad t \in \mathbb{Z} \tag{2.7}$$

where $(\varepsilon_t, t \in \mathbb{Z})$ is a real strong white noise, $\varphi_p \gamma_q \neq 0$ and the polynomials $1 - \sum_{j=1}^{p} \varphi_j z^j$ and $1 - \sum_{j=1}^{q} \gamma_j z^j$ have no common zeros and do not vanish for $|z| \leq 1$. Then (2.7) has a stationary solution

$$X_t = \sum_{j=1}^{\infty} \pi_j X_{t-j} + \varepsilon_t, \quad t \in \mathbb{Z}$$

where $\pi_j = \pi_j(\varphi_1, \ldots, \varphi_p; \gamma_1, \ldots, \gamma_q) := \pi_j(\theta)$, $j \geq 1$ and $(\pi_j, j \geq 1)$ decreases at an exponential rate. Thus we clearly have

$$X_{T+1}^* = \sum_{j=1}^{\infty} \pi_j(\theta) X_{T+1-j}$$

and if $\widehat{\theta}_T$ is the classical empirical estimator of θ, see Brockwell and Davis (1991), one may apply the above results.

Details are left to the reader. ◇

Example 1.14 (*continued*)
Consider the Ornstein–Uhlenbeck process

$$X_t = \int_{-\infty}^{t} \mathrm{e}^{-\theta(t-s)} dW(s), \quad t \in \mathbb{R}, \quad (\theta > 0);$$

in order to study quadratic error of the predictor it is convenient to choose the parameter

$$\beta = \frac{1}{2\theta} = \mathrm{Var}_\theta X_0,$$

and its natural estimator

$$\widehat{\beta}_T = \frac{1}{T} \int_0^T X_t^2 \, \mathrm{d}t.$$

Here

$$r_{T,h}(\beta, Y_T) = \mathrm{e}^{-\frac{h}{2\beta}} X_T,$$

and since

$$\left| \frac{\partial r_{T,h}(\beta, Y_T)}{\partial \eta} \right| \leq \frac{2}{h \mathrm{e}^2} |X_T|,$$

one may take $\phi(T) = T$ and $Z_T = (2/h)\mathrm{e}^{-2}X_T$. Now (X_t) is Gaussian stationary and GSM (see Doukhan (1994)), then conditions in Theorem 2.2 hold with $a(T) = 1$,

$$T \cdot \mathrm{E}_\beta|\widehat{\beta}_T - \eta|^2 \to \ell_\eta > 0$$

and

$$\overline{\lim_{T\to\infty}}\, T \cdot \mathrm{E}_\beta|\widehat{\beta}_T - \eta|^{2r} < \infty, \; r > 1.$$

Finally choosing $\psi(T) = T - \ln T \cdot \ln\ln T$ one obtains

$$\overline{\lim_{T\to\infty}}\, T \cdot \mathrm{E}_\beta|\widehat{X}_{T+h} - X^*_{T+h}|^2 < \infty.$$

Concerning limit in distribution we come back to θ and take the maximum likelihood estimator

$$\widehat{\theta}_T = \frac{1 - \dfrac{X_T^2}{T} + \dfrac{X_0^2}{T}}{\dfrac{2}{T}\displaystyle\int_0^T X_t^2\,\mathrm{d}t},$$

then (see Kutoyants (2004)),

$$\widehat{\theta}_T \to \theta \text{ a.s.}$$

and

$$\mathcal{L}(\sqrt{T}\,(\widehat{\theta}_T - \theta)) \xrightarrow{w} \mathcal{N}(0, 2\theta).$$

In order to apply Theorem 2.3 one may use $\psi(T) = T - \ln\ln T$ and note that

$$\frac{\partial r_{T,h}}{\partial\theta}(\theta, X_T) \overset{d}{=} \mathcal{N}\left(0, \frac{h^2\mathrm{e}^{-2\theta h}}{2\theta}\right)$$

hence

$$\sqrt{T}\,(\widehat{X}_{T+h} - X^*_{T+h}) \xrightarrow{\mathcal{D}} U \cdot V$$

where $(U, V) \overset{d}{=} \mathcal{N}(0, 2\theta) \otimes \mathcal{N}(0, (h^2\mathrm{e}^{-2\theta h})/2\theta)$.

The distribution of $U \cdot V$ has characteristic function

$$\phi_\theta(u) = (1 + u^2h^2\mathrm{e}^{-2\theta h})^{-1/2}, \quad u \in \mathbb{R}$$

and we have

$$\mathrm{E}_\theta(N_\theta \cdot N'_\theta) = 0$$

and

$$\mathrm{Var}_\theta(N_\theta \cdot N'_\theta) = h^2 \mathrm{e}^{-2\theta h}.$$

As an application it is possible to construct a prediction interval. First

$$D_T(\theta) = \sqrt{T}\,\frac{\widehat{X}_{T+h} - X^*_{T+h}}{h\mathrm{e}^{-\theta h}} \xrightarrow{\mathcal{D}} N_1 \cdot N_2$$

where $(N_1, N_2) \stackrel{d}{=} \mathcal{N}(0,1)^{\otimes 2}$.

Now $\widehat{\theta}_{\psi(T)} \to \theta$ a.s. yields (see Billingsley 1968)

$$D_T(\widehat{\theta}_{\psi(T)}) \xrightarrow{\mathcal{D}} N_1 \cdot N_2;$$

then if

$$P(|N_1 N_2| \leq v_{\alpha/2}) = 1 - \frac{\alpha}{2} \quad (0 < \alpha < 1)$$

and

$$I_{T,h} = \left[\mathrm{e}^{-\widehat{\theta}_{\psi(T)} h}\left(X_T - \frac{h v_{\alpha/2}}{\sqrt{T}}\right),\ \mathrm{e}^{-\widehat{\theta}_{\psi(T)} h}\left(X_T + \frac{h v_{\alpha/2}}{\sqrt{T}}\right)\right],$$

we have

$$P_\theta(X^*_{T+h} \in I_{T,h}) \to 1 - \frac{\alpha}{2}.$$

Finally a prediction interval for X_{T+h} may be obtained by setting

$$D'_{T,h}(\theta) = \left(\frac{2\theta}{1 - \mathrm{e}^{-2\theta h}}\right)^{1/2} (X_{T+h} - X^*_{T+h}) \stackrel{d}{=} \mathcal{N}(0,1),$$

therefore

$$\mathcal{L}(D'_{T,h}(\widehat{\theta}_{\psi(T)})) \xrightarrow{w} \mathcal{N}(0,1).$$

Now define

$$a_\alpha^{(T)} = -v_{\alpha/2}\left(\frac{1 - \mathrm{e}^{-2\widehat{\theta}_{\psi(T)} h}}{2\widehat{\theta}_{\psi(T)}}\right)^{1/2} + \mathrm{e}^{-\widehat{\theta}_{\psi(T)} h}\left(X_T - \frac{h v_{\alpha/2}}{\sqrt{T}}\right)$$

and

$$b_\alpha^{(T)} = -v_{\alpha/2}\left(\frac{1-\mathrm{e}^{-2\widehat{\theta}_{\psi(T)}h}}{2\widehat{\theta}_{\psi(T)}}\right)^{1/2} + \mathrm{e}^{-\widehat{\theta}_{\psi(T)}h}\left(X_T + \frac{hv_{\alpha/2}}{\sqrt{T}}\right)$$

where

$$P(|N| \leq v_{\alpha/2}) = 1 - \frac{\alpha}{2} \quad (N \stackrel{d}{=} \mathcal{N}(0,1)),$$

then

$$\lim_{T\to\infty} P_\theta(X_{T+h} \in [a_\alpha^{(T)}, b_\alpha^{(T)}]) \geq 1 - \alpha.$$

◇

Example 2.2 *(Ergodic diffusion process)*
Consider a *diffusion process*, defined by

$$X_t = X_0 + \int_0^t S(X_s,\theta)\,\mathrm{d}s + \int_0^t \sigma(X_s)\mathrm{d}W_s, \quad t \geq 0.$$

Under some regularity conditions, in particular if

$$\overline{\lim_{|x|\to\infty}} \sup_{\theta\in\Theta} \operatorname{sgn} x \cdot \frac{S(x,\theta)}{\sigma^2(x)} < 0$$

and if X_0 has density

$$f_S(x,\theta) = \frac{1}{G\sigma^2(x)}\exp\left(2\int_0^x \frac{S(v,\theta)}{\sigma^2(v)}\,\mathrm{d}v\right), \quad x \in \mathbb{R}$$

where

$$G = \int_{-\infty}^{+\infty} \sigma^{-2}(y)\exp\left(2\int_0^y \frac{S(v,\theta)}{\sigma^2(v)}\,\mathrm{d}v\right)\mathrm{d}y,$$

then $(X_t, t \geq 0)$ is strictly stationary, ergodic, Markovian and the maximum likelihood and Bayesian estimators are asymptotically efficient. For details we refer to Kutoyants (2004).

Now, we have

$$X_{T+h} - X_T = \int_T^{T+h} S(X_s,\theta)\,\mathrm{d}s + \int_T^{T+h} \sigma(X_s)\,\mathrm{d}W_s$$

and, since (W_t) has independent increments and is adapted to $(\mathcal{F}_o^t)$,

$$X_{T+h}^* = X_T + \mathrm{E}_\theta^{\mathcal{F}_o^T}\left[\int_T^{T+h} S(X_s,\theta)\,\mathrm{d}s\right]$$
$$= X_T + \int_T^{T+h} M_s(X_T,\theta)\,\mathrm{d}s = r_{T,h}(\theta, X_T)$$

where

$$M_s(X_T,\theta) = \mathrm{E}_\theta^{\mathcal{F}_o^T} S(X_s,\theta), \quad T \le s \le T+h.$$

Making the additional assumption

$$|S(X_s,\theta'') - S(X_s,\theta')| \le U_s|\theta'' - \theta'|;\ \theta',\ \theta'' \in \Theta,$$

where $\mathrm{E}_\theta(U_s^{2q}) < \infty (q > 1)$, we may apply Theorem 2.2 to $\widehat{X}_{T+h}$ where

$$\widehat{X}_{T+h} = r_{T,h}(\widehat{\theta}_{\psi(T)}, X_T)$$

to obtain

$$\overline{\lim_{T\to\infty}}\, T \cdot \mathrm{E}_\theta(\widehat{X}_{T+h} - X_{T+h}^*)^2 < \infty;$$

here $\widehat{\theta}_T$ is the MLE or the Bayesian estimator and $\psi(T) = T - \ln T \cdot \ln\ln T$.

Concerning limit in distribution, first note that

$$Q_\theta = \mathcal{L}\left(\frac{\partial}{\partial\theta}\mathrm{E}_\theta^{\mathcal{F}_o^T} \int_T^{T+h} S(X_s,\theta)\,\mathrm{d}s\right)$$

does not depend on T. Now, from

$$\mathcal{L}(\sqrt{T}\,(\widehat{\theta}_T - \theta)) \xrightarrow{w} \mathcal{N}(0, (I(\theta))^{-1})$$

where $I(\theta)$ is Fisher information, it follows that

$$\sqrt{T}\,(\widehat{X}_{T+h} - X_{T+h}^*) \xrightarrow{\mathcal{D}} U \cdot V$$

where $(U, V) \stackrel{d}{=} Q_\theta \otimes \mathcal{N}(0, I(\theta)^{-1})$. ◇

In the next two examples the mixing condition is not satisfied but straightforward arguments give asymptotic results.

Example 1.13 (*continued*)
The efficient predictor $\widehat{X}_{T+h} = (T+h)N_t/T$ is such that

$$T \cdot \mathrm{E}\,(X_{T+h}^* - \widehat{X}_{T+h})^2 = \lambda h^2$$

and, since N_T/T satisfies the CLT,

$$\mathcal{L}\left(\frac{\sqrt{T}}{h\sqrt{\lambda}}(\widehat{X}_{T+h} - X^*_{T+h})\right) \xrightarrow{w} \mathcal{N}(0,1). \qquad \diamond$$

Example 1.17 *(continued)*
We have seen that the efficient predictor is

$$\widehat{X}_{T+h} = X_T + \int_T^{T+h} f(t)\,\mathrm{d}t \cdot \frac{\int_0^T f(s)\,\mathrm{d}s}{\int_0^T f^2(s)\,\mathrm{d}s}$$

with quadratic error

$$\mathrm{E}_\theta(\widehat{X}_{T+h} - X^*_{T+h})^2 = \left(\int_T^{T+h} f(t)\,\mathrm{d}t\right)^2 \Big/ \int_0^T f^2(t)\,\mathrm{d}t;$$

thus, consistency depends on f.

If $f(t) = t^\gamma (\gamma \geq 0)$ one obtains a $1/T$-rate; if $f(t) = \mathrm{e}^{-t}$ the predictor is not consistent. $\diamond$

2.5 Equivalent risks

Until now we have studied the asymptotic behaviour of quadratic prediction error. In this section we consider an alternative risk, the integrated quadratic error.

For simplicity we suppose that $(X_n, n \in \mathbb{Z})$ is a strictly stationary real Markov process such that $(X_1, \ldots, X_n)$ has a density $f_{(X_1,\ldots,X_n)}$ with respect to a σ-finite measure $\mu^{\otimes n}$, $n \geq 1$.

One observes $(X_1, \ldots, X_n) = X_{(n)}$ and intends to predict $m(X_{n+1})$, where m is a real Borel function satisfying $\mathrm{E}m^2(X_0) < \infty$.

Let us set

$$r(x) = \mathrm{E}(m(X_{n+1})|X_n = x), \quad x \in \mathbb{R},$$

and consider an estimator r_n of r, based on $X_1, \ldots, X_{n-k_n}$ where $1 < k_n < n$. Our goal is to compare

$$I_n = \mathrm{E}\left(r_n(X_n) - r(X_n)\right)^2$$

and

$$J_n = \int_{\mathbb{R}} \mathrm{E}\left(r_n(x) - r(x)\right)^2 f_{X_0}(x)\,\mathrm{d}\mu(x).$$

For this purpose we use the following mixing coefficients:

$$\beta_{k_n} = \beta(\sigma(X_{n-k_n}), \sigma(X_n)) = \mathrm{E} \sup_{C \in \sigma(X_n)} |P(C) - P(C|\sigma(X_{n-k_n}))|,$$

and

$$\varphi_{k_n} = \varphi(\sigma(X_{n-k_n}), \sigma(X_n)) = \sup_{\substack{B \in \sigma(X_{n-k_n}), P(B)>0 \\ C \in \sigma(X_n)}} |P(C) - P(C|B)|.$$

Details concerning these coefficients appear in the Appendix. Now, we have the following bounds:

Lemma 2.1
Suppose that $R_n = \sup_{x \in \mathbb{R}} |r_n(x) - r(x)|^2 < \infty$, *then*

(a) *If* $\mathrm{E}\, R_n^p < \infty (p > 1)$,

$$|I_n - J_n| \leq 2 \parallel R_n \parallel_p \beta_{k_n}^{1-\frac{1}{p}}, \tag{2.8}$$

(b) *if* $\mathrm{E}\, R_n < \infty$,

$$|I_n - J_n| \leq 2 \parallel R_n \parallel_1 \varphi_{k_n}, \tag{2.9}$$

(c) *if* $\parallel R_n \parallel < \infty$,

$$|I_n - J_n| \leq 2 \parallel R_n \parallel_\infty \beta_{k_n}. \tag{2.10}$$

PROOF:
Note that

$$\begin{aligned} I_n &= \int (r_n(x_{(n-k_n)}, x) - r(x))^2 f_{X_{(n-k_n)}, X_n}(x_{(n-k_n)}, x) \, \mathrm{d}\mu^{\otimes(n-k_n+1)}(x_{(n-k_n)}, x) \\ &= \int (r_n - r)^2 f_{X_{(n-k_n)}} f_{X_n | X_{(n-k_n)}} \, \mathrm{d}\mu^{\otimes(n-k_n+1)}. \end{aligned}$$

Using the Markov property one obtains

$$|I_n - J_n| = \int (r_n - r)^2 f_{(X_{n-k_n})} [f_{X_n | X_{(n-k_n)}} - f_{X_n}] \, \mathrm{d}\mu^{\otimes(n-k_n+1)}.$$

Now, from a Delyon inequality (see Viennet 1997, p. 478), it follows that there exists $b(x_{(n-k_n)})$ such that

$$|I_n - J_n| \leq 2 \int R_n \cdot b \cdot f_{X_{(n-k_n)}} \, \mathrm{d}\mu^{\otimes(n-k_n)},$$

and $\| b \|_\infty = \varphi_{k_n}$, $\| b \|_1 = \beta_{k_n}$, $\| b \|_q \leq \beta_{k_n}^{1/q} (q > 1)$, hence (2.8), (2.9) and (2.10) by using the Hölder inequality. ■

As an application we consider the case where

$$r(x) = r_o(x, \theta), \quad \theta \in \Theta \subset \mathbb{R}$$

with r_o known and such that

$$|r_o(x, \theta') - r_o(x, \theta)| \leq c|\theta' - \theta|; \ \theta', \theta \in \Theta. \tag{2.11}$$

As in Section 2.3 we consider a predictor of the form

$$r_n(X_n) = r_o(X_n, \widehat{\theta}_{n-k_n})$$

where $\widehat{\theta}_{n-k_n}$ is an estimator of θ based on $X_{(n-k_n)}$. If $\mathrm{E}(\widehat{\theta}_{n-k_n} - \theta)^4 < \infty$, one may apply (2.11) and (2.8) to obtain

$$|I_n - J_n| \leq 2c^2 [\mathrm{E}(\widehat{\theta}_{n-k_n} - \theta)^4]^{1/2} \beta_{k_n}^{1/2}.$$

Thus, if $I_n \simeq v_n$, one has $J_n \simeq v_n$ provided

$$v_n^{-1} [\mathrm{E}(\widehat{\theta}_{n-k_n} - \theta)^4]^{1/2} \beta_{k_n}^{1/2} \to 0. \tag{2.12}$$

If, for example, Θ is compact, $\widehat{\theta}_n \in \theta$ a.s., $n \geq 1$, $\mathrm{E}(\widehat{\theta}_n - \theta)^2 \simeq 1/n$ and $\beta_{k_n} = \mathcal{O}(k_n^{-\beta})$, then the choice $k_n \simeq n^{-\delta}$ $(1 > \delta > 1/\beta)$ yields (2.12).

More generally, since β-mixing implies α-mixing, one may show equivalence of I_n and J_n in the context of Theorem 2.2. Details are left to the reader.

2.6 Prediction for small time lags

Suppose that $X = (X_t, \, t \in \mathbb{R})$ is observed on $[0, T]$ and one wants to predict X_{T+h} for small h.

Here X is a zero-mean, nondeterministic, weakly stationary process with the rational spectral density

$$f(\lambda) = \left| \frac{P_\alpha(\lambda)}{Q_\beta(\lambda)} \right|^2$$

where α and $\beta > \alpha$ are the degrees of the polynomials P_α and Q_β. Then, from Cramér and Leadbetter (1967), it follows that X has exactly $m = \beta - \alpha - 1$ derivatives in mean-square $(m \geq 0)$.

We now consider the suboptimal linear predictor

$$\widetilde{X}_{T+h} = \sum_{j=0}^{m} \frac{h^j}{j!} X^{(j)}(T)$$

where $X^{(j)}$ denotes the jth derivative of X.

Then one may show (see Bucklew (1985)), that it is locally asymptotically optimal as $h \to 0\,(+)$. More precisely the best linear predictor X^{**}_{T+h} satisfies

$$\begin{aligned} \mathrm{E}_\theta(X^{**}_{T+h} - X_{T+h})^2 &= \int_0^h f^2(\lambda)\mathrm{d}\lambda \\ &\sim_{h\to 0\,(+)} \frac{2(-1)^{m+1} h^{2m+1}}{(2m+1)(m!)^2} \gamma^{(2m+1)}(0) \end{aligned}$$

where γ is the ⋆autocovariance⋆ of X, and we have

$$\overline{\lim_{h\to 0\,(+)}} \frac{\mathrm{E}_\theta(\widetilde{X}_{T+h} - X_{T+h})^2}{\mathrm{E}_\theta(X^{**}_{T+h} - X_{T+h})^2} = 1.$$

Thus, if m is known, $\widetilde{X}_{T+h}$ is a locally asymptotically optimal linear statistical predictor. Moreover it is determined by $(X_t,\ T-\varepsilon \le t \le T)$ for every strictly positive ε. Then, in order to make local prediction, it is not necessary to estimate θ. For example, if X is an Ornstein–Uhlenbeck process we have $\widetilde{X}_{T+h} = X_T$ and $X^{**}_{T+h} = \mathrm{e}^{-\theta h} X_T$.

We now consider a process with regular sample paths:

Example 2.3 (*Wong process*)
Consider the strictly stationary Gaussian process

$$X_t = \sqrt{3}\,\exp(-\sqrt{3}t) \int_0^{\exp(2t/\sqrt{3})} W_s\,\mathrm{d}s, \quad t \in \mathbb{R}_+,$$

where W is a standard Brownian motion. Its spectral density is given by

$$f(\lambda) = \frac{4\sqrt{3}}{\pi} \frac{1}{(1+3\lambda^2)(3+\lambda^2)}, \quad \lambda \in \mathbb{R},$$

thus $m = 1$ and the asymptotically optimal linear statistical predictor is

$$\widetilde{X}_{T+h} = X_T + hX'_T. \qquad \diamond$$

Finally, since rational spectra are dense in the class of all spectra, it is possible to extend the above property. If $m = 0$ the extension holds as soon as $\gamma(t) = g(t)g(-t)$ where g is an absolutely continuous function, see Bucklew (1985).

2.7 Prediction for large time lags

Let $X = (X_t,\ t \in \mathbb{Z})$ be a discrete time, real, *nondeterministic* and weakly stationary process. Then

$$X_t = m + \sum_{j=0}^{\infty} a_j \varepsilon_{t-j}, \quad t \in \mathbb{Z}, \quad (m \in \mathbb{R}),$$

where $\sum a_j^2 < \infty$, $a_o = 1$ and (ε_t) is a white noise, and

$$X_{T+h}^{**} = \sum_{j=0}^{\infty} a_j \varepsilon_{t+h-j},$$

hence

$$\mathrm{E}(X_{T+h}^{**} - X_{T+h})^2 = \sigma^2 \sum_{j=0}^{h-1} a_j^2, \quad h \geq 1.$$

We are interested in prediction for large h. The best asymptotic linear error of prediction is

$$\gamma(0) = \lim_{h \to \infty} \mathrm{E}\,(X_{T+h}^{**} - X_{T+h})^2$$

where

$$\gamma(u) = \mathrm{Cov}\,(X_0, X_u), \quad u \in \mathbb{R}.$$

Thus a linear statistical predictor $\widetilde{X}_{T+h}$ based on $X_0, \ldots, X_T$ is asymptotically optimal (AO) if

$$\lim_{h \to \infty} \mathrm{E}(\widetilde{X}_{T+h} - X_{T+h})^2 = \gamma(0).$$

The naive predictor X_T is not AO since

$$\mathrm{E}\,(X_T - X_{T+h})^2 = 2(\gamma(0) - \gamma(h)) \underset{h \to \infty}{\longrightarrow} 2\gamma(0);$$

actually $\overline{X}_T = \left(\sum_{t=0}^{T} X_t\right)/(T+1)$ is AO and

$$T \cdot [\mathrm{E}(\overline{X}_T - X_{T+h})^2 - \gamma(0)] \underset{h\to\infty}{\longrightarrow} \sum_{u\in\mathbb{Z}} \gamma(u).$$

Similar results are achievable in continuous time: if $X = (X_t,\, t \in \mathbb{R})$ is real measurable, weakly stationary and nondeterministic, it admits the representation

$$X_t = \int_{-\infty}^{t} g(t-u)\,\mathrm{d}V(u), \quad t \in \mathbb{R},$$

where g is a nonrandom square integrable function and V a process with orthogonal increments (see Cramér and Leadbetter 1967). Then

$$X_{T+h}^{**} = \int_{-\infty}^{t} g(t+h-u)\,\mathrm{d}V(u)$$

and

$$\lim_{h\to\infty} \mathrm{E}(X_{T+h}^{**} - X_{T+h})^2 = \int_0^{\infty} g^2(v)\,\mathrm{d}v = \gamma(0).$$

Now the naive predictor has asymptotic quadratic error $2\gamma(0)$ when $\overline{X}_T = \left(\int_0^T X_t\,\mathrm{d}t\right)/T$ is AO with

$$\lim_{h\to\infty} T[\mathrm{E}\,(\overline{X}_T - X_{T+h})^2 - \gamma(0)] = \int_{\infty}^{+\infty} \gamma(u)\,\mathrm{d}u.$$

Notes

The Blackwell algorithm (1956) has been studied in detail by Lerche and Sarkar (1993); they established Theorem 2.1. Theorems 2.2 and 2.3 and their applications are new.

Lemma 2.1 comes from a discussion with J. Dedecker (Personal communication, 2006).

Section 2.6 is due to Bucklew (1985); for a discussion we refer to Stein (1988).

Finally results concerning prediction for large time lags are well known or easy to check.

Part II

Inference by Projection

3

Estimation by adaptive projection

3.1 Introduction

In this chapter we study an estimation method that applies to a large class of functional parameters. This class contains discrete probability distributions, density, spectral density, covariance operators, among others.

For estimating a functional parameter a general procedure consists in approximating it by suitable functions that can be easily estimated. Various methods of approximation are conceivable: regularization by convolution, sieves, wavelets, neural networks,... No approach may be considered as systematically better than another one. Here we use the basic orthogonal projection method: the principle is to estimate the projection of the unknown functional parameter on a space with finite dimension k_n that increases with the size n of the observed sample.

Since a suitable choice of k_n depends on the true value of the parameter, it is convenient to modify the estimator by replacing k_n with an integer $\hat{k}_n$ which is a function of the data and appears to be a good approximation of the best k_n. In this way one obtains an *adaptive projection estimator* (APE). In some special situations the APE reaches a $1/n$-rate.

3.2 A class of functional parameters

Let $\mathcal{P}$ be a family of discrete time stochastic processes with values in some measurable space $(E, \mathcal{B})$. A member X of $\mathcal{P}$ has the form $X = (X_t,\ t \in \mathbb{Z})$. Let φ

Inference and Prediction in Large Dimensions D. Bosq and D. Blanke

(or φ_X) be an unknown parameter that depends on the distribution P_X of X:

$$\varphi_X = g(P_X), \quad X \in \mathcal{P}$$

where g is not necessarily injective.

We suppose that $\varphi \in \phi \subset H$, where H is a separable real Hilbert space, equipped with its scalar product $\langle \cdot, \cdot \rangle$ and its norm $\| \cdot \|$.

We will say that φ is (e, h)*-adapted*, or simply e*-adapted*, if there exists a fixed orthonormal system $e = (e_j, j \geq 0)$ of H and a family $h = (h_j, j \geq 0)$ of applications $h_j : E^{\nu(j)+1} \to \mathbb{R}, j \geq 0$ such that

$$\varphi = \sum_{j=0}^{\infty} \varphi_j e_j$$

with

$$\varphi_j = \langle \varphi, e_j \rangle = \mathrm{E}\big[h_j(X_0, \ldots, X_{\nu(j)})\big], \quad j \geq 0, \varphi \in \phi.$$

Finally we suppose that $\mathrm{E}[h_j^2(X_0, \ldots, X_{\nu_{(j)}})] < \infty$ and

$$\nu(j) \leq j, \quad j \geq 0.$$

We indicate a typical example of such a parameter. Other examples will appear in Sections 3.4 and 3.5.

Example 3.1 (*Spectral density*)
Let $X = (X_t, t \in \mathbb{Z})$ be a real zero-mean stationary process with ⋆autocovariance⋆

$$\gamma_j = \mathrm{E}(X_0 X_j), \quad j \geq 0, \quad \text{such that} \sum_{j=0}^{\infty} |\gamma_j| < \infty;$$

its spectral density is defined as

$$\varphi(\lambda) = \frac{1}{2\pi} \sum_{j \in \mathbb{Z}} \gamma_j \cos \lambda j, \quad \lambda \in [-\pi, +\pi].$$

Then

$$e_o = \frac{1}{\sqrt{2\pi}}, \; e_j(\lambda) = \frac{\cos \lambda j}{\sqrt{\pi}}, \quad j \geq 0, \; \nu(j) = j$$

and

$$h_o(X_0) = \frac{X_0^2}{\sqrt{2\pi}}, \; h_j(X_0, \ldots, X_j) = \frac{X_0 X_j}{\sqrt{\pi}}, \quad j \geq 1. \qquad \diamond$$

Now the problem is to specify estimation rates for some subclasses of ϕ. In order to define them it is convenient to consider a function $b : \mathbb{R}_+ \to \mathbb{R}_+^*$ that is integrable, continuously differentiable and such that $x \mapsto xb(x)$ is decreasing (large sense) for x large enough $(x > x_0)$. Note that this implies the following facts:

- b is strictly decreasing for $x > x_0$,
- $\sum_{j=0}^{\infty} b(j) < \infty$,
- $jb(j) \searrow 0$ as $j \uparrow \infty$.

Functions of this type are powers, negative exponentials, combinations of powers and logarithms, . . .

Now we set $b_j = b(j), j \geq 0$ and $\phi_b = \{\varphi : \varphi \in \phi, \ |\varphi_j| \leq b_j, j \geq 0\}$. We assume that there exists $\varphi^* \in \phi_b$ whose Fourier coefficients decrease at the same rate as (b_j). More precisely:

Assumptions 3.1 (A3.1)
There exists $\varphi^* \in \phi_b$ such that $\varphi_j^* \neq 0, \ j \geq 0$ and

$$|\varphi_j^*| = \delta\, b_j, \quad j \geq j_0 \ \ (0 < \delta \leq 1).$$

In order to estimate φ from the data $X_1, \ldots, X_n$ we use estimators of the form

$$\overline{\varphi}_n = \sum_{j=0}^{\infty} \lambda_{jn} \overline{\overline{\varphi}}_{jn} e_j \tag{3.1}$$

where

$$\overline{\overline{\varphi}}_{jn} =: \overline{\varphi}_{jn} = \frac{1}{n - \nu(j)} \sum_{i=1}^{n-\nu(j)} h_j(X_i, \ldots, X_{i+\nu(j)}) \ \text{ if } n > \nu(j) \tag{3.2}$$

$$= 0, \text{ if not.}$$

The estimator $\overline{\varphi}_n$ is well defined as soon as

$$\sum_{j: n > \nu(j)} \lambda_{jn}^2 \overline{\varphi}_{jn}^2 < \infty \ \text{ a.s.} \tag{3.3}$$

Now let $\mathcal{T}_\ell$ be the family of estimators of φ such that (3.1), (3.2) and (3.3) hold. We want to optimize the asymptotic behaviour of $\inf_{\overline{\varphi}_n \in \mathcal{T}_\ell} \sup_{X \in \mathcal{P}} \mathrm{E}_X \parallel \overline{\varphi}_n - \varphi_X \parallel^2$, in order to obtain the *minimax rate* for the family $\mathcal{T}_\ell$.

For this purpose, we will use the *oracle* associated with φ^*, assuming that $\overline{\varphi}_{jn}$ is an unbiased estimator of φ_j.

3.3 Oracle

The oracle is the best estimator, belonging to $\mathcal{T}_\ell$, at the point $\varphi = \varphi^*$, for some process X^* such that $\varphi^* = g(P_{X^*})$. Its construction uses the following elementary result.

Lemma 3.1
Let Z be an unbiased estimator of a real parameter θ such that $\theta \neq 0$. Let $V_\theta Z$ be the variance of Z. Then, the minimizer of $\mathrm{E}_\theta(\lambda Z - \theta)^2$ with respect to λ is

$$\lambda^* = \frac{\theta^2}{\theta^2 + V_\theta Z},$$

and the corresponding quadratic error is given by

$$\mathrm{E}_\theta(\lambda^* Z - \theta)^2 = \frac{\theta^2 \cdot V_\theta Z}{\theta^2 + V_\theta Z}.$$

PROOF:
Straightforward and therefore omitted. ■

Using Lemma 3.1 we obtain the oracle

$$\varphi_n^0 = \sum_{j \geq 0} \lambda_{jn}^* \overline{\overline{\varphi}}_{jn} e_j$$

with

$$\lambda_{jn}^* = \frac{\varphi_j^{*^2}}{\varphi_j^{*^2} + V_{jn}^*}, \quad j \geq 0 \tag{3.4}$$

where V_{jn}^* denotes variance of $\overline{\overline{\varphi}}_{jn}$ for some X^* such that $\varphi^* = g(P_{X^*})$.

Actually, if $n > \nu(j)$ the choice $Z = \overline{\overline{\varphi}}_{jn}$ in Lemma 3.1 leads to (3.4); if $n \leq \nu(j)$ one has $\overline{\overline{\varphi}}_{jn} = 0$, thus $V_{jn}^* = 0$ and $\lambda_{jn}^* = 1$ is suitable.

Note that φ_n^0 is well defined since

$$\mathrm{E}\left(\sum_{j \geq 0} \lambda_{jn}^{*^2} \overline{\overline{\varphi}}_{jn}^2\right) = \sum_{n > \nu(j)} \lambda_{jn}^{*^2} (V_{jn}^* + \varphi_j^{*^2})$$

$$\leq \sum_{n > \nu(j)} \frac{\varphi_j^{*^4}}{\varphi_j^{*^2} + V_{jn}^*} \leq \sum_{n > \nu(j)} \varphi_j^{*^2} \leq \| \varphi^* \|^2 < \infty.$$

Now, in order to study $\mathrm{E}_{X^*} ||\varphi_n^0 - \varphi^*||^2$ we make the general assumption:

Assumptions 3.2 (A3.2)
$E_{X^*}\overline{\varphi}_{jn} = \varphi_j^*$, $j \geq 0$, $n > \nu(j)$ and

$$0 \leq \frac{\alpha_j}{u_n} \leq V_{jn}^* \leq \frac{\beta_j}{u_n}, \quad j \geq 0$$

where (α_j) and (β_j) do not depend on n, and (u_n) is a sequence of integers such that $(u_n) \nearrow \infty$ and $u_n \leq n$, $n \geq n_o$.
We shall see that, under mild conditions, $u_n = n$. The case $u_n = o(n)$ corresponds to long memory processes.

Lemma 3.2
If A3.2 holds and $n \to \infty$

(1) *If $\sum_j \beta_j < \infty$, then*

$$0 \leq \sum_j \alpha_j \leq \underline{\lim}_{n\to\infty} u_n E_{X^*} \parallel \varphi_n^0 - \varphi^* \parallel^2 \tag{3.5}$$

and

$$\overline{\lim}_{n\to\infty} u_n E_{X^*} \parallel \varphi_n^0 - \varphi^* \parallel^2 \leq 2B^2\delta^2 + \sum_j \beta_j < \infty \tag{3.6}$$

where $B = \max_{j\geq 1} jb_j$.
(2) *If $\sum_j \alpha_j = \infty$, then*

$$\lim_{n\to\infty} u_n E_{X^*} \parallel \varphi_n^0 - \varphi^* \parallel^2 = \infty. \tag{3.7}$$

PROOF:

(1) Since $jb_j \to 0$ one has $B < \infty$ and $b_j^2 \leq B^2/j^2, j \geq 1$. Now, $n \leq \nu(j)$ implies $n \leq j$, thus, if $n > j_o + 1$,

$$\sum_{j:n\leq\nu(j)} \varphi_j^{*^2} = \delta^2 \sum_{j:n\leq\nu(j)} b_j^2 \leq \delta^2 \sum_{n\leq j} b_j^2$$
$$\leq B^2\delta^2 \sum_{j\geq n} \frac{1}{j^2} \leq \frac{B^2\delta^2}{n-1} \leq \frac{2B^2\delta^2}{u_n},$$

moreover

$$u_n \sum_{j:n>\nu(j)} \frac{\varphi_j^{*^2} \cdot V_{jn}^*}{\varphi_j^{*^2} + V_{jn}^*} \leq \sum_{n>\nu(j)} u_n V_{jn}^* \leq \sum_{j\geq 0} \beta_j,$$

therefore

$$\mathrm{E}_{X^*} \parallel \varphi_n^0 - \varphi^* \parallel^2 = \sum_{j:n>\nu(j)} \frac{\varphi_j^{*^2} \cdot V_{jn}^*}{\varphi_j^{*^2} + V_{jn}^*} + \sum_{j:n\leq\nu(j)} \varphi_j^{*^2} \leq \frac{2\beta^2\delta^2}{u_n} + \frac{\sum \beta_j}{u_n} \tag{3.8}$$

hence (3.6).

In order to get a lower bound, we write

$$u_n \mathrm{E}_{X^*} \parallel \varphi_n^0 - \varphi^* \parallel^2 \geq \sum_{j:n>\nu(j)} \frac{\varphi_j^{*^2} \cdot u_n V_{jn}^*}{\varphi_j^{*^2} + V_{jn}^*}$$

$$\geq \sum_{j\geq 0} \frac{\varphi_j^{*^2} \cdot \alpha_j}{\varphi_j^{*^2} + V_{jn}^*} \mathbb{1}_{\{n>\nu(j)\}}(j) := m_n,$$

all terms in m_n belong to $[0, \alpha_j]$ and tend to α_j (respectively); then the dominated convergence theorem gives $m_n \to \sum_{j\geq 0} \alpha_j$ which yields (3.5).

(2) If $\sum_j \alpha_j = \infty$, the Fatou lemma entails

$$\underline{\lim}_{n\to\infty} u_n \mathrm{E}_{X^*} \parallel \varphi_n^0 - \varphi^* \parallel^2 \geq \sum_{j\geq 0} \underline{\lim}_{n\to\infty} u_n \psi_{jn}$$

where ψ_{jn} is the term of order j in (3.8).

Since, for n large enough, one has

$$u_n \psi_{jn} = \frac{\varphi_j^{*^2} \cdot u_n V_{jn}^*}{\varphi_j^{*^2} + V_{jn}^*}$$

it follows that $\underline{\lim}_{n\to\infty} u_n \psi_{jn} \geq \alpha_j$, hence (3.7). ■

3.4 Parametric rate

If $\nu(j) = 0, j \geq 0$ and $\sum_j \beta_j < \infty$ the rate u_n^{-1} is reached provided the following assumption holds.

Assumptions 3.3 (A3.3)
$\overline{\varphi}_{jn}$ is an unbiased estimator of $\varphi_j, j \geq 0$ and there exists $c > 0$ such that

$$\sup_{X\in\mathcal{P}} \mathrm{Var}_X \overline{\varphi}_{jn} \leq c\frac{\beta_j}{u_n}, \quad j \geq 0, \quad n \geq n_o.$$

To show this we define two estimators of φ:

$$\widetilde{\varphi}_{n,0} = \sum_{j=0}^{\infty} \overline{\varphi_{jn}} e_j,$$

and

$$\widetilde{\varphi}_n = \sum_{j=0}^{u_n} \overline{\varphi_{jn}} e_j.$$

Note that explicit calculation of $\widetilde{\varphi}_{n,0}$ is not possible in general, thus it is preferable to employ $\widetilde{\varphi}_n$.

Theorem 3.1
If A3.2 and A3.3 hold with $0 < \sum_j \alpha_j \leq \sum_j \beta_j < \infty$ *then*

$$\varliminf_{n\to\infty} \inf_{\overline{\varphi}_n \in \mathcal{T}_\ell} \sup_{X\in\mathcal{P}} u_n \, \mathrm{E}_X \parallel \overline{\varphi}_n - \varphi_X \parallel^2 \geq \sum_j \alpha_j \tag{3.9}$$

when

$$\sup_{X\in\mathcal{P}} u_n \, \mathrm{E}_X \parallel \widetilde{\varphi}_{n,0} - \varphi_X \parallel^2 \leq c \sum_j \beta_j \tag{3.10}$$

and

$$\sup_{X\in\mathcal{P}} u_n \mathrm{E}_X \parallel \widetilde{\varphi}_n - \varphi_X \parallel^2 \leq B^2 + c \sum_j \beta_j. \tag{3.11}$$

PROOF:

For each $\overline{\varphi}_n$ in $\mathcal{T}_\ell$ we have

$$\begin{aligned}\sup_{X\in\mathcal{P}} u_n \, \mathrm{E}_X \parallel \overline{\varphi}_n - \varphi_X \parallel^2 &\geq u_n \mathrm{E}_{X^*} \parallel \overline{\varphi}_n - \varphi^* \parallel^2 \\ &\geq u_n \mathrm{E}_{X^*} \parallel \varphi_n^0 - \varphi^* \parallel^2,\end{aligned}$$

therefore

$$\inf_{\overline{\varphi}_n} \sup_{X\in\mathcal{P}} u_n \, \mathrm{E}_{X^*} \parallel \overline{\varphi}_n - \varphi_X \parallel^2 \geq u_n \mathrm{E}_{X^*} \parallel \varphi_n^0 - \varphi^* \parallel^2$$

and (3.5) yields (3.9).

Now, from A3.3 it follows that

$$\mathrm{E}_X \parallel \widetilde{\varphi}_{n,0} - \varphi_X \parallel^2 = \sum_{j=0}^{\infty} \mathrm{E}_X (\overline{\varphi_{jn}} - \varphi_j)^2 \leq c \frac{\sum_{j=0}^{\infty} \beta_j}{u_n}$$

and

$$u_n \mathrm{E}_X \| \widetilde{\varphi}_n - \varphi_X \|^2 = \sum_{j=0}^{u_n} \mathrm{E}_X(\overline{\varphi}_{jn} - \varphi_j)^2 + u_n \sum_{j>u_n} \varphi_j^2$$
$$\leq c \sum_{j=0}^{u_n} \beta_j + u_n \sum_{j>u_n} b_j^2$$
$$\leq c \sum_{j=0}^{\infty} \beta_j + B^2 u_n \sum_{j>u_n} \frac{1}{j^2}$$

hence (3.10) and (3.11). ■

Example 3.2 (*Discrete distributions*)
Let $\mathcal{P}$ be a family of $\mathbb{N}$-valued processes, defined on a rich enough probability space, and such that

(a) $P_{X_t} = P_{X_0}, t \in \mathbb{Z}$

(b) $P(X_t = j) \leq b_j, j \geq 0$ where $b_j \downarrow 0$ and $1 < \sum_j b_j < \infty$

(c) $P_{(X_s, X_t)} = P_{(X_s+h, X_t+h)}$; $s,\ t,\ h \in \mathbb{Z}$

(d) $\sum_{h\in\mathbb{Z}} \sup_{j\geq 0} |P(X_h = j|X_0 = j) - P(X_h = j)| \leq \gamma < \infty$ where $P(X_h = j\ |X_0 = j) = P(X_h = j)$ if $P(X_0 = j) = 0$ and γ does not depend on X.

Here we set $\varphi = \varphi_X = \sum_{j=0}^{\infty} P(X_0 = j) \mathbb{1}_{\{j\}}$, then $\varphi \in L^2(\mu)$ where μ is the counting measure on $\mathbb{N}$ and φ is (e, h)-adapted with $e_j = h_j = \mathbb{1}_{\{j\}}$, and $\nu(j) = 0, j \geq 0$.

In this situation one may choose

$$\varphi^* = \sum_{j=0}^{\infty} \frac{b_j}{\sum_j b_j} \mathbb{1}_{\{j\}}$$

and $X^* = (X_t^*,\ t \in \mathbb{Z})$ where the X_t^*'s are i.i.d. with distribution φ^*. Thus we have

$$V_{jn}^* = \frac{\varphi_j^*(1 - \varphi_j^*)}{n} = \frac{\alpha_j}{n} < \frac{\beta_j}{n} = \frac{\varphi_j^*}{n}$$

and $\sum \beta_j < \infty$. Moreover

$$n \mathrm{Var}\, \overline{\varphi}_{jn} = \varphi_j(1 - \varphi_j) + \sum_{1\leq|h|\leq n-1} \left(1 - \frac{|h|}{n}\right) \big[P(X_0 = j, X_h = j)$$
$$- P(X_0 = j)P(X_h = j)\big]$$
$$< b_j + 2 \sum_{h=1}^{\infty} P(X_0 = j) \sup_{j'\geq 0} |P(X_h = j'|X_0 = j') - P(X_h = j')|$$
$$\leq (1 + 2\gamma) b_j := c\beta_j$$

where $c = (1 + 2\gamma) \sum_{j'} b_{j'}$.

Therefore A3.2 and A3.3 hold and, from Theorem 3.1, it follows that the minimax rate is $1/n$. The estimator

$$\widetilde{\varphi}_n = \sum_{j=1}^{n} \left(\frac{1}{n} \sum_{t=1}^{n} \mathbb{1}_{\{j\}}(X_t) \right) \mathbb{1}_{\{j\}}$$

reaches this rate. ◇

Example 3.3 (*Distribution functions*)
Consider a sequence $(X_t,\ t \in \mathbb{Z})$ of i.i.d. [0, 1]-valued random variables. Suppose that their common distribution function F possesses a continuous derivative f on [0, 1]. Let (e_j) be an orthonormal basis of $L^2(\mu)$ where μ is Lebesgue measure over [0, 1]. Then, if e_j has a continuous version, we have

$$\langle F, e_j \rangle = -\int_0^1 E_j(x) f(x) \mathrm{d}x = \mathrm{E}(-E_j(X_0))$$

where E_j is the primitive of e_j such that $E_j(1) = 0$. We see that F is (e, h)-adapted with $h_j = -E_j$ and $\nu(j) = 0,\ j \geq 0$.

If, for example, (e_j) is the trigonometric system, it is easy to prove that Var $\overline{\varphi}_{jn} = \mathcal{O}(1/nj^2)$; it follows that the minimax rate is $1/n$. Details are left to the reader. ◇

Example 3.4 (*Covariance operator*)
Let $(X_t,\ t \in \mathbb{Z})$ be a sequence of G-valued i.i.d. random variables, where G is a separable Hilbert space with scalar product $\langle \cdot, \cdot \rangle_G$ and norm $\| \cdot \|_G$.

Assume that $\mathrm{E} \parallel X_0 \parallel_G^4 < \infty$ and $\mathrm{E}X_0 = 0$. One wants to estimate the covariance operator of X_0 defined as

$$C_o(x) = \mathrm{E}(\langle X_0, x \rangle_G X_0), \quad x \in G.$$

C_o belongs to the Hilbert space $H = \mathcal{S}_G$ of Hilbert–Schmidt operators on G (see Chapter 11) and has the spectral decomposition $C_o = \sum_{j=0}^{\infty} \lambda_j v_j \otimes v_j$ with $\lambda_j \geq 0, \sum_j \lambda_j < \infty$. We suppose here that the eigenvectors v_j are known. Since $e_j = v_j \otimes v_j, j \geq 0$ is an orthonormal system in $\mathcal{S}_G$ and $\lambda_j = \mathrm{E}(\langle X_0, v_j \rangle_G^2),\ j \geq 0$, we see that C_o is (e, h)-adapted.

Now we have

$$\overline{\varphi}_{jn} = \frac{1}{n} \sum_{i=1}^{n} \langle X_i, v_j \rangle_G^2$$

and

$$\mathrm{Var}\ \overline{\varphi}_{jn} = \frac{1}{n} (\mathrm{E}(\langle X_0, v_j \rangle_G^4) - \lambda_j^2), \quad j \geq 0.$$

Noting that

$$\mathrm{E}\left(\sum_j \langle X_i, v_j\rangle_G^4\right) \le \mathrm{E}\left[\left(\sum_j \langle X_0, v_j\rangle_G^2\right)^2\right] = \mathrm{E} \parallel X_0 \parallel_G^4$$

it is easy to see that the minimax rate is n^{-1}. Again details are left to the reader. This parametric rate can be reached in the more general framework where (X_t) is an autoregressive Hilbertian process and the v_j's are unknown. We refer to Chapter 11 for a detailed study of that case. ◇

3.5 Nonparametric rates

If $\sum_j \alpha_j = \infty$, Lemma 3.2 shows that the rate of oracle is less than u_n^{-1}. In order to specify this rate we are going to define a truncation index k_n^* that indicates when 0 is a better approximation of φ_j^* than $\overline{\varphi}_{jn}$.

First we need a definition:

Definition 3.1
A strictly positive real sequence (γ_j) *is said to be of* moderate variation *(MV) if, as* $k \to \infty$

$$\overline{\gamma}_k \simeq \gamma_k$$

where $\overline{\gamma}_k = (\sum_0^k \gamma_j)/k$.
A typical example of such a sequence is

$$\gamma_j = c\, j^a (\ln(j+1))^b; \quad (c > 0,\ a > -1,\ b \text{ integer} \ge 0),$$

in particular a constant sequence is MV.

The following properties of an MV sequence are easy to check.

$$\sum_{j=0}^{\infty} \gamma_j = \infty$$

$$\gamma_k \simeq \gamma_{k+1} \tag{3.12}$$

$$\delta_k^{-1} \gamma_k \to \infty \tag{3.13}$$

for each positive sequence (δ_j) such that $k\delta_k \to 0$.

Now we make the following assumption:

Assumptions 3.4 (A3.4)
(β_j) is MV and $\sum_0^k \alpha_j \simeq \sum_0^k \beta_j$.
This allows us to define k_n^*. Set

$$d = \frac{|\varphi_{j_0}^*| u_{n_0}^{1/2}}{\beta_{j_0}^{1/2}},$$

where j_o and n_o are specified in A3.1 and A3.2 respectively. Note that it is always possible to suppose that $\beta_{j_o} \neq 0$.

Now, if $n \geq n_o$, k_n^* is the greatest $j \geq j_o$ such that

$$\varphi_j^{*^2} \geq d^2 \frac{\beta_j}{u_n},$$

that is

$$\varphi_{k_n^*}^{*^2} \geq d^2 \frac{\beta_{k_n^*}}{u_n}, \tag{3.14}$$

$$\varphi_j^{*^2} < d^2 \frac{\beta_j}{u_n}, \quad j > k_n^*. \tag{3.15}$$

k_n^* is well defined since

$$\varphi_{j_o}^{*^2} = d^2 \frac{\beta_{j_o}}{u_{n_o}} \geq d^2 \frac{\beta_{j_o}}{u_n}, \quad (n \geq n_o)$$

and, from (3.13), it follows that $\lim_{j\to\infty} \beta_j/\varphi_j^{*^2} \to \infty$; thus, for j large enough, we have (3.15).

The following statement furnishes some useful properties of k_n^*.

Lemma 3.3

$$k_n^* \to \infty, \tag{3.16}$$

$$k_n^* = o(u_n), \tag{3.17}$$

$$j_o \leq j \leq k_n^* \Rightarrow n > \nu(j); \quad n \geq n_1. \tag{3.18}$$

PROOF:
If $k_n^* \not\to \infty$ there exists $K > 0$ and a sequence S of integers such that $k_{n'}^* \leq K$, $n' \in S$. Thus $k_{n'}^* + 1 \in \{1, \ldots, K+1\}$, $n' \in S$, which implies that $\beta_{k_{n'}^*+1}\varphi_{k_{n'}^*+1}^{-2}$ is bounded, which is impossible from (3.15) and $u_{n'} \to \infty$, hence (3.16).
Now, since $\sum_{j=0}^{k_n^*} \beta_j \simeq k_n^* \beta_{k_n^*}$ and $k_n^* \to \infty$ we have

$$\beta_{k_n^*}^{-1} \simeq \frac{k_n^*}{\sum_0^{k_n^*} \beta_j} = o(k_n^*) \tag{3.19}$$

because $\sum_0^{k_n^*} \beta_j \to \infty$. Moreover, for each $\varepsilon > 0$, we have, for n large enough, $|k_n^* \varphi_{k_n^*}^*| < \varepsilon$, but $|\varphi_{k_n^*}^*| \geq d(\beta_{k_n^*}/u_n)^{1/2}$, therefore $k_n^* < \varepsilon d^{-1} u_n^{1/2} \beta_{k_n^*}^{-1/2}$ and consequently

$$k_n^* = o\left(u_n^{1/2} \beta_{k_n^*}^{-1/2}\right).$$

Now, from (3.19) it follows that $u_n^{1/2}\beta_{k_n^*}^{-1/2} = o(k_n^{*^{1/2}} u_n^{1/2})$ hence (3.17) since $k_n^* = o(k_n^{*^{1/2}} u_n^{1/2})$.

Concerning (3.18) note that $j_o \leq j \leq k_n^*$ yields $\nu(j) \leq j \leq k_n^*$, and since $k_n^* = o(u_n) = o(n)$ we have $n > k_n^*$ for n large enough, hence $n > \nu(j)$. ■

Lemma 3.4 (*Bias of oracle*)
If A3.2 and A3.4 hold then, for n large enough,

$$\sum_{j>k_n^*} \varphi_j^{*^2} < d^2 \frac{(\gamma + 1 + k_n^*)\beta_{k_n^*+1}}{u_n},$$

where $\gamma > 0$ is such that $\beta_{k_n^*} \geq \gamma\beta_{k_n^*+1}$, hence

$$\sum_{j>k_n^*} \varphi_j^{*^2} = \mathcal{O}\left(\frac{\sum_{j=0}^{k_n^*} \beta_j}{u_n}\right). \tag{3.20}$$

PROOF:
We set $\varphi^*(x) = \delta b(x)$, $x \geq j_o$. Then, monotony of φ^* yields

$$\sum_{j>k_n^*} \varphi_j^{*^2} < \varphi_{k_n^*+1}^2 + \int_{k_n^*}^{\infty} \varphi^{*^2}(x)\mathrm{d}x, \tag{3.21}$$

now we put

$$v_n = \frac{1}{d}\sqrt{\frac{u_n}{\beta_{k_n^*+1}}},$$

since $\beta_{k_n^*+1} \simeq \beta_{k_n^*}$ (cf. (3.12) and A3.4) it follows from (3.14) that $v_n \to \infty$ and, if $\beta_{k_n^*} \geq \gamma\beta_{k_n^*+1}, \varphi^*(k_n^*) \geq \gamma v_n^{-1} > 0$.

Denoting by ψ the inverse function of φ^* one obtains

$$k_n^* \leq \psi(\gamma v_n^{-1})$$

and

$$k_n^* + 1 > \psi(v_n^{-1}),$$

hence

$$\int_{k_n^*+1}^{\infty} \varphi^{*^2}(x)\mathrm{d}x < \int_{\psi(v_n^{-1})}^{\infty} \varphi^{*^2}(x)\mathrm{d}x = \int_{v_n^{-1}}^{0} u^2\psi'(u)\mathrm{d}u := J_n;$$

now an integration by parts gives

$$J_n = -v_n^{-2}\psi(v_n^{-1}) - 2\int_{v_n^{-1}}^{0} u\psi(u)\,\mathrm{d}u, \tag{3.22}$$

and the last integral may be written as

$$H_n = 2\int_{\psi(v_n^{-1})}^{\infty} x\varphi^*(x)\left[-\varphi^{*'}(x)\right]\mathrm{d}x.$$

Noting that $-\varphi^{*'}$ is positive and that $x\varphi^*(x)$ is decreasing for $x \geq j_0$ we get

$$H_n \leq 2\psi(v_n^{-1})\varphi^*\left[\psi(v_n^{-1})\right]\int_{\psi(v_n^{-1})}^{\infty}\left[-\varphi^{*'}(x)\right]\mathrm{d}x$$

thus

$$H_n \leq 2v_n^{-2}\psi(v_n^{-1}). \tag{3.23}$$

Collecting (3.21), (3.22) and (3.23) one obtains

$$\sum_{j>k_n^*}\varphi_j^{*^2} < v_n^{-2}(1+\gamma+k_n^*) = \frac{d^2(1+\gamma+k_n^*)\beta_{k_n^*+1}}{u_n}$$

which implies (3.20) since $\sum_{j=0}^{k_n^*}\beta_j \simeq k_n^*\beta_{k_n^*+1}$. ■

Lemma 3.5
If A3.2 and A3.4 hold and if $\beta_j/\varphi_j^{*^2} \nearrow$ then

$$\mathrm{E}_{X^*} \parallel \varphi_n^0 - \varphi^* \parallel^2 \simeq \frac{\sum_0^{k_n^*}\beta_j}{u_n}.$$

PROOF:
Using (3.18) in Lemma 3.3 we may write

$$\begin{aligned}\mathrm{E}_{X^*} \parallel \varphi_n^0 - \varphi^* \parallel^2 &= \sum_0^{k_n^*}\frac{\varphi_j^{*^2}\cdot V_{jn}^*}{\varphi_j^{*^2}+V_{jn}^*} + \sum_{j>k_n^*}\mathrm{E}\left(\varphi_{n,j}^0 - \varphi_j^*\right)^2 \\ &\leq \sum_0^{k_n^*} V_{jn}^* + \sum_{j>k_n^*}\varphi_j^{*^2},\end{aligned} \tag{3.24}$$

then A3.2 and Lemma 3.4 give

$$\mathrm{E}_{X^*} \parallel \varphi_n^0 - \varphi^* \parallel^2 = \mathcal{O}\left(\frac{\sum_0^{k_n^*} \beta_j}{u_n}\right).$$

Now (3.24) implies

$$\mathrm{E}_{X^*} \parallel \varphi_n^0 - \varphi^* \parallel^2 \geq \sum_{j_\circ}^{k_n^*} \frac{\varphi_j^{*^2} \cdot V_{jn}^*}{\varphi_j^{*^2} + V_{jn}^*}$$

and monotony of $(\beta_j \varphi_j^{*^{-2}})$ and (3.14) yield

$$\varphi_j^{*^2} \geq d^2 \frac{\beta_j}{u_n} \geq d^2 V_{jn}^*, \quad j_\circ \leq j \leq k_n^*,$$

therefore

$$\sum_0^{k_n^*} \frac{\varphi_j^{*^2} \cdot V_{jn}^*}{\varphi_j^{*^2} + V_{jn}^*} \geq \sum_{j_\circ}^{k_n^*} \frac{\varphi_j^{*^2} \cdot \frac{\alpha_j}{u_n}}{\varphi_j^{*^2} + \frac{\varphi_j^{*^2}}{d^2}} \geq \frac{1}{1 + d^{-2}} \left(\sum_{j_\circ}^{k_n^*} \alpha_j\right) u_n^{-1}$$

and, since $\sum_{j_\circ}^{k_n^*} \alpha_j \simeq \sum_{j_\circ}^{k_n^*} \beta_j$ it follows that $\left(\sum_0^{k_n^*} \beta_j\right) u_n^{-1} = \mathcal{O}\left(\mathrm{E}_{X^*} \parallel \varphi_n^0 - \varphi^* \parallel^2\right)$. ■

Finally we study risk of the *projection estimator* $\widetilde{\varphi}_n = \sum_{j=0}^{k_n^*} \overline{\varphi}_{jn} e_j$.

Lemma 3.6
Under A3.3 and A3.4 we have

$$\sup_{X \in \mathcal{P}} \mathrm{E}_X \parallel \widetilde{\varphi}_n - \varphi_X \parallel^2 = \mathcal{O}\left(\frac{\sum_0^{k_n^*} \beta_j}{u_n}\right).$$

PROOF:

$$\begin{aligned} \mathrm{E}_X \parallel \widetilde{\varphi}_n - \varphi \parallel^2 &= \sum_0^{k_n^*} \mathrm{Var}_X\, \overline{\varphi}_{jn} + \sum_{j > k_n^*} \varphi_j^2 \\ &\leq c \frac{\sum_0^{k_n^*} \beta_j}{u_n} + \frac{1}{\delta^2} \sum_{j > k_n^*} \varphi_j^{*^2}, \quad X \in \mathcal{P} \end{aligned}$$

and the desired result follows from Lemma 3.4. ■

Now we are in a position to state the final result.

Theorem 3.2
If A3.2, A3.3 and A3.4 hold and if $\beta_j \varphi_j^{*^{-2}} \nearrow$ *then the minimax rate for* $\mathcal{T}_\ell$ *is* $v_n = \sum_0^{k_n^*} \beta_j / u_n$ *and* $\widetilde{\varphi}_n$ *reaches this rate.*

PROOF:
Similarly to the proof of Theorem 3.1, Lemma 3.5 entails

$$\varliminf_{n\to\infty} \inf_{\overline{\varphi}_n \in \mathcal{T}_\ell} \sup_{X\in\mathcal{P}} \frac{u_n}{\sum_0^{k_n^*} \beta_j} \mathrm{E}_X \| \overline{\varphi}_n - \varphi \|^2 > 0$$

and, by Lemma 3.6, the proof is complete. ■

Discussion
It is noteworthy that existence of k_n^* is not linked to $\sum_j \beta_j = \infty$ as shown by the following simple example: if $\varphi_j^* = j^{-a}$ $(a > 1)$ and $\alpha_j = \beta_j = j^{-b}$ $(b \in \mathbb{R})$, then, for $1 < b < 2a$ we have $\sum_j \beta_j < \infty$ but $\beta_j \varphi_j^{*^{-2}} = j^{2a-b} \to \infty$, thus $k_n^* \simeq u_n^{1/(2a-b)}$, and the estimator $\sum_0^{k_n^*} \overline{\varphi}_{jn} e_j$ reaches the best rate. However the estimator $\sum_0^{u_n} \overline{\varphi}_{jn} e_j$ reaches the best rate for all $a > 1$, thus it works even if a is unknown.

Finally note that, if $b \geq 2a$ one may take $k_n^* = \infty$ and write the parametric rate under the form $\left(\sum_0^{k_n^*} \beta_j\right)/u_n$ used for the nonparametric rate. ◇

Now in order to give applications of Theorem 3.2, we indicate a useful lemma:

Lemma 3.7
If X is a strictly stationary, α-mixing process such that $\sum_\ell \ell[\alpha(\ell)]^{(q-2)/q} < \infty$ *where* $2 < q \leq \infty$ *and if* $\sup_{j\geq 0} \mathrm{E}\left(|h_j^q(X_0, \ldots, X_{\nu(j)})|^{1/q}\right) < \infty$ *then*

$$\sup_{0\leq j\leq k_n^*} |\Gamma_j - n\mathrm{Var}_X \overline{\varphi}_{jn}| \to 0, \quad n \to \infty$$

where $\Gamma_j = \gamma_{0j} + 2\sum_{\ell=1}^{\infty} \gamma_{\ell j}, \quad j \geq 0,$
with $\gamma_{\ell j} = \mathrm{Cov}\left(h_j(X_0, \ldots, X_{\nu(j)}), h_j(X_\ell, \ldots, X_{\ell+\nu(j)})\right), \ell \geq 0, j \geq 0.$

PROOF:
The proof uses *Davydov inequality* and the relation

$$(n - \nu(j))\mathrm{Var}_X \overline{\varphi}_{jn} = \gamma_{0j} + 2 \sum_{\ell=1}^{n-\nu(j)-1} \left[1 - \frac{\ell}{n - \nu(j)}\right] \gamma_{\ell j}, \quad n > \nu(j);$$

details are left to the reader. ■

Example 3.5 (*Density*)
Suppose that $X = (X_t,\ t \in \mathbb{Z})$ takes its values in a measure space $(E, \mathcal{B}, \mu)$ and that the X_t's have a common density f, with respect to μ, with $f \in L^2(\mu)$.

For convenience we assume that μ is a probability and that $(e_j, j \geq 0)$ is an orthonormal basis of $L^2(\mu)$ such that $e_o = 1$ and $M = \sup_{j\geq 0} \parallel e_j \parallel_\infty < \infty$. Then $f = 1 + \sum_{j=1}^{\infty} \varphi_j e_j$ with $\varphi_j = \int_E f e_j \, d\mu = \mathrm{E}(e_j(X_0)), j \geq 1$.

In order to construct $\varphi^* = f^*$ we set $f^* = 1 + \delta \sum_{j=1}^{\infty} b_j e_j$ and take $\delta \in]0, \min\left(1/2, (2M \sum_{j=1}^{\infty} b_j)^{-1}\right)[$, consequently $1/2 \leq f^*(x) \leq 1 + \delta M \sum_{j=1}^{\infty} b_j$, $x \in E$.

If the X_t's are i.i.d. with density f^* it follows that $V_{jn}^* = \left[\int e_j^2 f^* \mathrm{d}\mu - \varphi_j^{*^2}\right]/n$. Putting $\alpha_j = \int e_j^2 f^* \mathrm{d}\mu - \varphi_j^{*^2}$ and $\beta_j = \int e_j^2 f^* \mathrm{d}\mu$ and noting that $\lim_{k\to\infty} \left(\sum_1^k \varphi_j^{*^2}\right) /k = 0$ we see that A3.4 is satisfied with $1/2 \leq \beta_j \leq M^2, j \geq 1$. Then $b(k_n^*) \simeq 1/\sqrt{n}$ and the rate of oracle is k_n^*/n.

Now, if $\widetilde{f}_n = \sum_{j=0}^{k_n} \overline{\varphi}_{jn} e_j$ and if each X in $\mathcal{P}$ is strictly stationary with

$$\sup_{X \in \mathcal{P}} \sup_{j \geq 1} \sum_{\ell=1}^{\infty} |\mathrm{Cov}(e_j(X_0),\ e_j(X_\ell))| < \infty$$

it follows that

$$\sup_{X \in \mathcal{P}} \mathrm{E}_X \parallel \widetilde{f}_n - f \parallel^2 = \mathcal{O}\left(\frac{k_n^*}{n}\right).$$

The above condition is satisfied if X is α_X-mixing with $\sup_{X\in\mathcal{P}} \sum_{\ell=1}^{\infty} \alpha_X(\ell) < \infty$. ◇

Example 3.6 (*Derivatives of density*)
In Example 3.3 if F is of class C_2, e_j of class C_1 and if $f(0)e_j(0) = f(1)e_j(1) = 0$ then $\langle f', e_j \rangle = \mathrm{E}[-e_j'(X_0)]$ thus f' is $(e',\ e)$-adapted.

If (e_j) is the trigonometrical system and the X_t's are i.i.d. with density f^* defined in Example 3.5, then $V_{jn}^* \simeq j^2/n$ and Theorem 3.2 applies under mild conditions. Clearly similar results can be obtained for derivatives of order greater than 1.

Example 3.1 (*Spectral density, continued*)
Here

$$\overline{\varphi}_{0n} = \frac{1}{n\sqrt{2\pi}} \sum_{t=1}^{n} X_t^2,$$

$$\overline{\varphi}_{jn} = \frac{1}{(n-j)\sqrt{\pi}} \sum_{t=1}^{n-j} X_t X_{t+j}, \quad 1 \leq j \leq n-1.$$

Now if $b_j = ar^j, j \geq 0$ $(a \geq 1,\ 0 < r < 1)$ then $\mathcal{P}$ contains linear processes of the form

$$X_t = \sum_{j=0}^{\infty} a_j \varepsilon_{t-j}, \quad t \in \mathbb{Z}$$

where (ε_t) is a white noise and $|a_j| \leq c\,\rho^j$, $j \geq 0$ with $0 < \rho \leq r$ and c small enough. Actually we have

$$|\gamma_j| \leq \frac{c\sigma^2}{1-\rho^2}\rho^j, \quad j \geq 0 \text{ where } \sigma^2 = \mathrm{E}\varepsilon_n^2.$$

In particular $\mathcal{P}$ contains the ARMA processes. Now we set

$$f^*(\lambda) = \frac{1}{2\pi}\sum_{j=0}^{\infty} r^j \cos\lambda j, \quad \lambda \in [-\pi, +\pi]$$

which is associated with an AR(1) process:

$$X_t^* = rX_{t-1}^* + \varepsilon_t, \quad t \in \mathbb{Z}$$

where $\operatorname{Var} X_t^* = 1$. In this case we have (see Brockwell and Davis 1991, p. 225)

$$nV_{jn}^* \simeq \frac{(1-r^{2j})(1+r^2)}{1-r^2} - 2jr^{2j} \simeq \frac{1+r^2}{1-r^2}.$$

If (X_t^*) satisfies the conditions in Lemma 3.7, we have

$$\sup_{0\leq j\leq k_n^*}\left|\frac{1+r^2}{1-r^2} - nV_{jn}^*\right| \underset{n\to\infty}{\longrightarrow} 0;$$

then one may choose

$$\alpha_j = \frac{1}{2}\frac{1+r^2}{1-r^2}, \quad \beta_j = 2\frac{1+r^2}{1-r^2}, \quad j \geq 1$$

and Theorem 3.2 applies with $k_n^* \simeq \ln n$ and optimal rate $(\ln n)/n$. ◇

3.6 Rate in uniform norm

If φ has a preferred functional norm it is interesting to use the uniform norm in order to have an idea about the form of its graph. In the follows we will say that φ is 'functional' if H is a space of functions or of equivalence classes of functions. The following simple exponential type inequality is useful.

Lemma 3.8
Let $Y_1, \ldots, Y_n$ be a finite sequence of bounded zero-mean real random variables and let $(\alpha(k),\ 0 \leq k \leq n-1)$ be the associated α-mixing coefficients. Then

$$P(|\overline{Y}_n| > \varepsilon) \leq 4\exp\left(-\frac{\varepsilon^2}{8B^2}q\right) + 22\left(1+\frac{4B}{\varepsilon}\right)^{1/2} q\alpha\left(\left[\frac{n}{2q}\right]\right), \tag{3.25}$$

$\varepsilon > 0$, $q \in \{1, 2, \ldots, [n/2]\}$; $(B = \sup_{1\leq t\leq n} \| Y_t \|_\infty)$.

PROOF:
See Bosq (1998). ■

We now suppose that $b_j = ar^j$ ($a \geq 1$, $0 < r \leq r_o < 1$ with r_o known), that $H = \sup_{j\geq 0} \| h_j \|_\infty, M = \sup_{j\geq 0} \| e_j \|_\infty < \infty$ and that X is *GSM* with $\alpha(k) \leq ce^{-dk}(c \geq 1/4,\ d > 0)$. We study the behaviour of the estimator $\widetilde{\varphi}_n = \sum_{j=0}^{[\delta \ln n]} \overline{\varphi}_{jn} e_j$, $\delta > 1/(2\ln(1/r_o))$.

Theorem 3.3
If φ is functional, X is strictly stationary and GSM, φ is (e, h)-adapted with $(e_j), (h_j)$ uniformly bounded and $(b_j) \to 0$ at an exponential rate, then

$$\| \widetilde{\varphi}_n - \varphi \|_\infty = \mathcal{O}\left(\frac{(\ln n)^2}{\sqrt{n}}\right) \quad \textit{almost surely.} \tag{3.26}$$

PROOF:
We have

$$\| \widetilde{\varphi}_n - \varphi \|_\infty \leq M \sum_{j=0}^{[\delta \ln n]} |\overline{\varphi}_{jn} - \varphi_j| + \sum_{j>\delta[\ln n]} |\varphi_j|$$

and

$$\sum_{j>[\delta \ln n]} |\varphi_j| \leq a\frac{r^{[\delta \ln n]+1}}{1-r} \leq \frac{ar^{\delta \ln n}}{1-r} = \frac{a}{1-r}\frac{1}{n^{\delta \ln \frac{1}{r}}}.$$

Now we set $\varepsilon_n = \gamma(\ln n)^2/\sqrt{n}, (\gamma > 0)$, then, since $\delta > 1/(2\ln(1/r))$, we have, for n large enough, $\varepsilon_n/2 > \sum_{j>[\delta \ln n]} |\varphi_j|$, thus

$$\begin{aligned} P(\| \widetilde{\varphi}_n - \varphi \|_\infty > \varepsilon_n) &\leq \sum_{j=0}^{[\delta \ln n]} P\left(|\overline{\varphi}_{jn} - \varphi_j| > \frac{\varepsilon_n}{2M[\delta \ln n]}\right) \\ &\leq \sum_{j=0}^{[\delta \ln n]} P\left(|\overline{\varphi}_{jn} - \varphi_j| > \frac{\gamma}{2M\delta}\frac{\ln n}{\sqrt{n}}\right). \end{aligned}$$

Using inequality (3.25) with $Y_i = h_j(X_0, \ldots, X_{\nu(j)}) - \mathrm{E}h_j(X_0, \ldots, X_{\nu(j)})$ one obtains

$$\begin{aligned} P\left(|\overline{\varphi}_{jn} - \varphi_j| > \frac{\gamma}{2M\delta}\frac{\ln n}{\sqrt{n}}\right) &\leq 4\exp\left(-\frac{\gamma^2}{32M^2\delta^2H^2}\frac{(\ln n)^2}{n}q\right) \\ &\quad + 22\left(1 + \frac{4H\sqrt{n}}{\gamma(\ln n)^2}\right)^{1/2} q\alpha_Y\left(\left[\frac{n-\nu(j)}{2q}\right]\right) \end{aligned}$$

with $1 \leq q \leq (n - \nu(j))/2$ and $\alpha_Y(k) \leq c\exp\Big(-d\max(0, k - \nu(j))\Big)$.

Now, choosing $q = \left[c' \frac{n}{\ln n}\right]$, $(c' > 0)$, yields

$$P(\| \widetilde{\varphi}_n - \varphi \|_\infty > \varepsilon_n) = \mathcal{O}\left(\frac{\ln n}{n^{c'\gamma^2/32M^2\delta^2H^2}}\right) + \mathcal{O}\left(\frac{1}{(\ln n)^2 n^{\frac{d}{2c'}-\frac{5}{4}}}\right)$$

hence, choosing $0 < c' < 2d/9, \gamma^2 > 32M^2\delta^2H^2/c'$ and using the Borel–Cantelli lemma one obtains (3.26). ■

3.7 Adaptive projection

The best truncation index, say $k_{n,\varphi}$, for estimating φ by projection, depends on (φ_j) and (β_j). If these sequences are unknown, one tries to approximate $k_{n,\varphi}$.

The principle of such an approximation is to determine an index $\hat{k}_n$, characterized by smallness of the empirical Fourier coefficients for $j > \hat{k}_n$. More precisely one sets

$$\hat{k}_n = \max\{j : 0 \leq j \leq k_n, \ |\overline{\varphi}_{jn}| \geq \gamma_n\}$$

where (k_n, γ_n) is chosen by the statistician, and with the convention $\hat{k}_n = k_n$ if $\{\cdot\} = \varnothing$.

In the following we always suppose that $k_n < n, (k_n) \to \infty$ and $(k_n/n) \to 0$, thus

$$\overline{\varphi}_{jn} = \frac{1}{n - \nu(j)} \sum_{j=1}^{n-\nu(j)} h_j(X_i, \ldots, X_{i+\nu(j)})$$

is well defined for $j \leq k_n$, since $\nu(j) \leq j \leq k_n < n$.

Concerning the threshold, we take

$$\gamma_n = \frac{\ln n \cdot \log_2 n}{\sqrt{n}}, \quad n \geq 3. \tag{3.27}$$

This choice appears as suitable in various situations, in particular if $|\varphi_j| \downarrow$ and the observed variables are i.i.d. since, in this case, we have $|\varphi_{k_{n,\varphi}}| \simeq 1/\sqrt{n}$.

The logarithm term in (3.27) allows as to compensate variability of $\overline{\varphi}_{jn}$, $1 \leq j \leq k_n$, and takes into account correlation between the observed variables.

The associated estimator is defined as

$$\hat{\varphi}_n = \sum_{j=0}^{\hat{k}_n} \overline{\varphi}_{jn} e_j.$$

Now, in order to study asymptotic properties of $\hat{\varphi}_n$ we make the following assumption.

Assumptions 3.5(A3.5)
$X = (X_t,\, t \in \mathbb{Z})$ is a strictly stationary, geometrically strongly mixing process and $(h_j,\, j \geq 0)$ is uniformly bounded.

In the following we put $h = \sup_{j\geq 0} \| h_j \|_\infty$ and write $\alpha_k \leq c\exp(-dk)$, $k \geq 0$, $(c \geq 1/4,\ d > 0)$, where α_k is the sequence of strongly mixing coefficients of X.

3.7.1 Behaviour of truncation index

Set $B_n = \bigcup_{j=0}^{k_n}\{|\overline{\varphi}_{jn}| \geq \gamma_n\}$, the first statement asserts asymptotic negligibility of B_n^c.

Lemma 3.9
If A3.5 holds, and if j_1 is an index such that $\varphi_{j_1} \neq 0$, then conditions $k_n \geq j_1$ and $\gamma_n < |\varphi_{j_1}/2|$ yield

$$P(B_n^c) \leq a\exp(-b\sqrt{n}) \quad (a > 0,\, b > 0). \tag{3.28}$$

PROOF:
If $k_n \geq j_1$ and $\gamma_n < |\varphi_{j_1}/2|$, one has

$$B_n^c \Rightarrow |\overline{\varphi}_{j_1 n}| < \gamma_n \Rightarrow |\overline{\varphi}_{j_1 n} - \varphi_{j_1}| \geq \frac{|\varphi_{j_1}|}{2},$$

then inequality (3.25) gives

$$P(B_n^c) \leq 4\exp\left(-\frac{\varphi_{j_1}^2}{32h^2}q\right) + 22\left(1 + \frac{8h}{|\varphi_{j_1}|}\right)^{1/2} qc\exp\left(-d\left[\frac{n}{2q}\right]\right),$$

choosing $q \simeq \sqrt{n}$ one obtains (3.28). ■

Now let us set $\phi_o(K) = \{\varphi : \varphi \in \phi,\ \varphi_K \neq 0; \varphi_j = 0,\, j > K\}$ and $\phi_o = \bigcup_{K=0}^{\infty}\phi_o(K)$. The next theorem shows that, if $\varphi \in \phi_o$, $\hat{k}_n$ is a strongly consistent estimator of $K = K_\varphi$.

Theorem 3.4
If A3.5 holds, then

$$P(\hat{k}_n \neq K_\varphi) = \mathcal{O}(n^{-\delta\log_2 n}), \quad (\delta > 0), \tag{3.29}$$

hence, almost surely, $\hat{k}_n = K_\varphi$ for n large enough.

PROOF:
Applying Lemma 3.9 with $j_1 = K_\varphi = K$ one obtains $P(\{\hat{k}_n \neq K_\varphi\} \cap B_n^c) \leq P(B_n^c) \leq a\exp(-b\sqrt{n})$. Now, for n large enough,

$$C_n := \{\{\hat{k}_n > K\} \cap B_n\} \cup \{\{\hat{k}_n < K\} \cap B_n\}$$
$$\Rightarrow \bigcup_{j=K+1}^{k_n} \left\{|\overline{\varphi_{jn}}| > \frac{\gamma_n}{2}\right\} \cup \left\{|\overline{\varphi_{Kn}} - \varphi_K| > \frac{|\varphi_K|}{2}\right\},$$

then (3.25) gives

$$P(C_n) \leq 4\sum_{j=K+1}^{k_n} \exp\left(-\frac{\gamma_n^2 q}{32h^2}\right) + 22\sum_{j=K+1}^{k_n}\left(1+\frac{8h}{\gamma_n}\right)^{1/2} qc\exp\left(-d\left[\frac{n}{2q}\right]\right)$$
$$+ 4\exp\left(-\frac{\varphi_K^2}{32h^2}q'\right) + 22\left(1+\frac{8h}{|\varphi_K|}\right)^{1/2}$$
$$q'c\exp\left(-d\left[\frac{n}{2q'}\right]\right),$$

and the choice $q \sim n/(\ln n \cdot \log_2 n), q' \simeq \sqrt{n}$ and the fact that $k_n < n$ entail (3.29).

We now consider the general case where

$$\varphi \in \phi_1 = \{\varphi,\ \varphi \in \phi,\ \varphi_j \neq 0 \text{ for an infinity of } j\text{'s}\}.$$

First, similarly to the proof of Theorem 3.4, one may note that, if $\varphi_{j_1} \neq 0$, $P(\hat{k}_n > j_o)$ tends to zero at an exponential rate, hence

$$\hat{k}_n \to \infty \quad \text{a.s.}$$

In order to specify the asymptotic behaviour of $\hat{k}_n$ we set

$$q(\eta) = \min\{q \in \mathbb{N},\ |\varphi_j| \leq \eta,\ j > q\}, \quad \eta > 0$$

and

$$q_n(\varepsilon) = q((1+\varepsilon)\gamma_n), \quad \varepsilon > 0$$
$$q_n'(\varepsilon') = q((1-\varepsilon')\gamma_n), \quad 0 < \varepsilon' < 1.$$

Then we have the following asymptotic bounds.

Theorem 3.5
Under A3.5, if $\varphi \in \phi_1$ and $k_n > q_n'(\varepsilon')$, we have

$$P\left(\hat{k}_n \notin [q_n(\varepsilon), q_n'(\varepsilon')]\right) = \mathcal{O}\left(n^{-\delta' \log_2 n}\right), \quad (\delta' > 0), \tag{3.30}$$

therefore, a.s. for n large enough

$$\hat{k}_n \in \left[q_n(\varepsilon), q'_n(\varepsilon')\right], \tag{3.31}$$

in particular, if $q_n(\varepsilon) \simeq q'_n(\varepsilon')$, *then*

$$\hat{k}_n \simeq q(\gamma_n) \quad a.s. \tag{3.32}$$

PROOF:
First, $k_n > q'_n(\varepsilon')$ yields

$$\{\hat{k}_n > q'_n(\varepsilon')\} \cap B_n \Rightarrow \bigcup_{j=q'_n(\varepsilon')+1}^{k_n} \{\overline{\varphi}_{jn} \geq \gamma_n\},$$

but

$$j > q'_n(\varepsilon') \Rightarrow |\varphi_j| < (1-\varepsilon')\gamma_n,$$

thus

$$|\varphi_{jn}| \geq \gamma_n \Rightarrow |\overline{\varphi}_{jn} - \varphi_j| \geq |\overline{\varphi}_{jn}| - |\varphi_j| > \varepsilon'\gamma_n,$$

using again (3.25), one arrives at

$$P\big(\{\hat{k}_n > q'_n(\varepsilon')\} \cap B_n\big) \leq \big(k_n - q'_n(\varepsilon')\big)\Delta_n$$

where

$$\Delta_n = 4\exp\left(-\frac{\varepsilon'^2\gamma_n^2}{8h^2}q\right) + 22\left(1+\frac{4h}{\varepsilon'\gamma_n}\right)^{1/2} qc\exp\left(-d\left[\frac{n}{2q}\right]\right);$$

note that this bound remains valid if $k_n = q'_n(\varepsilon')$.

Now, the choice $q \simeq n/(\ln n \cdot \log_2 n)$ and (3.28) yield

$$P\big(\hat{k}_n > q'_n(\varepsilon')\big) = \mathcal{O}\big(n^{-\delta_1 \log_2 n}\big) \quad (\delta_1 > 0). \tag{3.33}$$

On the other hand, if $q_n(\varepsilon) > 0$, we have $|\varphi_{q_n(\varepsilon)}| > (1+\varepsilon)\gamma_n$ thus

$$\begin{aligned}\{\hat{k}_n < q_n(\varepsilon)\} &\Rightarrow \left|\overline{\varphi}_{q_n(\varepsilon),n}\right| \leq \gamma_n \\ &\Rightarrow \left|\overline{\varphi}_{q_n(\varepsilon),n} - \varphi_{q_n(\varepsilon)}\right| > \varepsilon\gamma_n\end{aligned}$$

and (3.25) gives

$$P\big(\hat{k}_n < q_n(\varepsilon)\big) = \mathcal{O}\big(n^{-\delta_2 \log_2 n}\big). \tag{3.34}$$

If $q_n(\varepsilon) = 0$ the same bound holds, and (3.30) follows from (3.33) and (3.34).
Finally (3.31) and (3.32) are clear. ■

Example 3.7
If $|\varphi_j| \simeq j^{-b}, j \geq 0$ $(b > 1/2)$, one has $k_{n,\varphi} \simeq n^{1/(2b)}$ and for $k_n \simeq n/(\ln n)$ one obtains

$$\hat{k}_n \simeq \left(\frac{n}{\ln n}\right)^{1/(2b)} \quad a.s.$$

◇

Example 3.8
If $|\varphi_j| = \alpha\rho^j (\alpha > 0,\ 0 < \rho < 1)$ and $k_n > \ln n/\ln(1/\rho)$ one has $k_{n,\varphi} \sim q_n(\varepsilon) \sim q'_n(\varepsilon') \sim \ln\ n/(2\ln(1/\rho))$, hence

$$\frac{\hat{k}_n}{\ln n} \rightarrow \frac{1}{2\ln(1/\rho)} \quad \text{a.s.}$$

Note that $\hat{\rho}_n = n^{-1/2(\hat{k}_n+1)}$ is then a strongly consistent estimator of ρ:

$$\hat{\rho}_n \rightarrow \rho \quad \text{a.s.}$$

◇

3.7.2 Superoptimal rate

Theorem 3.4 shows that, in the special case where $\varphi \in \phi_o$, $\hat{\varphi}_n$ has the same asymptotic behaviour as the pseudo-estimator

$$\varphi_{n,K_\varphi} = \sum_{j=0}^{K_\varphi} \overline{\varphi}_{jn} e_j, \quad n > K_\varphi.$$

Before specifying this fact, let us set $S_\varphi = \sum_{j=0}^{K_\varphi}\big(\sum_{\ell\in\mathbb{Z}} \gamma_{j\ell}\big)$, where

$$\gamma_{j\ell} = \operatorname{Cov}\big(h_j(X_0,\ldots,X_{\nu(j)}),\ h_j\big(X_\ell,\ldots,X_{\ell+\nu(j)}\big)\big).$$

Note that S_φ is well defined if A3.5 holds since, by using the Schwarz and *Billingsley* inequalities, it is easy to prove that $\sum_{\ell\in\mathbb{Z}} |\gamma_{j\ell}| = \mathcal{O}\big(\sum_{k\geq 0}\alpha_k\big)$.
Now we have the following:

Theorem 3.6
If $\varphi \in \phi_o$ and A3.5 holds, then, as $n \to \infty$,

$$n \cdot \mathrm{E} \parallel \hat{\varphi}_n - \varphi \parallel^2 \to S_\varphi \tag{3.35}$$

and

$$\sqrt{n}\,(\hat{\varphi}_n - \varphi) \xrightarrow{\mathcal{D}} \sum_{j=0}^{K_\varphi} N_j e_j \tag{3.36}$$

where $N = (N_o, \ldots, N_{K_\varphi})$ is a $\mathbb{R}^{K_\varphi+1}$-dimensional zero-mean Gaussian vector and $\mathcal{D}$ denotes convergence in distribution in the space H.

If, in addition, φ is functional and $M = \sup_{j\geq 0} \parallel e_j \parallel_\infty < \infty$, then

$$\overline{\lim}_{n\to\infty} \left(\frac{n}{\log_2 n}\right)^{1/2} \parallel \hat{\varphi}_n - \varphi \parallel_\infty < \infty \quad a.s. \tag{3.37}$$

PROOF:
The quadratic error of $\hat{\varphi}_n$ has the form

$$\mathrm{E} \parallel \hat{\varphi}_n - \varphi \parallel^2 = \mathrm{E}\left(\sum_{j=0}^{\hat{k}_n} (\overline{\varphi}_{jn} - \varphi_j)^2\right) + \mathrm{E}\left(\sum_{j>\hat{k}_n} \varphi_j^2\right).$$

From $\mathbb{1}_{\{\hat{k}_n = K_\varphi\}} + \mathbb{1}_{\{\hat{k}_n \neq K_\varphi\}} = 1$ it follows that

$$\mathrm{E} \parallel \hat{\varphi}_n - \varphi \parallel^2 = \sum_{j=0}^{K_\varphi} \mathrm{Var}\, \overline{\varphi}_{jn} - \mathrm{E}\left(\left(\sum_{j=0}^{K_\varphi} (\overline{\varphi}_{jn} - \varphi_j)^2 - \parallel \hat{\varphi}_n - \varphi \parallel^2\right) \mathbb{1}_{\{\hat{k}_n \neq K_\varphi\}}\right)$$
$$:= V_n - V_n'.$$

Using A3.5 one easily obtains $n \cdot V_n \to S_\varphi$, and Theorem 3.4 yields

$$n \cdot V_n' \leq n\left[4(1 + K_\varphi)h^2 + 2(k_n + 1)h^2 + 2 \parallel \varphi \parallel^2\right] P(\hat{k}_n \neq K_\varphi)$$
$$= \mathcal{O}\left(n^{-c \log_2 n + 2}\right)$$

hence (3.35).

Concerning (3.36), the CLT for mixing variables and the Cramér–Wold device (see Billingsley 1968) entail

$$U_n := \sqrt{n} \begin{pmatrix} \overline{\varphi}_{0n} - \varphi_o \\ \vdots \\ \overline{\varphi}_{K_\varphi,n} - \varphi_{K_\varphi} \end{pmatrix} \xrightarrow{\mathcal{D}} N = \begin{pmatrix} N_o \\ \vdots \\ N_{K_\varphi} \end{pmatrix}.$$

Now consider the linear operator defined as $s(x_0, \ldots, x_{K_\varphi}) = \sum_{j=0}^{K_\varphi} x_j e_j$, $(x_0, \ldots, x_{K_\varphi})' \in \mathbb{R}^{1+K_\varphi}$; since it is continuous, we obtain

$$\sqrt{n}\,(\varphi_{n,K_\varphi} - \varphi) = s(U_n) \xrightarrow{\mathcal{D}} s(N) = \sum_{j=0}^{K_\varphi} N_j e_j.$$

On the other hand, since

$$\{\hat{k}_n = K_\varphi\} \Rightarrow \{\hat{\varphi}_n = \varphi_{n,K_\varphi}\},$$

Theorem 3.4 yields $P\Big(\sqrt{n}\,(\hat{\varphi}_n - \varphi_{n,K_\varphi}) = 0\Big) \to 1$, and (3.36) comes from $\sqrt{n}\,(\hat{\varphi}_n - \varphi) = \sqrt{n}\,(\hat{\varphi}_n - \varphi_{n,K_\varphi}) + \sqrt{n}\,(\varphi_{n,K_\varphi} - \varphi)$.

Finally the LIL for stationary GSM processes (see Rio 2000) implies that, with probability one,

$$\overline{\lim_{n\to\infty}} \left(\frac{n}{\log_2 n}\right)^{1/2} |\overline{\varphi}_{jn} - \varphi_j| < c_j < \infty; \quad j = 0, \ldots, K_\varphi,$$

thus

$$\overline{\lim_{n\to\infty}} \left(\frac{n}{\log_2 n}\right)^{1/2} \| \varphi_{n,K_\varphi} - \varphi \|_\infty \leq M \sum_{j=0}^{K_\varphi} c_j$$

and, since $\hat{\varphi}_n = \varphi_{n,K_\varphi}$, a.s. for n large enough, (3.37) holds. ■

Example 3.1 (*Spectral density, continued*)
Suppose that X is a MA(K) process:

$$X_t = \sum_{j=0}^{K} a_j \varepsilon_{t-j}, \quad t \in \mathbb{Z}$$

where $a_o = 1$, $a_K \neq 0$, $\sum_{j=0}^{K} a_j z^j \neq 0$ if $|z| \leq 1$, and (ε_t) is a strong white noise. Then

$$\varphi(\lambda) = \frac{1}{2\pi} \sum_{|j|\leq K} \gamma_j \cos \lambda j, \quad \lambda \in [-\pi, +\pi],$$

thus $\varphi \in \phi_o(K)$. Since A3.5 holds it follows that $\hat{k}_n$ is a strongly consistent estimator of K (Theorem 3.4) and $\hat{\varphi}_n$ satisfies (3.35), (3.36) and (3.37). ◇

Example 3.5 (*Density, continued*)
Let $(g_j, j \geq 0)$ be a sequence of square integrable densities with respect to μ and $(e_j, j \geq 0)$ an orthonormal system of $L^2(\mu)$ such that $e_j = \sum_{\ell=0}^{j} b_\ell g_\ell$, $j \geq 0$; then, if the density f of X_t is a finite mixture: $f = \sum_{j=0}^{K} p_j g_j$ ($p_j \geq 0, \sum_0^K p_j = 1$) then

$f = \sum_{j=0}^{K} \varphi_j e_j$. This implies that $f \in \phi_o(K)$ and, if A3.5 holds, conclusions in Theorems 3.4 and 3.6 hold. ◇

Example 3.9 (*Regression*)
If $X_t = (Y_t, Z_t), t \in \mathbb{Z}$ is an $E_o = E_1 \times \mathbb{R}$ valued strictly stationary process, with $\mathrm{E}Z_o^2 < \infty$ and $P_{Y_o} = \mu$ assumed to be known, one may set:

$$r(y) = \mathrm{E}(Z_o | Y_o = y), \quad y \in E_1$$

and, if (e_j) is an arbitrary orthonormal basis of $L^2(\mu)$, one has

$$\begin{aligned}\langle r, e_j \rangle &= \int r e_j \,\mathrm{d}\mu = \mathrm{E}\big(\mathrm{E}(Z_o | Y_o) e_j(Y_o)\big)\\ &= \mathrm{E}\Big(Z_o e_j(Y_o)\Big), \quad j \geq 0,\end{aligned}$$

hence r is (e_j)-adapted.

In the particular case where $E_1 = [0, 1]$ and r is a polynomial of degree K, one may use the Legendre polynomials (i.e. the orthonormal basis of $L^2\big([0, 1], \mathcal{B}_{[0,1]}, \lambda\big)$ generated by the monomials) for estimating r at a parametric rate, and K, by using the above results. ◇

3.7.3 The general case

We now consider the case where $\varphi \in \phi_1$. The first statement gives consistency.

Theorem 3.7

(1) Under A3.5 we have

$$\mathrm{E} \parallel \hat{\varphi}_n - \varphi \parallel^2 \to 0, \quad \varphi \in \phi. \tag{3.38}$$

(2) If, in addition φ is functional and $M = \sup_{j \geq 0} \parallel e_j \parallel_\infty < \infty$, then, conditions $\sum |\varphi_j| < \infty$ and $k_n \simeq n^\delta (0 < \delta < 1)$ yield

$$\parallel \hat{\varphi}_n - \varphi \parallel \to 0 \quad a.s. \tag{3.39}$$

PROOF:
(1) The *Billingsley inequality* implies

$$\mathrm{Var}\, \overline{\varphi}_{jn} \leq \frac{4h^2}{n - \nu(j)} \left(1 + 2\sum_{\ell=1}^{\infty} \alpha(\ell)\right), \quad 0 \leq j \leq k_n, \tag{3.40}$$

thus

$$\sum_{j=0}^{k_n} \operatorname{Var} \overline{\varphi}_{jn} = \mathcal{O}\left(\frac{k_n}{n}\right) \to 0, \tag{3.41}$$

on the other hand

$$\mathrm{E}\left(\sum_{j>\hat{k}_n} \varphi_j^2\right) = \sum_{j>k_n} \varphi_j^2 + \sum_{\hat{k}_n<j\leq k_n} \varphi_j^2 \mathbb{1}_{\{\hat{k}_n<k_n\}}.$$

Now, given $\eta > 0$, there exists k_η such that $\sum_{j>k_\eta} \varphi_j^2 < \eta/2$, hence, for n large enough,

$$\mathrm{E}\left(\sum_{j=\hat{k}_n}^{k_n} \varphi_j^2 \mathbb{1}_{\{\hat{k}_n>k_\eta\}}\right) + \mathrm{E}\left(\sum_{j=\hat{k}_n}^{k_n} \varphi_j^2 \mathbb{1}_{\{\hat{k}_n\leq k_\eta\}}\right)$$
$$\leq \sum_{j>k_\eta} \varphi_j^2 + \| \varphi \|^2 P(\hat{k}_n \leq k_\eta) < \eta.$$

This clearly implies,

$$\mathrm{E}\left(\sum_{j>\hat{k}_n} \varphi_j^2\right) \to 0, \tag{3.42}$$

and (3.38) follows from (3.35), (3.41) and (3.42).

(2) Since $\hat{k}_n \leq k_n$ we have the bound

$$\| \hat{\varphi}_n - \varphi \|_\infty \leq M \sum_{j=0}^{k_n} |\overline{\varphi}_{jn} - \varphi_j| + M \sum_{j>\hat{k}_n} |\varphi_j|.$$

First $\hat{k}_n \to \infty$ a.s. gives $M \sum_{j>\hat{k}_n} |\varphi_j| \to 0$ a.s. Now (3.25) yields

$$P\left(\sum_{j=0}^{k_n} |\overline{\varphi}_{jn} - \varphi_j| > \eta\right) \leq \sum_{j=0}^{k_n} P\left(|\overline{\varphi}_{jn} - \varphi_j| > \frac{\eta}{k_n+1}\right)$$
$$\leq (k_n+1)\left[4\exp\left(-\frac{\eta^2 q}{32h^2(k_n+1)^2}\right)\right.$$
$$\left.+22\left(1+\frac{8h(k_n+1)}{\eta}\right)^{1/2} qc\exp\left(-d\left[\frac{n}{2q}\right]\right)\right],$$

$\eta > 0$. The choice $q \simeq n^\gamma$ $(2\delta < \gamma < 1)$ and Borel–Cantelli lemma give (3.39). ■

We now specify rates of convergence.

Theorem 3.8
If A3.5 holds and $k_n \geq q'_n(\varepsilon')$, then

$$\mathrm{E} \parallel \hat{\varphi}_n - \varphi \parallel^2 = \mathcal{O}\left(\frac{q'_n(\varepsilon')}{n} + \sum_{j>q_n(\varepsilon)} \varphi_j^2 \right). \tag{3.43}$$

PROOF:
Set $A_n = \{q_n(\varepsilon) \leq \hat{k}_n \leq q'_n(\varepsilon')\}$, then

$$\mathrm{E}\big(\parallel \hat{\varphi}_n - \varphi \parallel^2 \mathbb{1}_{A_n}\big) \leq \sum_{j=0}^{q'_n(\varepsilon')} \mathrm{Var}\, \overline{\varphi}_{jn} + \sum_{j>q_n(\varepsilon)} \varphi_j^2,$$

from (3.40), and n large enough, it follows that

$$\mathrm{E}\big(\parallel \hat{\varphi}_n - \varphi \parallel^2 \mathbb{1}_{A_n}\big) \leq \frac{1 + q'_n(\varepsilon')}{n} 8h^2 \left(1 + 2\sum_{\ell=1}^{\infty} \alpha(\ell) \right) + \sum_{j>q_n(\varepsilon)} \varphi_j^2.$$

Moreover,

$$\mathrm{E}\big(\parallel \hat{\varphi}_n - \varphi \parallel^2 \mathbb{1}_{A_n^c}\big) \leq 2\big((1 + k_n)h^2 + \parallel \varphi \parallel^2\big)P(A_n^c),$$

and Theorem 3.5 gives

$$\mathrm{E}\left(\parallel \hat{\varphi}_n - \varphi \parallel^2 \mathbb{1}_{A_n^C}\right) = \mathcal{O}\left(k_n n^{-\delta' \log_2 n}\right) = o\left(\frac{q'_n(\varepsilon')}{n}\right),$$

hence (3.43). ■

Example 3.7 (*continued*)
Here the best rate at φ is $n^{-(2b-1)/2b}$ when

$$\mathrm{E} \parallel \hat{\varphi}_n - \varphi \parallel^2 = \mathcal{O}\left(n^{-\frac{2b-1}{2b}} (\ln n \cdot \log_2 n)^{1/2b}\right).$$ ◇

Example 3.8 (*continued*)
The best rate at φ is $\ln n/n$ and $\hat{\varphi}_n$ reaches this rate. One may take $k_n \simeq \ln n \cdot \log_2 n$. ◇

We now give a general but less sharp result, since it only shows that the possible loss of rate is at most $(\ln n \cdot \log_2 n)^2$.

Theorem 3.9
If conditions in Theorem 3.2 and A3.5 hold, and if $k_n = k_n^* + 1$ then

$$\sup_{\varphi\in\phi_b} \mathrm{E} \parallel \hat{\varphi}_n - \varphi \parallel^2 = \mathcal{O}\Big(v_n(\ln n \cdot \log_2 n)^2\Big). \tag{3.44}$$

PROOF:
From (3.15) and A3.1 it follows that

$$|\varphi_{k_n^*+1}| \leq b_{k_n^*+1} = \frac{1}{\delta}\left|\varphi^*_{k_n^*+1}\right| < \frac{d}{\delta}\left(\frac{\beta_{k_n^*+1}}{u_n}\right)^{1/2},$$

but, since X is GSM, one may choose $u_n = n$ and $\beta_{k_n^*+1} := \beta = 8h^2 \sum_{k\geq 0}\alpha_k$. Therefore, as soon as $\ln n \cdot \log_2 n \geq 1$, $|\varphi^*_{k_n^*+1}| < d\beta^{1/2}\gamma_n/\delta$, thus, $k_n \geq k_n^* + 1 \geq q_{\varphi^*}(d\beta^{1/2}\gamma_n/\delta)$, and since

$$d = \frac{|\varphi^*_{j_o}|u_{n_o}^{1/2}}{\beta_{j_o}^{1/2}} = \frac{|\varphi^*_{j_o}|}{\beta^{1/2}},$$

we have $k_n \geq q_{\varphi^*}\left(\frac{|\varphi^*_{j_o}|}{\delta}\gamma_n\right) = q_{\varphi^*}(b_{j_o}\gamma_n)$.

Now we may always assume that $b_{j_o} < 1$ (if not we choose a new model associated with the sequence $(b_j/(2b_{j_o}),\ j\geq 0)$, with the parameter $\varphi/(2b_{j_o})$ instead of φ), so one may set $b_{j_o} = 1-\varepsilon'$, and finally

$$k_n \geq q_\varphi((1-\varepsilon')\gamma_n) \geq q_\varphi((1+\varepsilon)\gamma_n),$$

that is

$$k_n \geq q((1-\varepsilon')\gamma_n) \geq q((1+\varepsilon)\gamma_n).$$

We are now in a position to use Theorem 3.8 for obtaining

$$\begin{aligned}
\mathrm{E} \parallel \hat{\varphi}_n - \varphi \parallel^2 &= \mathcal{O}\left(\frac{k_n^*}{n} + \sum_{q_n(\varepsilon)<j\leq k_n^*} \varphi_j^2 + \sum_{j>k_n^*}\varphi_j^2\right)\\
&= \mathcal{O}\Big(v_n + k_n^*(1+\varepsilon)^2\gamma_n^2\Big)\\
&= \mathcal{O}\Big(v_n(\ln n \cdot \log_2 n)^2\Big)
\end{aligned}$$

and (3.44) follows since the bound is uniform with respect to $\varphi \in \phi_b$. ■

We now turn to uniform rate.

Theorem 3.10
If A3.5 holds, $|\varphi_j| \leq c\rho^j$, $j \geq 0$ $(c > 0,\ 0 < \rho < 1)$, φ *is functional and* $k_n \simeq \ln n \cdot \log_2 n$ *then*

$$\| \hat{\varphi}_n - \varphi \|_\infty = \mathcal{O}\left(\frac{(\ln n)^3}{\sqrt{n}}\right) \quad a.s.$$

PROOF:
Similar to the proof of Theorem 3.7(2) and is therefore omitted. ■

3.7.4 Discussion and implementation

For the practical construction of $\hat{\varphi}_n$, two questions arise: the choice of an orthonormal system (e_j) and determination of k_n.

In various situations there is a natural orthonormal system associated with the estimation problem. If, for example, φ is the spectral density, the cosine system appears as natural. Otherwise one may try to use an attractive system. If, for example $\varphi = f$ is the marginal density (Example 3.5), φ is adapted to every orthonormal basis in $L^2(\mu)$. In such a situation one may choose a family of preferred densities $(g_j, j \geq 0)$ and use the associated orthonormal system:

$$e_o = \frac{g_o}{\| g_o \|},\ e_j = \frac{g_j - \sum_{\ell=0}^{j-1} \langle g_j, e_\ell \rangle e_\ell}{\left\| g_j - \sum_{\ell=0}^{j-1} \langle g_j, e_\ell \rangle e_\ell \right\|},\ j \geq 1,$$

possibly completed.

For example, if $E_o = [0, 1]$ the Legendre polynomials are generated by the monomial densities and the trigonometric functions by a family of periodic densities. The Haar functions, see Neveu (1972), are associated with simple densities. Conversely the choice of a particular orthonormal system is associated with the selection (possibly involuntary!) of a sequence of preferred densities.

Finally, note that it can be interesting to used 'mixed' systems, for example presence of trigonometrical functions and of polynomials in the same system reinforce its efficiency for approximating various densities.

Concerning k_n, a generally-applicable choice would be $k_n \simeq n/(\ln n)^2$ since it implies $k_n \geq q_n'(\varepsilon')$, for n large enough, under the mild condition $|\varphi_j| \leq A/\sqrt{j}$ $(A > 0)$. However a practical choice of k_n must be more accurate because a too large k_n might disturb the estimator by introducing isolated j's such that $|\overline{\varphi}_{jn}| \geq \gamma_n$.

3.8 Adaptive estimation in continuous time

Estimation by adaptive projection can be developed in the framework of continuous time processes. For convenience we restrict our study to estimation of the marginal density of an observed strictly stationary continuous time process, but extension to more general functional parameters is not difficult.

So, let $X = (X_t, t \in \mathbb{R})$ be a measurable strictly stationary continuous time process, with values in the measure space $(E_o, \mathcal{B}_o, \mu)$ where μ is a σ-finite measure. X_0 has an unknown density, say f, with respect to μ, and there exists an orthonormal system $(e_j, j \geq 0)$ in $L^2(\mu)$ such that:

$$M_X = \sup_{j\geq 0} \| e_j(X_0) \|_\infty < \infty, \text{ and } f = \sum_{j=0}^{\infty} a_j(f) e_j,$$

where $a_j = a_j(f) = \int e_j f \mathrm{d}\mu$, $j \geq 0$, and $\sum_j a_j^2 < \infty$.

The adaptive projection estimator of f is defined as $\hat{f}_T = \sum_{j=0}^{\hat{k}_T} a_{jT} e_j$, $T > 0$, where

$$\hat{a}_{jT} = \frac{1}{T} \int_0^T e_j(X_t) \mathrm{d}t,$$

$j \geq 0$ and $\hat{k}_T = \max\{j : 0 \leq j \leq k_T, |\hat{a}_{jT}| \geq \gamma_T\}$ with the convention $\hat{k}_T = k_T$ if $\{\cdot\} = \varnothing$. The family of integers (k_T) is such that $k_T \to \infty$ and $k_T/T \to 0$ as $T \to \infty$, and the threshold γ_T is

$$\gamma_T = \left(\frac{\ln T \cdot \log_2 T}{T} \right)^{1/2},$$

$T > \mathrm{e}^1$.

In the following we use similar notation to above, namely $\phi_o, \phi_o(K)$, $\phi_1, q_T(\varepsilon), q_T'(\varepsilon')$.

Now (3.25) is replaced by the following:

Lemma 3.10

Let $Y = (Y_t, t \geq 0)$ be a real measurable bounded strictly stationary process with a strong mixing coefficient $(\alpha_Y(u), u \geq 0)$ such that $\int_0^\infty \alpha_Y(u)\mathrm{d}u < \infty$, then if $1 \leq \tau \leq T/2, \eta > 0, \zeta > 0$, we have

$$P\left(\left| \frac{1}{T} \int_0^T (Y_t - \mathrm{E}Y_t)\mathrm{d}t \right| \geq \eta \right) \leq 4 \exp\left(- \frac{\eta^2 \| Y_o \|_\infty^{-2} T}{c_1 + c_2 \tau T^{-1} + c_3 \| Y_o \|_\infty^{-1} \eta\tau} \right) + \frac{c_4}{\eta} \| Y_o \|_\infty \alpha_Y(\tau)$$

with $c_1 = 32(1+\zeta)^2 \int_0^\infty \alpha_Y(u)\mathrm{d}u$, $c_2 = 4c_1$, $c_3 = 16(1+\zeta)/3, c_4 = 16(1+\zeta)/\zeta$.

PROOF:

The proof appears in Bosq and Blanke (2004). ■

Let $\mathcal{X}$ be the family of processes that satisfy the above assumptions and that are geometrically strongly mixing. We now state results concerning the asymptotic behaviour of $\hat{k}_T$ and $\hat{f}_T$. The proofs use the same method as in the discrete case and are, therefore, left to the reader.

Theorem 3.11
If $X \in \mathcal{X}$, then

(1) *if $f \in \phi_o$, one has, for all $\delta > 0$,*

$$P(\hat{k}_T \neq K(f)) = \mathcal{O}(T^{-\delta}),$$

(2) *if $f \in \phi_1$, for all $A > 0$,*

$$P(\hat{k}_T < A) = \mathcal{O}\Big(\exp(-c_A\sqrt{T})\Big), \quad (c_A > 0),$$

(3) *if $q_T(\varepsilon) \leq k_T$,*

$$P\big(\hat{k}_T \notin [q'_T(\varepsilon'), q_T(\varepsilon)]\big) = \mathcal{O}(T^{-\delta}), \quad \delta > 0.$$

Example 3.8 (*continued*)
If $|a_j| = \alpha\rho^j$ $(0 < \rho < 1,\ \alpha > 0)$ and if $k_T > (\ln T)/\ln(1/\rho)$, since $q_T(\varepsilon) \simeq \ln T/[2\ln(1/\rho)]$ we have

$$P\left(\left|\frac{\hat{k}_T}{\ln T} - \frac{1}{2\ln(1/\rho)}\right| \geq \xi\right) = \mathcal{O}(T^{-\delta}), \quad \xi > 0,\ \delta > 0.$$

As before $\hat{\rho}_T = T^{-1/(2\hat{k}_T+1)}$ is a strongly consistent estimator of ρ. ◇

The next statement specifies asymptotic behaviour of $\hat{f}_T$ over ϕ_o.

Theorem 3.12
If $X \in \mathcal{X}$ and $f \in \phi_o$ then

(1) $T \cdot \mathrm{E}\| \hat{f}_T - f \|^2 \underset{T\to\infty}{\longrightarrow} 2\sum_{j=0}^{K(f)} \int_o^\infty \mathrm{Cov}\big(e_j(X_0), e_j(X_u)\big)\, du.$

(2) *Moreover*

$$P(\| \hat{f}_T - f \|_\infty \geq \xi) = \mathcal{O}(T^{-\delta}), \quad \delta > 0,\ \xi > 0.$$

(3) *Finally, if $T = nh$ $(h > 0)$,*

$$\sqrt{T}\,(\hat{f}_T - f) \xrightarrow{\mathcal{D}} N$$

where convergence in distribution takes place in $L^2(\mu)$ and N is a zero-mean Gaussian $L^2(\mu)$-valued random variable with a $1 + K(f)$-dimensional support.

Finally we have:

Theorem 3.13
If $X \in \mathcal{X}, f \in \phi_1, k_T \geq q_T(\varepsilon)$, then

(1) $\mathrm{E}\| \hat{f}_T - f \|^2 = \mathcal{O}(q'_T(\varepsilon')/T) + \sum_{j>q_T(\varepsilon)} a_j^2$.

(2) *If, in particular,* $|a_j| = \alpha\, \rho^j (0 < \rho < 1,\ \alpha > 0)$ *and* $\ln T = \mathcal{O}(k_T)$ *it follows that*

$$\mathrm{E}\| \hat{f}_T - f \|^2 = \mathcal{O}\left(\frac{\ln T \cdot \log_2 T}{T}\right).$$

(3) *Moreover, if* $\ln T = o(k_T), T = T_n$ *with* $\sum_n T_n^{-\delta_o} \ln T_n < \infty\ (\delta_o > 0)$ *then, if* $\alpha_X(u) \leq ae^{-bu},\ u > 0$

$$\overline{\lim_{T_n \uparrow \infty}} \frac{T_n^{1/2}}{(\ln T_n)^{1/2}} \| \hat{f}_{T_n} - f \|_\infty \leq 2\sqrt{\frac{2a\delta_o}{b}\ \frac{M_X^2}{\ln(1/\rho)}}$$

almost surely.

In Chapter 9 we will see that f_T and $\hat{f}_T$ may reach a $1/T$-rate on ϕ_1 if X admits an empirical density.

Notes

Carbon (1982) has studied our class of functional parameters in the context of pointwise and uniform convergence.

Results presented in this chapter are new or very recent. Results concerning density appear in Bosq (2005a); Bosq and Blanke (2004). The case of regression is studied in Aubin (2005); J.B. Aubin and R. Ignaccolo (Adaptive projection estimation for a wide class of functional parameters, personal communication 2006).

Donoho *et al.* (1996) have studied a density estimator by wavelet thresholding with a slightly different method.

4

Functional tests of fit

4.1 Generalized chi-square tests

In order to construct a nonparametric test of fit, it is natural to use an estimator of the observed variables' distribution.

A typical example of this procedure is the famous χ^2-test of Karl Pearson (1900) which is based on an elementary density estimator, namely the histogram.

More precisely, let $X_1, \ldots, X_n$ be a sample of $(E_0, \mathcal{B}_0, \mu)$ valued random variables, where μ is a probability on $(E_0, \mathcal{B}_0)$. For testing $P_{X_i} = \mu$ one considers a partition $A_0, A_1, \ldots, A_k$ of E_0 such that $p_j = \mu(A_j) > 0, 0 \leq j \leq k$, and sets

$$f_{n,0}(x) = \sum_{j=0}^{k} \left[\frac{1}{np_j} \sum_{i=1}^{n} \mathbb{1}_{A_j}(X_i) \right] \mathbb{1}_{A_j}(x), \quad x \in E_0.$$

This histogram may be interpreted as the projection density estimator associated with the orthonormal system of $L^2(\mu)$ defined by $(p_j^{-1/2}\mathbb{1}_{A_j}, 0 \leq j \leq k)$. This system is an orthonormal basis of $\mathrm{sp}(\mathbb{1}_{A_0}, \ldots, \mathbb{1}_{A_k})$, a $(k+1)$-dimensional subspace of $L^2(\mu)$, which contains the constants, and has the following reproducing kernel (see Berlinet and Thomas-Agnan 2004):

$$K_0 = \sum_{j=0}^{k} p_j^{-1} \mathbb{1}_{A_j} \otimes \mathbb{1}_{A_j}.$$

Thus, the histogram takes the form

$$f_{n,0}(\cdot) = \frac{1}{n} \sum_{i=1}^{n} K_0(X_i, \cdot).$$

Inference and Prediction in Large Dimensions D. Bosq and D. Blanke

Now one may construct a test statistic based on $\| f_{n,0} - 1 \|$ where $\| \cdot \|$ is the $L^2(\mu)$-norm:

$$Q_{n,0} = n \| f_{n,0} - 1 \|^2 = n \sum_{j=0}^{k} p_j^{-1} \left[\frac{1}{n} \sum_{j=1}^{n} \mathbb{1}_{A_j}(X_i) - p_j \right]^2 ,$$

that is the classical χ^2-test statistic.

Since the histogram is in general a suboptimal density estimator it is natural to employ others estimators for constructing tests of fit. Here we consider tests based on the linear estimators studied in Chapter 3. We begin with the projection estimator:

$$f_n = \sum_{j=0}^{k_n} \hat{a}_{jn} e_j$$

where (e_j) is an orthonormal system in $L^2(\mu)$ and

$$\hat{a}_{jn} = \frac{1}{n} \sum_{i=1}^{n} e_j(X_i).$$

In the current section we take $k_n = k \geq 1$, choose $e_0 \equiv 1$ and set

$$K = 1 + \sum_{j=1}^{k} e_j \otimes e_j;$$

we will say that K is a *kernel of order* k. The associated test statistic is

$$Q_n = n \sum_{j=1}^{k} \hat{a}_{jn}^2.$$

In order to study Q_n we will use a multidimensional Berry–Esséen type inequality:

Lemma 4.1 *(Sazonov inequality)*
Let $(U_n, n \geq 1)$ be a sequence of k-dimensional random vectors, i.i.d. and such that $\mathrm{E} \| U_n \|^2 < \infty$ and $\mathrm{E}U_n = 0$. Let $\mathbb{C}$ be the class of convex measurable sets in $\mathbb{R}^k$. Finally let $t = \{t_1, \ldots, t_k\}$ be a finite subset of $\mathbb{R}^k$ such that the scalar products $\langle U_n, t_\ell \rangle; \ell = 1, \ldots, k$ are noncorrelated real random variables such that $\mathrm{E}|\langle U_n, t_\ell \rangle|^3 < \infty$. Then

$$\sup_{C \in \mathbb{C}} |P_n(C) - N(C)| \leq c_0 k^3 \left[\sum_{\ell=1}^{k} \rho_\ell^{(t)} \right] n^{-1/2}, n \geq 1,$$

where P_n is the distribution of $\left(\sum_{i=1}^{n} U_i\right)/\sqrt{n}$, N the normal distribution with the same moments of order 1 and 2 as P_{U_n}, and

$$\rho_\ell^{(t)} = \frac{\mathrm{E}|\langle U_n, t_\ell\rangle|^3}{(\mathrm{E}(\langle U_n, t_\ell\rangle^2))^{3/2}}, \quad 1 \leq \ell \leq k.$$

Finally c_0 is a universal constant.

Proof:
See Sazonov (1968a,b).

We now study the asymptotic behaviour of Q_n.

Theorem 4.1

(a) *If $P_{X_1} = \mu$ and $n \to \infty$ then*

$$Q_n \xrightarrow{\mathcal{D}} Q$$

where $Q \approx \chi^2(k)$.

(b) *If $P_{X_1} = \nu$ such that $\int |e_{j_0}| \mathrm{d}\nu < \infty$ and $\int e_{j_0} \mathrm{d}\nu \neq 0$ for some $j_0 \in \{1, \ldots, k\}$, then*

$$Q_n \to \infty \quad a.s.$$

(c) *If $P_{X_1} = \mu$ and $s_3 = \int \left(\sum_{j=1}^{k} |e_j|^3\right) \mathrm{d}\mu < \infty$, then*

$$\sup_{C \in \mathbb{C}} |P(Q_n \in C) - P(Q \in C)| \leq c_0 k^3 s_3 n^{-1/2}, \quad n \geq 1 \tag{4.1}$$

where c_0 is a universal constant.

Proof:

(a) Consider the random vector

$$Z_n = n^{1/2}(\hat{a}_{1n}, \ldots, \hat{a}_{k_n})$$

and apply the central limit theorem in $\mathbb{R}^k$ (see Rao 1984) to obtain $Z_n \xrightarrow{\mathcal{D}} Z$ where Z is a standard Gaussian vector. By continuity it follows that

$$Q_n = \| Z_n \|_k^2 \xrightarrow{\mathcal{D}} \| Z \|_k^2 \approx \chi^2(k),$$

where $\| \cdot \|$ is euclidian norm in $\mathbb{R}^k$.

(b) The strong law of large numbers (see Rao 1984) implies

$$\hat{a}_{j_0 n} \to \int e_{j_0} \mathrm{d}\nu \quad \text{a.s.}$$

therefore

$$Q_n = n \sum_{j=1}^{k} \hat{a}_{jn}^2 \geq n\hat{a}_{j_0 n}^2 \to \infty \text{ a.s.}$$

(c) Inequality (4.1) is a direct consequence of the Sazonov inequality where $U_i = (e_1(X_i), \ldots, e_k(X_i)), 1 \leq i \leq n$ and $\{t_1, \ldots, t_k\}$ is the canonical basis of $\mathbb{R}^k$. ■

Notice that one may obtain the limit in distribution and a Berry–Esséen inequality if $P_{X_1} = \nu$, see Bosq (1980).

Now in order to test

$$H_0 \colon P_{X_1} = \mu$$

against

$$H_1(k) \colon P_{X_1} = \nu$$

such that $\int |e_j| \mathrm{d}\nu < \infty$ and $\int e_j \mathrm{d}\nu \neq 0$ for at least one $j \in \{1, \ldots, k\}$, we choose the test whose critical region is defined by

$$Q_n > \chi_\alpha^2(k)$$

with

$$P(Q > \chi_\alpha^2(k)) = \alpha \ (\alpha \in]0, 1[).$$

Theorem 4.1 shows that this test has asymptotic level α and asymptotic power 1. Moreover, if $s_3 < \infty$,

$$|P(Q > \chi_\alpha^2(k)) - \alpha| = \mathcal{O}(n^{-1/2}).$$

Example 4.1 *(Application: testing absence of mixture)*
Suppose that X_1 has a density f with respect to μ. One wants to test $f = 1$ against $f = b_0 + \sum_{j=1}^{k} b_j f_j$ where $f_1, \ldots, f_k$ are densities which belong to $L^2(\mu)$ and $\max_{1 \leq j \leq k} |b_j| \neq 0$. To this aim one may construct an orthonormal basis of $\mathrm{sp}\{1, f_1, \ldots, f_k\}$ and then apply the above test. This example shows that the choice of (e_j) depends on the alternative hypothesis. We will return to this choice in the next sections. ◇

Finally simulations indicate that the 'smooth test' defined above is in general better than the classical χ^2-test (see Bosq 1989; Neyman 1937).

4.2 Tests based on linear estimators

We now consider the more general framework where one uses estimators of the form

$$g_n = \frac{1}{n}\sum_{i=1}^{n} K_n(X_i, \cdot)$$

with

$$K_n = \sum_{j=0}^{\infty} \lambda_{jn} e_{jn} \otimes e_{jn}, \quad n \geq 1$$

where $(e_{jn}, j \geq 0)$ is an orthonormal system in $L^2(\mu)$ such that $e_{0n} \equiv 1$, $\sum \lambda_{jn}^2 < \infty$ and $\lambda_{0n} = 1$ for each $n \geq 1$. Moreover we suppose that $1 \geq |\lambda_{jn}| \geq |\lambda_{j+1,n}|$, $j \geq 0$, $n \geq 1$.

The test is based on the Hilbertian statistic

$$T_n = n^{1/2}(g_n - 1),$$

which rejects H_0 for large values of $\| T_n \|$.

In order to state results concerning T_n, we introduce the following assumptions and notations

$$\begin{aligned}
\ell_n &= \frac{1}{n}(1 + |\lambda_{3n}|^{-3} 2^{-1/2}), \inf_{n\geq 1} |\lambda_{3n}| > 0, \\
M_n &= \sup_{j\geq 0} \| e_j \|_\infty < \infty, \\
\Lambda_{r,n} &= \sum_{j>r} \lambda_{jn}^2, \\
U_n &= \sum_{j=1}^{\infty} \lambda_{jn} N_j e_{jn}
\end{aligned}$$

where $(N_j, j \geq 1)$ is an auxiliary sequence of independent random variables with common law $\mathcal{N}(0, 1)$.

Then, we have the following bound.

Theorem 4.2
If $P_{X_1} = \mu$,

$$\sup_{a\geq 0}(|P(\| T_n \|^2 \leq a) - P(\| U_n \|^2 \leq a)|) \leq C_n \quad (4.2)$$

with

$$C_n = (6 + 2\ell_n + M_n^2)\Lambda_{r,n}^{2/3} + 3c_0 M_n \frac{r^4}{\sqrt{n}}, \quad n \geq 1, r \geq 3.$$

PROOF:
We divide this into five parts.

(1) Set $Z_{r,n} = \sum_{j=1}^{r} \lambda_{jn}^2 N_j^2, \quad r \geq 3,\ n \geq 1.$

The characteristic function of $Z_{r,n}$ being integrable, its density is given by

$$f_{r,n}(x) = \frac{1}{2\pi} \int_{-\infty}^{+\infty} \prod_{j=1}^{r} (1 - 2\mathrm{i}\lambda_{jn}^2 t)^{-1/2} \mathrm{e}^{-\mathrm{i}tx} \mathrm{d}t, \quad x \in \mathbb{R}$$

hence

$$\sup_{x\in\mathbb{R}} f_{r,n}(x) \leq \frac{1}{2\pi} \left[2 + \int_{|t|\geq 1} \prod_{j=1}^{3} (1 + 4\lambda_{jn}^4 t^2)^{-1/4} \mathrm{d}t \right]$$
$$\leq \ell_n.$$

(2) Consider the event

$$\Delta_a = \{Z_{r,n} \leq a, \| U_n \|^2 > a\}, \quad a \in \mathbb{R}_+,$$

then, if $0 < \gamma < a$,

$$\Delta_a \Rightarrow \{\| U_n \|^2 - Z_{r,n} > \gamma\} \cup \{a - \gamma \leq Z_{r,n} \leq a\},$$

therefore

$$P(\Delta_a) \leq \frac{\mathrm{E}\left(\sum_{j>n} \lambda_{jn}^2 N_j^2 \right)^2}{\gamma^2} + \gamma \sup_{x\in\mathbb{R}} f_{r,n}(x)$$
$$\leq \frac{3\Lambda_{r,n}^2}{\gamma^2} + \ln \gamma,$$

for $a > \Lambda_{r,n}^{2/3}$ and $\gamma = \Lambda_{r,n}^{2/3}$ one obtains

$$P(\Delta_a) \leq (3 + \ell_n)\Lambda_{r,n}^{2/3},$$

for $a \leq \Lambda_{r,n}^{2/3}$ we have

$$P(\Delta_a) \leq P(Z_{r,n} \leq a) \leq \ell_n \Lambda_n^{2/3}.$$

Consequently

$$|P(Z_{r,n} \leq a) - P(\| U_n \|^2 \leq a)| \leq P(\Delta_a) \leq (3 + \ell_n)\Lambda_{r,n}^{2/3} \qquad (4.3)$$
$$a \geq 0, r \geq 3, n \geq 1.$$

(3) Now, let us set

$$R_{r,n} = n \sum_{r+1}^{\infty} \lambda_{jn} \hat{a}_{jn}^2$$

where

$$\hat{a}_{jn} = \frac{1}{n} \sum_{i=1}^{n} e_{jn}(X_i), \quad j \geq 0.$$

After some tedious but easy calculations one gets

$$\begin{aligned} \mathrm{E}R_{r,n}^2 &= \frac{1}{n} \sum_{r+1}^{\infty} \lambda_{jn}^4 \int e_{jn}^4 \mathrm{d}\mu + 3 \frac{n-1}{n} \sum_{r+1}^{\infty} \lambda_{jn}^4 \\ &+ \frac{1}{n} \sum_{\substack{j,j'>r \\ j \neq j'}} \lambda_{jn}^2 \lambda_{j'n}^2 \int e_{jn}^2 e_{j'n}^2 \mathrm{d}\mu + \frac{n-1}{n} \sum_{\substack{j,j'>r \\ j \neq j'}} \lambda_{jn}^2 \lambda_{j'n}^2, \end{aligned}$$

which yields

$$\mathrm{E}R_{r,n}^2 \leq (M_n^2 + 3)\Lambda_{r,n}^2.$$

(4) Consider the random variables

$$Z'_{r,n} = n \sum_{1}^{n} \lambda_{jn}^2 \hat{a}_{jn}^2, \quad r \geq 3, \quad n \geq 1.$$

As in part (2) we may bound $|P(\| S_n \|^2 \leq a) - P(Z'_{r,n} \leq a)|$ by putting

$$\Delta'_a = \{Z'_{r,n} \leq a, \| S_n \|^2 > a\} \subseteq \{R_{r,n} > \gamma\} \cup \{a - \gamma \leq Z'_{r,n} \leq a\}, \qquad (4.4)$$
$$0 < \gamma < a.$$

Then the Sazonov inequality (Lemma 4.1) entails

$$|P(Z_{r,n} \leq a) - P(Z'_{r,n} \leq a)| \leq c_0 M_n \frac{r^4}{\sqrt{n}},$$

hence

$$|P(Z_{r,n} \in]a-\gamma, a]) - P(Z'_{r,n} \in]a-\gamma, a])| \leq 2c_0 M_n \frac{r^4}{\sqrt{n}}, \tag{4.5}$$

and from (4.4) it follows that

$$P(\Delta'_a) \leq \frac{M_n^2+3}{\gamma^2}\Lambda_{r,n}^2 + \ell_n\gamma + 2c_0 M_n \frac{r^4}{\sqrt{n}},$$

now, if $\lambda = \Lambda_{r,n}^{2/3} < a$, we have

$$P(\Delta'_a) \leq (M_n^2+3)\Lambda_{r,n}^{2/3} + \ell_n\Lambda_{r,n}^{2/3} + 2c_0 M_n \frac{r^4}{\sqrt{n}},$$

when, if $a \leq \Lambda_{r,n}^{2/3}$,

$$P(\Delta'_a) \leq \ell_n\Lambda_{r,n}^{2/3} + c_0 M_n \frac{r^4}{\sqrt{n}},$$

consequently

$$|P(Z'_{r,n} \leq a) - P(\| S_n \|^2 \leq a)| \leq (\ell_n + M_n^2 + 3)\Lambda_{r,n}^{2/3} + 2c_0 M_n \frac{r^4}{\sqrt{n}}. \tag{4.6}$$

(5) Finally (4.2) follows from (4.3), (4.5) and (4.6). ■

From (4.2) one derives limits in distribution for T_n. Two cases are considered:

Assumptions 4.1 (A4.1)
$\lambda_{jn} \xrightarrow[n\to\infty]{} \lambda_j$, $j \geq 1$ with $|\lambda_{jn}| \leq \lambda'_j$, $j \geq 1$ where $\sum_j \lambda'^2_j < \infty$.

Assumptions 4.2 (A4.2)
There exists (r_n) such that

$$\frac{r_n^4}{\sqrt{n}} \to 0, \sum_{j>r_n} \lambda_{jn}^2 \to 0, \sum_j \lambda_{jn}^4 < \infty.$$

Corollary 4.1
If $\sup_{n\geq 1} M_n < \infty$, *then*

(*a*) *If A4.1 holds,*

$$\| T_n \|^2 \xrightarrow{\mathcal{D}} \| U \|^2 = \sum_{j=1}^{\infty} \lambda_j N_j^2,$$

(*b*) *If A4.2 holds,*

$$\frac{\| T_n \|^2 - \sum_1^\infty \lambda_{jn}^2}{\sqrt{2}\left(\sum_1^\infty \lambda_{jn}^4\right)^{1/2}} \xrightarrow{\mathcal{D}} N \sim \mathcal{N}(0,1).$$

PROOF:
Clear. ■

Part (a) of Corollary 4.1 applies to fixed kernels of the form $K = \sum_{j=0}^\infty \lambda_j e_j \otimes e_j$, for example:

- $K = \sum_{j=0}^k e_j \otimes e_j$ (see Section 4.1).
- $K = 1 + \sum_{j=1}^\infty j^{-1} e_j \otimes e_j$; for this kernel the limit is the Cramér–von Mises distribution (see Nikitin 1995).
- $K = \sum_{j=0}^\infty \rho^j e_j \otimes e_j (0 < \rho < 1)$, here the limit in distribution has characteristic function

$$\varphi(t) = \prod_{\ell=1}^\infty (1 - 2i\rho^\ell t)^{-1/2}, \quad t \in \mathbb{R}$$

and the choice $r_n = [(3/8)\ln(1/\rho) \cdot \ln n]$ leads to the bound $c(\ln n)^4/\sqrt{n}$ where c is constant.

Concerning part (b), it holds for kernels of the form $K_n = \sum_{j=0}^{k_n} e_j \otimes e_j$, with $k_n \to \infty$, $k_n^4/\sqrt{n} \to 0$.

4.2.1 Consistency of the test

Here the hypotheses of the test of fit are

$$H_0\colon P_{X_1} = \mu$$
$$H_1\colon P_{X_1} = \nu$$

where ν is such that there exists $(j_n(\nu), n \geq 1)$ for which $\underline{\lim}_{n\to\infty} |\lambda_{j_n(\nu),n} \int e_{j_n(\nu),n} \mathrm{d}\nu| > 0$.

If $e_{jn} = e_j$, $n \geq 1$ and $\underline{\lim}_{n\to\infty} \lambda_{j,n} > 0$, $j \geq 1$ the above condition is equivalent to $\int e_{j(\nu)} \mathrm{d}\nu \neq 0$ for some $j(\nu) \geq 1$. Then, if (e_j) is an orthonormal basis of $L^2(\mu)$, $H_0 + H_1$ contains all distributions of the form $\nu = f\cdot\mu + \gamma$ where $0 \neq f \in L^2(\mu)$ and γ is ⋆orthogonal⋆ to μ.

Recall that the test $\| T_n \|^2 > c_n$ is said to be consistent if, as $n \to \infty$,

$$\alpha_n = P_\mu(\| T_n \|^2 > c_n) \to 0$$

and

$$\beta_n(\nu) = P_\nu(\| T_n \|^2 > c_n) \to 1, \quad \nu \in H_1.$$

Then, we have:

Theorem 4.3
If A4.2 is satisfied the test $\| T_n \|^2 > c_n$ *is consistent, provided*

$$\frac{c_n}{n} \to 0, \quad \frac{c_n - \sum_1^\infty \lambda_{jn}^2}{\left(\sum \lambda_{jn}^4\right)^{1/2}} \to \infty. \tag{4.7}$$

PROOF:
Clear. ■

If $\forall \eta > 0$, $\exists \nu \in H_1$: $\overline{\lim}_{n\to\infty} \sum_{j=1}^\infty \lambda_{jn}^2 (\int e_{jn}\, d\nu)^2 \le \eta$ one can show that (4.7) is necessary for consistency (see Bosq 1980).

For obtaining a rate of convergence we need an exponential Hilbertian inequality

Lemma 4.2 *(Pinelis–Sakhanenko inequality)*
Let $Y_1, \ldots, Y_n$ *be H-random variables that are independent, zero-mean and bounded:* $\| Y_i \| \le b$, $1 \le i \le n$, *where b is a constant, then*

$$P\left(\| \sum_{i=1}^n Y_i \| \ge t \right) \le 2 \exp\left(-\frac{t^2}{2s_n^2 + (2/3)bt} \right), \quad t > 0,$$

where $s_n^2 = \sum_{i=1}^n \mathrm{E} \| Y_i \|^2$.

PROOF:
See Pinelis and Sakhanenko (1985). ■

4.2.2 Application

We write the test under the form

$$\| \frac{1}{n} \sum_{i=1}^n K_n'(X_i, \cdot) \| > d_n$$

where $K_n' = K_n - 1$ and $d_n = (c_n/n)^{1/2}$.

Applying Lemma 4.2 one obtains

$$\alpha_n \le 2 \exp\left(-\frac{n d_n^2}{2\mathrm{E}_\mu \| K_n' \|^2 + (2/3) \| K_n' \|_\infty d_n} \right),$$

and, if $\gamma_n = \mathrm{E}_\nu \parallel K_n'(X_1, \cdot) \parallel -d_n > 0$,

$$\beta_n(\nu) \geq 1 - 2\exp\left(-\frac{n\gamma_n^2}{2\mathrm{E}_\nu \parallel \bar{K}_n' \parallel^2 + (2/3) \parallel K_n' \parallel_\infty \gamma_n}\right),$$

where $\bar{K}_n' = K_n'(X_1, \cdot) - \mathrm{E}_\nu K_n'(X_1, \cdot)$.

In particular if

$$K_n = K = \sum_{j=0}^{\infty} \lambda_j e_j \otimes e_j$$

and $d_n = d > 0$, we have

$$\alpha_n \leq 2\exp\left(-\frac{nd^2}{2\mathrm{E}_\nu \parallel K' \parallel^2 + (2/3) \parallel K' \parallel_\infty}\right)$$

and

$$\beta_n \geq 1 - 2\exp\left(-\frac{n\gamma^2}{2\mathrm{E}_\nu \parallel K' \parallel^2 + (2/3) \parallel K' \parallel_\infty}\right)$$

provided $\gamma = \mathrm{E}_\nu \parallel K' \parallel -d^2 > 0$.

Thus (α_n) and (β_n) converge at an exponential rate. This rate is optimal since the Neyman–Pearson test has the same property, see Tusnády (1973). Notice that the choice $c_n = nd$ does not contradict condition $c_n/n \to 0$ in Theorem 4.3, since this condition is necessary for consistency only if $\mathrm{E}_\nu \parallel \bar{K}'(X_1, \cdot) \parallel^2 \leq d$.

4.3 Efficiency of functional tests of fit

In this section we study efficiency of functional tests of fit under local hypotheses and in Bahadur's sense.

4.3.1 Adjacent hypotheses

Suppose that we have i.i.d. data $X_{1n}, \ldots, X_{nn}$ with distribution ν_n close to μ as n is large enough; if K is a kernel of order k, ν_n is said to be *adjacent* to μ if it fulfills Assumptions A4.3.

Assumptions 4.3 (A4.3)

(*i*) $\sqrt{n} \int K'(x, \cdot)\, \mathrm{d}\nu_n(x) \underset{n\to\infty}{\longrightarrow} g(\cdot)$, with $\parallel g \parallel^2 = \lambda^2 \neq 0$ where convergence takes place in $\mathcal{E}' = \mathrm{sp}(e_1, \ldots, e_k)$,

(*ii*) $\Gamma_{\nu_n}(j, \ell) = \int e_j e_\ell \mathrm{d}\nu_n - \int e_j \mathrm{d}\nu_n \int e_\ell \mathrm{d}\nu_n \underset{n\to\infty}{\longrightarrow} \delta_{j\ell}, 1 \leq j, \ell \leq k$.

For interpreting these conditions one may note that if

$$\nu_n = (1 + h_n)\mu, \quad (h_n \in \mathcal{E}'), \quad n \geq 1,$$

Assumptions A4.3 can be replaced by:

Assumptions 4.4 (A4.4)

$$\sqrt{n}\left\|1 - \frac{\mathrm{d}\nu_n}{\mathrm{d}\mu}\right\|_{L^2(\mu)} \to \lambda \neq 0.$$

Limit in distribution under adjacency appears in the following statement.

Theorem 4.4
If the test is based on a kernel of order k, then

(1) if A4.3 holds, one has

$$Q_n \xrightarrow{\mathcal{D}} Q(k, \lambda)$$

where $Q(k, \lambda) \approx \chi^2(k, \lambda)$;
(2) if A4.4 holds with $h_n = g/\sqrt{n}$ then

$$\sup_{a \geq 0} |P(Q_n \leq a) - P(Q \leq a)| \leq \frac{c}{\sqrt{n}}$$

where c is constant.

PROOF:
Similar to that of Theorem 4.1, see Bosq (1983a). ■

From Theorem 4.4 it follows that the test $Q_n > \chi^2_\alpha(k)$ where $P(Q(k, \lambda) > \chi^2_\alpha(k)) = \alpha$ has asymptotic level α and asymptotic power

$$\begin{aligned}\beta_k &= P(Q(k, \lambda) > \chi^2_\alpha(k)) \\ &= P(Q(k, 0) > \chi^2_\alpha(k) - \lambda^2)\end{aligned}$$

with a rate of convergence toward β_k which is a $\mathcal{O}(1/\sqrt{n})$.

Conversely, consider alternatives of the form

$$\nu_{n,j} = \left(1 + \frac{g_{n,j}}{\sqrt{n}}\right)\mu, \quad 1 \leq j \leq k$$

where $g_{n,j} \to g_j$ weakly in $L^2(\mu)$ and the g_j are mutually orthogonal. Then, since $\int(1 + (g_{n,j}/\sqrt{n}))\mathrm{d}\mu = 1$ one deduces that $g_j \perp 1$, $1 \leq j \leq k$ and it is easy to see that the test of asymptotic level α, based on the kernel

$$K = 1 + \sum_{j=1}^{k} \frac{g_j \otimes g_j}{\| g_j \|^2}$$

has maximal asymptotic power among functional tests based on kernel of order k.

The obtained asymptotic power is

$$P(Q(k,0) > \chi^2_\alpha(k) - \| g_j \|^2), 1 \leq j \leq k.$$

Now in order to determine local efficiency of this test we first investigate asymptotic behaviour of the Neyman–Pearson (NP) test under general adjacent hypotheses.

Lemma 4.3

If $\nu_n = (1 + (c_n g_n/\sqrt{n}))\mu$, where $\| g_n \|^2 \to \lambda^2$ and (c_n) is a sequence of real numbers such that $n^{-1/2}\max(1, c_n^2)\max(1, \| g_n \|^3) \to 0$, then

(1) if $P_{X_{1n}} = \mu$, $n \geq 1$,

$$c_n^{-1} \sum_{i=1}^{n} \ln\left(1 + \frac{c_n}{\sqrt{n}} g_n(X_{in})\right) + \frac{c_n}{2} \| g_n \|^2 \xrightarrow{\mathcal{D}} N \approx \mathcal{N}(0, \lambda^2). \qquad (4.8)$$

(2) If $P_{X_{1n}} = \nu_n$, $n \geq 1$,

$$c_n^{-1} \sum_{i=1}^{n} \ln\left(1 + \frac{c_n}{\sqrt{n}} g_n(X_{in})\right) - \frac{c_n}{2} \| g_n \|^2 \xrightarrow{\mathcal{D}} N \approx \mathcal{N}(0, \lambda^2). \qquad (4.9)$$

PROOF:

The proof can be carried out by using characteristic functions, see Bosq (1983a). If $c_n = c$ and $g_n = g$ one may also establish the results from contiguity properties, cf. Roussas (1972). ■

Now, in order to compare asymptotic powers we take $c_n = c$, then (4.8) and (4.9) respectively become

$$\sum_{i=1}^{n} \ln\left(1 + \frac{c}{\sqrt{n}} g_n(X_{in})\right) \xrightarrow{\mathcal{D}} N_1 \approx \mathcal{N}\left(-\frac{\lambda^2}{2}, \lambda^2\right),$$

and

$$\sum_{i=1}^{n} \ln\left(1 + \frac{c}{\sqrt{n}} g_n(X_{in})\right) \xrightarrow{\mathcal{D}} N_2 \approx \mathcal{N}\left(\frac{\lambda^2}{2}, \lambda^2\right).$$

Then the asymptotic power of the NP test of asymptotic level α is

$$\beta_0 = P(N_{(0)} > N_\alpha - \lambda)$$

where $N_{(0)} \approx \mathcal{N}(0, 1)$ and $P(N_{(0)} > N_\alpha) = \alpha$.

Consequently, the asymptotic efficiency of the test associated with the optimal kernel of order k is given by

$$e_k = \frac{P(Q_k(0) > \chi^2_\alpha(k) - \lambda^2)}{P(N_{(0)} > N_\alpha - \lambda)}.$$

If $k = 1$, it takes the simple form

$$e_1 = \frac{P(|N_{(0)} - \lambda| > N_{\alpha/2})}{P(N_{(0)} > N_\alpha - \lambda)}.$$

4.3.2 Bahadur efficiency

Bahadur efficiency is based on the required number of observations for obtaining a given power.

Definition 4.1
Let (U_n) and (V_n) be two sequences of statistics used for testing H_0 against H_1.

One denotes as $N_U(\alpha, \beta, \theta)$ the required sample size for the test of level α associated with (U_n), to obtain at least a power β for θ fixed in H_1. Similarly one defines $N_V(\alpha, \beta, \theta)$.

Then, Bahadur efficiency of (U_n) with respect to (V_n) is given by

$$e_{U,V}(\beta, \theta) = \lim_{\alpha \downarrow 0} \frac{N_U(\alpha, \beta, \theta)}{N_V(\alpha, \beta, \theta)}.$$

Details concerning theory of Bahadur efficiency appear in Nikitin (1995). Now we have the following result.

Theorem 4.5
Consider the test problem

$$H_0 : P_{X_1} = \mu$$

against

$$H_1 : P_{X_1} = (1 + h)\mu \quad \left(|h| < 1, \int h^2 \mathrm{d}\mu > 0, \int h \mathrm{d}\mu = 0\right).$$

Let Q_n be the statistic associated with the kernel

$$K = 1 + \frac{h}{\| h \|} \otimes \frac{h}{\| h \|}$$

then, its Bahadur efficiency with respect to the NP test is

$$e_{Q,NP}^{(h)} = \frac{\int h^2 \mathrm{d}\mu}{2\int (1+h)\ln(1+h)\mathrm{d}\mu}.$$

PROOF:
See Bosq (2002b). ■

Note that, since $\lim_{\|h\|_\infty \to 0} e_{Q,NP}^{(h)} = 1$, the test based on (Q_n) is asymptotically locally efficient

4.4 Tests based on the uniform norm

Let K be a bounded kernel of order k and

$$f_n(\cdot) = \frac{1}{n}\sum_{i=1}^{n} K(X_i, \cdot).$$

In this section we consider tests of fit of the form

$$\sqrt{n} \| f_n - 1 \|_\infty > c.$$

Such a test is more discriminating than a Hilbertian test, but its sensitivity may produce an important rate of rejection.

In order to simplify exposition we suppose here that $E_0 = [0,1]^d$, but other spaces could be considered, in particular those that satisfy a 'majorizing measure condition' (see Ledoux and Talagrand 1991).

Now, limit in distribution appears in the following statement.

Theorem 4.6
If $K = 1 + \sum_{j=1}^{k} e_j \otimes e_j$, where $\{1, e_1, \ldots, e_k\}$ is an orthonormal system of $L^2([0,1]^d, \mathcal{B}_{[0,1]^d}, \ell^d)$, $P_{X_1} = \ell^d$, and the e_j's are Lipschitzian, then

$$\sqrt{n}(f_n - 1) \xrightarrow{\mathcal{D}} Z = \sum_{j=1}^{k} N_j e_j \quad (4.10)$$

where $(N_1, \ldots, N_k) \approx \mathcal{N}(0,1)^{\otimes k}$ and convergence in distribution takes place in $C([0,1]^d)$.

PROOF:
$K(X_i, \cdot)$ is a random variable with values in $C([0,1]^d)$, thus (4.10) follows from the CLT for Lipschitz processes (cf. Ledoux and Talagrand 1991, p. 396). ■

The associated test of asymptotic level α and asymptotic power 1 is given by

$$\sqrt{n} \parallel f_n - 1 \parallel_\infty > c_\alpha$$

where

$$P(\parallel Z \parallel_\infty > c_\alpha) = \alpha, \quad (\alpha \in]0,1[).$$

However , determination of c_α is rather intricate since the exact distribution of $\parallel Z \parallel_\infty$ is in general unknown. An exception is $k = 1$ where $\parallel Z \parallel_\infty = |N_1| \parallel e_1 \parallel_\infty$.

For $k > 1$ one must use approximation. The most rudimentary comes from the bound $\parallel Z \parallel_\infty \leq M \sum_{j=1}^k |N_j|$ with $M = \max_{1 \leq j \leq k} \parallel e_j \parallel_\infty$. We have

$$P\left(\sum_{j=1}^{k} |N_j| > c_\alpha M^{-1}\right) \geq \alpha$$

hence $c_\alpha \leq M F_k^{-1}(1-\alpha)$ where F_k is the distribution function of $\sum_{j=1}^k |N_j|$.

If $d = 1$, improvement of that rough approximation uses the following lemma.

Lemma 4.4
Let $(Z_t, t \in [0,1])$ be a Gaussian process such that Var $Z_0 = \sigma^2 > 0$ and

$$\mathrm{E}|Z_t - Z_s|^2 \leq a|t-s|^b \quad (0 < b \leq 2)$$

then, there exists a strictly positive constant γ_b such that

$$P\left(\sup_{0 \leq t \leq 1} Z_t > \eta\right) \leq 4\exp\left(-\frac{\gamma_b}{a}\eta^2\right) + \frac{1}{2}\exp\left(-\frac{\eta^2}{8\sigma^2}\right), \quad \eta > 0.$$

PROOF:
See Leadbetter *et al.* (1983). ■

Here $b = 2$, therefore

$$\alpha = P(\parallel Z \parallel_\infty > c_\alpha) \leq 8\exp\left(-\frac{\gamma_2}{a}c_\alpha^2\right) + \frac{1}{2}\exp\left(-\frac{c_\alpha^2}{8\sigma^2}\right)$$

where $\sigma^2 = \sum_{j=1}^k e_j^2(0)$ is assumed to be strictly positive.

4.5 Extensions. Testing regression

Extensions of functional tests of fit are based on the following idea: let φ be an $(h - e)$-adapted parameter (cf. Chapter 3); in order to test $\varphi = \varphi_0$ one can employ projection estimators of φ or, more generally, linear estimators of φ.

If $\varphi_0 = \varphi_{\theta_0} e_0$ and $\bar{\varphi}_n = \sum_{j=0}^{\infty} \lambda_{jn} \bar{\bar{\varphi}}_{jn} e_j$, cf. (3.1), the test has critical region

$$\| \bar{\varphi}_n - \varphi_0 \|^2 > c_n$$

and one may derive results similar to those in Sections 4.2, 4.3 and 4.4.

Here we only consider the case of regression: let (X, Z) be a $E_0 \times \mathbb{R}$ valued random vector such that $P_X = \mu$ and $\mathrm{E}|Z| < \infty$. One wants to test

$$\mathrm{E}(Z|X) = r_0(X)$$

where r_0 is a fixed version of regression of Z with respect to X, from the i.i.d. sample (X_i, Z_i), $1 \leq i \leq n$.

To begin, one may reduce the problem to a noninfluence test by setting

$$Y_i = Z_i - r_0(X_i),$$

hence H_0 is given by

$$\mathrm{E}(Y|X) = r(X) = 0.$$

The general alternative has the form

$$H_1 : r(X) \in L^2(\mu) - \{0\}.$$

Now let $(e_j, j \geq 0)$ be an orthonormal basis of $L^2(\mu)$ (assumed to be separable), with $e_0 = 1$; r admits the decomposition

$$r = \sum_{j=0}^{\infty} b_j e_j, \quad \sum_j b_j^2 < \infty$$

where

$$b_j = \int r e_j \, \mathrm{d}\mu = \mathrm{E}(\mathrm{E}(Y|X) e_j(X)) = \mathrm{E}(Y e_j(X)), \quad j \geq 0.$$

Natural estimators of b_j are

$$\hat{b}_{jn} = \frac{1}{n} \sum_{i=1}^{n} Y_i e_j(X_i), \quad j \geq 0.$$

Now, we restrict the problem by considering the alternative

$$H_1(k) = \left\{ r : r = \sum_{j=0}^{k} b_j e_j \quad \text{with} \quad \sum_{0}^{k} |b_j| \neq 0 \right\},$$

and we use the statistic $R_n = \sqrt{n} \sum_{j=0}^{k} \hat{b}_{jn} e_j$, hence the test $\| R_n \|^2 > c_n$.

The next statement gives limits in distribution for $\| R_n \|^2$ and

$$\| R'_n \|^2 = n \sum_{j=0}^{k} (\hat{b}_{jn} - b_j)^2.$$

Theorem 4.7

(1) Under $H_1(k)$,

$$\| R'_n \|^2 \xrightarrow{\mathcal{D}} R'$$

where R' has characteristic function $\det(I_{k+1} - 2it\Sigma)^{-1/2}$ with

$$\Sigma = \text{Cov}(Ye_j(X), Ye_\ell(X))_{0 \leq j, \ell \leq k}.$$

(2) Under H_0 and if $\mathrm{E}(Y^2|X) = \gamma$ where $\gamma \neq 0$ is constant,

$$\| R'_n \|^2 = \| R_n \|^2 \xrightarrow{\mathcal{D}} \gamma Q_{k+1}$$

where $Q_{k+1} \approx \chi^2(k+1)$.

(3) Under $H_1(k)$,

$$\| R_n \|^2 \to \infty \quad a.s.$$

PROOF:
Analogous to the proof of Theorem 4.1 and therefore omitted (see Ignaccolo 2002). ■

As before the Sazonov inequality allows us to specify convergence rate; put

$$\Delta_n = \sup_{a \geq 0} |P(\| R'_n \|^2 \leq a) - P(\gamma Q_{k+1} \leq a)|$$

then, if $d_3 = \max_{0 \leq j \leq k} \mathrm{E}|Ye_j(X)|^3 < \infty$ and Σ is positive definite, we have

$$\Delta_n \leq c_0 d_3 \frac{(k+1)^4}{\min\limits_{0 \leq j \leq k} \lambda_j^{3/2}} n^{-1/2}$$

where $\lambda_0, \ldots, \lambda_k$ are eigenvalues of Σ. Note that, under H_0, $\lambda_0 = \cdots = \lambda_k = \gamma$.

It follows that the test

$$\| R_n \|^2 > \gamma \chi^2_\alpha(k+1)$$

has asymptotic level α and asymptotic power 1, with rate $\mathcal{O}(n^{-1/2})$.

If $\gamma = \mathrm{E}(Y^2|X) = \mathrm{E}(\mathrm{E}(Y^2|X)) = \mathrm{E}Y^2$ is unknown, it is replaced by the empirical estimator

$$\hat{\gamma}_n = \frac{1}{n}\sum_{i=1}^{n} Y_i^2,$$

and the limit in distribution of $\| R_n \|^2 \, \hat{\gamma}_n^{-1}$ is clearly a $\chi^2(k+1)$.

On the other hand, the test $\| R_n \|^2 > c_n$ will be consistent if and only if $c_n \to \infty$ and $c_n/n \to 0$ (see Gadiaga 2003).

Finally one may combine the tests of fit studied before and the regression test, to simultaneously test $P_X = \mu$ and $E(Y|X) = 0$.

4.6 Functional tests for stochastic processes

Suppose that data come from a strictly stationary process, then one may again use functional tests of fit.

As an example we consider a discrete strongly mixing process and a statistic based on a kernel of order k. Then the test takes the form

$$Q_n = n \sum_{j=1}^{k} \hat{a}_{jn}^2 > c.$$

If $\sum_\ell \alpha_\ell < \infty$, where (α_ℓ) is the sequence of strong mixing coefficients associated with the observed process $(X_t, t \in \mathbb{Z})$, then, under H_0,

$$Q_n \xrightarrow{\mathcal{D}} Q$$

where Q has characteristic function $\det(I_k - 2it\Gamma)^{-1}$ and

$$\Gamma = (\Gamma_{j\ell}) = \left(\sum_{m \in \mathbb{Z}} \mathrm{E}\Big(e_j(X_0) e_\ell(X_m) \Big) \right), \quad 1 \le j, \ell \le k.$$

In order to achieve construction of the test, estimation of Γ is required; one may set

$$\hat{\Gamma}_{j\ell n} = \sum_{m=-\nu_n}^{\nu_n} \frac{1}{n-m} \sum_{i=1}^{n-m} e_j(X_i) e_\ell(X_{i+m}),$$

$1 \leq j, \ell \leq k$, where $\nu_n < n$. Then, if $\alpha_\ell \leq a\rho^\ell$ $(a > 0,\ 0 < \rho < 1)$ and $\nu_n \simeq n^\gamma$ $(0 < \gamma < 1/2)$ it can be shown that

$$\hat{\Gamma}_{j\ell n} \longrightarrow \Gamma_{j\ell} \quad \text{a.s.}$$

If Γ is positive definite, it follows that eigenvalues $(\hat{\lambda}_{jn}, 1 \leq j \leq k)$ of $(\hat{\Gamma}_{j\ell n})$ converge to eigenvalues $(\lambda_j, 1 \leq j \leq k)$ of Γ, with probability 1.

Now, Q has representation

$$Q = \sum_{j=1}^{k} \lambda_j^2 N_j^2$$

where $(N_1, \ldots, N_k) \approx \mathcal{N}(0,1)^{\otimes k}$. Consequently one may approximate distribution of Q by distribution of

$$\hat{Q}_n = \sum_{j=1}^{k} \hat{\lambda}_{jn}^2 N_j^2$$

where $(N_1, \ldots, N_k)$ may be assumed to be independent of $(X_1, \ldots, X_n)$. This allows us to achieve construction of the test.

Notes

The ‘smooth’ χ^2-tests appear for the first time in Neyman (1937). Bosq (1978) has generalized these by interpreting them as tests based on projection density estimators.

As indicated above, results in Sections 4.1, 4.2 and 4.3 are due to Bosq (1980, 1983a, 2002b). Gadiaga (2003) has obtained results in Section 4.4. Extensions (Sections 4.5 and 4.6) come from Gadiaga and Ignaccolo (2005). Related results appear in Hart (1997).

Finally simulations are performed in Bosq (1989), Ignaccolo (2002) and Gadiaga (2003).

5

Prediction by projection

In Chapter 2 we have presented some asymptotic results about prediction when distribution of the observed process only depends on a finite-dimensional parameter.

We now study a class of nonparametric predictors based on projection methods. The general framework allows us to apply the obtained results to prediction of conditional distribution and to construction of prediction intervals.

5.1 A class of nonparametric predictors

Let $X = (X_t,\ t \in \mathbb{Z})$ be a strictly stationary Markov process with values in the measurable space $(E_0, \mathcal{B}_0)$, and observed at times $0, 1, \ldots, n$.

Suppose that $P_{X_0} = \mu$ is known. One wants to forecast $g(X_{n+1})$ where g is a measurable application from E_0 to H, a separable Hilbert space, with scalar product $\langle \cdot, \cdot \rangle$, norm $\| \cdot \|$ and Borel σ-field $\mathcal{B}_H$. Moreover suppose that $g \in L^2_H(\mu)$, that is

$$\| g \|_{L^2_H(\mu)} = \left(\int_{E_0} \| g(x) \|^2 \, \mathrm{d}\mu(x) \right)^{1/2} < \infty.$$

As noticed in Chapter 1, it is equivalent to consider prediction of

$$r(X_n) = \mathrm{E}(g(X_{n+1})|X_n), \quad r \in \mathcal{R}$$

where $\mathcal{R}$ is the family of possible regressions of $g(X_{n+1})$ with respect to X_n.

As in chapters 3 and 4 we use a method based on kernels of the form

$$K_n = \sum_{j=0}^{\infty} \lambda_{jn} e_{jn} \otimes e_{jn}, \quad n \geq 1, \tag{5.1}$$

Inference and Prediction in Large Dimensions D. Bosq and D. Blanke

where, for each n, $(e_{jn},\ j \geq 0)$ is an orthonormal system in $L^2(\mu)$ such that $\| e_{jn} \|_\infty = M_{jn} < \infty, j \geq 0$ and $\sum_{j=0}^{\infty} |\lambda_{jn}| M_{jn}^2 < \infty$.

K_n defines a continuous linear operator on $L_H^2(\mu)$ through the formula

$$(K_n\varphi)(y) = \int_{E_0} K_n(x,y)\varphi(x)\,\mathrm{d}\mu(x),\ y \in E_0; \varphi \in L_H^2(\mu).$$

The predictor associated with K_n is defined as

$$\begin{aligned} r_n(X_n) &= \frac{1}{n}\sum_{i=1}^{n} K_n(X_{i-1}, X_n) g(X_i) \\ &= \sum_{j=0}^{\infty} \lambda_{jn} e_{jn}(X_n)\widehat{b}_{jn} \end{aligned}$$

with

$$\widehat{b}_{jn} = \frac{1}{n}\sum_{i=1}^{n} e_{jn}(X_{i-1}) g(X_i) \in H, \quad j \geq 0; n \geq 1.$$

In order to study prediction error of $r_n(X_n)$ we shall suppose that (X_t) is strongly mixing with coefficients $(\alpha(\ell),\ \ell \geq 0)$ and use the following Hilbertian version of the Davydov inequality (cf. Merlevède *et al.* 1997).

Lemma 5.1
If Y and Z are H-valued random variables such that $\mathrm{E} \| Y \|^q < \infty$ *and* $\mathrm{E} \| Z \|^r < \infty$ *with* $1/r + 1/q = 1 - 1/p \geq 0$, *then*

$$|\mathrm{E}\langle Y, Z\rangle - \langle \mathrm{E}Y, \mathrm{E}Z\rangle| \leq 18(\alpha(\sigma(Y), \sigma(Z)))^{1/p}(\mathrm{E} \| Y \|^q)^{1/q} \cdot (\mathrm{E} \| Z \|^r)^{1/r} \quad (5.2)$$

From Lemma 5.1 one may obtain a bound for the quadratic error of prediction.

Lemma 5.2
If $g \in L_H^q(\mu)$ *where* $2 < q \leq \infty$ *and* $n \geq 2$, *then*

$$\mathrm{E} \| r_n(X_n) - r(X_n) \|^2 \leq A_n + B_n + C_n$$

where

$$\begin{aligned} A_n &= 2 \| K_n r - r \|_{L_H^2(\mu)}^2, \\ B_n &= \frac{2 \| g \|_q^2}{n} B_{n,q} \sum_{j \geq 0} \lambda_{jn}^2 M_{jn}^{2+\frac{4}{q}}, \end{aligned}$$

with

$$B_{n,q} = 1 + M_{jn}^{2-\frac{4}{q}} 18 \sum_{\ell=0}^{n-2} \alpha^{\frac{q-2}{q}}(\ell),$$

and where

$$C_n = \frac{8 \parallel g \parallel_q^2}{n^2} C_{n,q} \sum_{j \neq j'} |\lambda_{jn}\lambda_{j'n}| M_{jn}^2 M_{j'n}^2$$

with

$$C_{n,q} = 72 \sum_{\ell=0}^{[n/2]} (\ell + 2)\alpha^{\frac{q-2}{q}}(\ell) + 18 \sum_{\ell=0}^{n-1} (2\ell + 3)\alpha^{\frac{q-2}{q}}(\ell).$$

PROOF:
The proof distinguishes between various groupings of variables, in order to optimize the use of (5.2); the σ-fields that appear here taking the form $\sigma(X_s, \ s \leq t)$ and $\sigma(X_s, \ s \geq t + \ell)$. Computations are tedious since the presence of X_n creates intricate groupings. For details we refer to Bosq (1983b) where calculations are performed with a β-*mixing coefficient* which does not modify their principle. ■

The following statement is an immediate consequence of Lemma 5.2.

Theorem 5.1
If $g \in L_H^q(\mu)$, $(2 < q \leq \infty)$ and

(a) $\parallel K_n r - r \parallel_{L_H^2(\mu)} \underset{n\to\infty}{\longrightarrow} 0, \ r \in \mathcal{R},$
(b) $\frac{1}{n}\sum_{j\geq 0} |\lambda_{jn}|^{\gamma} M_{jn}^{2\gamma} \underset{n\to\infty}{\longrightarrow} 0; \ \gamma = 1, 2,$
(c) $\sum_{\ell\geq 1} \ell\alpha^{\frac{(q-2)}{q}}(\ell) < \infty,$
then

$$\mathrm{E} \parallel r_n(X_n) - r(X_n) \parallel^2 \underset{n\to\infty}{\longrightarrow} 0, \ r \in \mathcal{R}.$$

Now, if K_n is a *finite-dimensional kernel*, i.e.

$$K_n = \sum_{j=0}^{k_n} e_j \otimes e_j, \quad n \geq 1$$

where (e_j) is a uniformly bounded orthonormal system in $L^2(\mu)$ with $e_0 \equiv 1$, one may derive a more precise result.

Corollary 5.1

If K_n is finite-dimensional with $k_n \to \infty$, one may replace (b) by $k_n/n \to \infty$ to obtain

$$\mathrm{E} \parallel r_n(X_n) - r(X_n) \parallel^2 = \mathcal{O}\left(\sum_{j>k_n} \left\| \int e_j r \,\mathrm{d}\mu \right\|^2 + \frac{k_n+1}{n}\right), \quad r \in \mathcal{R}.$$

In particular, if $\parallel \int e_j r \mathrm{d}\mu \parallel = \mathcal{O}(j^{-\gamma})$, $(\gamma > 1/2)$ the choice $k_n \simeq n^{1/(2\gamma)}$ gives

$$\mathrm{E} \parallel r_n(X_n) - r(X_n) \parallel^2 = \mathcal{O}\left(n^{-\frac{2\gamma-1}{2\gamma}}\right).$$

For example, if $E_0 = [0,1]$, μ is uniform distribution on $[0,1]$, r has two continuous derivatives and $e_0 \equiv 1, e_j(x) = \sqrt{2}\cos 2\pi jx, 0 \le x \le 1; j \ge 1$ then $k_n \simeq n^{-1/4}$ gives

$$\mathrm{E} \parallel r_n(x) - r(x) \parallel^2 = \mathcal{O}(n^{-3/4}).$$

Now, if r is known to be a polynomial of degree less than or equal to k_0 over [0,1], one may use the orthonormal system associated with $1, x, \ldots, x^{k_0}$ (Legendre polynomials) and the rate is $\mathcal{O}(n^{-1})$.

Finally if magnitude of the Fourier coefficients of r is unknown one may use an estimation $\widehat{k_n}$ of the best k_n, similarly to Chapter 3.

Now it should be noticed that the assumption '$P_{X_0} = \mu$ known' is rather heavy. However we are going to see that $r_n(X_n)$ is, in some sense, robust with respect to μ:

Theorem 5.2

Let μ_0 be a probability on $(E_0, \mathcal{B}_0)$ such that $\mathrm{d}\mu/\mathrm{d}\mu_0$ does exist and is bounded. Then, if $\parallel K_n r - r \parallel_{L^2_H(\mu_0)} \to 0$, $r \in \mathcal{R}$, and $\sum_{j=0}^{\infty} |\lambda_{jn}| M_{jn}^2 < \infty$ (for the decomposition of K_n in $L^2(\mu)$), if $g \in L^q_H(\mu_0)$ and assumptions (b) *and* (c) *in Theorem 5.1 hold, we have*

$$\overline{\lim_{n\to\infty}} \mathrm{E} \parallel r_n(X_n) - r(X_n) \parallel^2_{L^2_H(\mu)} \le 2 \left\| \frac{\mathrm{d}\mu}{\mathrm{d}\mu_0} \right\|_\infty \left\| 1 - \frac{\mathrm{d}\mu}{\mathrm{d}\mu_0} \right\|^2_{L^2(\mu_0)}. \tag{5.3}$$

Therefore, $r_n(X_n)$ remains an acceptable predictor if μ_0 considered as an approximation of μ, is such that the bound in (5.3) is small with respect to the structural prediction error $\parallel g \parallel^2_{L^2_H(\mu)} - \parallel r \parallel^2_{L^2_H(\mu)}$.

Proof of Theorem 5.2 appears in Bosq (1983b).

Finally, if μ is completely unknown but has a density, one may construct an estimator μ_n of μ by using the projection density estimator studied in Chapter 3. Replacing $L^2_H(\mu)$ by $L^2_H(\mu_n)$ one obtains a statistical predictor close to $r_n(X_n)$ by using Theorem 5.2. We leave details to the reader.

5.2 Guilbart spaces

In order to apply the previous method to prediction of the conditional distribution $P_{X_{n+1}}^{X_n}$ we need a structure of Hilbert space for the family of bounded signed measures on $(E_0, \mathcal{B}_0)$. For this purpose we introduce the notion of Guilbart space.

Suppose that E_0 is a separable metric space, with its Borel σ-field $\mathcal{B}_{E_0}$ and consider $G : E_0 \times E_0 \mapsto \mathbb{R}$, a continuous symmetric bounded function of positive type, i.e.

$$(\forall k \geq 1),\ (\forall (a_1, \ldots, a_k) \in \mathbb{R}^k),\ (\forall (x_1, \ldots, x_k) \in E_0^k),\ \sum_{1 \leq i,j \leq k} a_i a_j G(x_i, x_j) \geq 0.$$

Then the *reproducing kernel Hilbert space* (RKHS) H_G generated by G is obtained by completing the space of functions of the form $\sum_{i=1}^{k} a_i G(x_i, \cdot)$, with respect to the scalar product,

$$\left(\sum_{i=1}^{k} a_i G(x_i, \cdot),\ \sum_{j=1}^{k'} b_j G(y_j, \cdot) \right)_G = \sum_{\substack{1 \leq i \leq k \\ 1 \leq j \leq k'}} a_i b_j G(x_i, y_j).$$

Then, G is a *reproducing kernel* in the sense that

$$f(x) = \langle f(\cdot),\ G(x, \cdot) \rangle_G,\quad x \in E_0,\ f \in H_G.$$

Now let $\mathcal{M}$ (respectively $\mathcal{M}_+$) be the family of bounded signed (respectively positive) measures on $(E_0, \mathcal{B}_0)$; one sets

$$f_\lambda(\cdot) = \int_{E_0} G(x, \cdot)\, \mathrm{d}\lambda(x),\quad \lambda \in \mathcal{M}.$$

Then, for a suitable choice of $G, \lambda \mapsto f_\lambda$ is injective and realizes a plug-in of $\mathcal{M}$ in H_G. Actually it can be proved that $\mathcal{M}$ is dense in H_G. The associated scalar product on $\mathcal{M}$ is defined as

$$\langle \lambda, \nu \rangle_G = \int_{E_0 \times E_0} G(x, y) \mathrm{d}\lambda(x)\, \mathrm{d}\nu(y);\quad \lambda, \nu \in \mathcal{M}.$$

A typical choice of G is

$$G(x, y) = \sum_{i=0}^{\infty} f_i(x) f_i(y)$$

where (f_i) determines the elements of $\mathcal{M}$, i.e.

$$\int f_i \mathrm{d}\lambda = 0,\quad i \geq 0 \Leftrightarrow \lambda = 0.$$

Under mild conditions the topology induced on $\mathcal{M}_+$ by $\langle\cdot,\cdot\rangle_G$ is the weak topology (see Guilbart 1979). In particular this property is always satisfied if E_0 is compact (provided $\langle\cdot,\cdot\rangle_G$ is not degenerate). If $\langle\cdot,\cdot\rangle_G$ induces weak topology on $\mathcal{M}_+$ we will say that H_G is a *Guilbart space*. Properties of Guilbart spaces appear in Guilbart (1979) and Berlinet and Thomas-Agnan (2004).

For example, if E_0 is compact in $\mathbb{R}$, $G(x,y) = \mathrm{e}^{xy}$ generates a Guilbart space.

5.3 Predicting the conditional distribution

We now study prediction of $P_{X_{n+1}}^{X_n}$, a more precise indicator of future than $\mathrm{E}^{X_n}(X_{n+1})$.

In order to apply results in Section 5.1 we take $H = H_G$ a Guilbart spaces space and define $g : E_0 \mapsto H_G$ by

$$g(x) = \delta_{(x)}, \quad x \in E_0$$

or equivalently

$$g(x) = G(x,\cdot), \quad x \in E_0.$$

The next statement shows that prediction of $g(X_{n+1})$ is equivalent to prediction of $P_{X_{n+1}}^{X_n}$.

Lemma 5.3
If H_G is a Guilbart space we have

$$E^{X_n}\delta_{(X_{n+1})} = P_{X_{n+1}}^{X_n}. \tag{5.4}$$

PROOF:
First, note that g is bounded since

$$\sup_{x\in E_0}\langle\delta_{(x)},\delta_{(x)}\rangle^{1/2} = \sup_{x\in E_0} G(x,x)^{1/2} = \| G \|_\infty^{1/2}.$$

Now, for every $\lambda \in \mathcal{M}$,

$$\begin{aligned}\langle \mathrm{E}^{X_n}\delta_{(\mathcal{X}_{n+1})},\lambda\rangle_G &= \mathrm{E}^{X_n}\langle\delta_{(\mathcal{X}_{n+1})},\lambda\rangle_G \\ &= \mathrm{E}^{X_n}\int G(X_{n+1},x)\,\mathrm{d}\lambda(x) \\ &= \int G(y,x)\mathrm{d}\lambda(x)\mathrm{d}P_{X_{n+1}}^{X_n}(y) = \langle P_{X_{n+1}}^{X_n},\lambda\rangle_G\end{aligned}$$

and, since $\mathcal{M}$ is dense in H_G, (5.4) follows. ■

We may now apply Theorem (5.1) with

$$r(X_n) = P_{X_{n+1}}^{X_n}$$

and our predictor takes the form of a random signed bounded measure:

$$\begin{aligned} r_n(X_n) &= \frac{1}{n}\sum_{i=1}^{n} K_n(X_{i-1}, X_n)\delta_{(X_i)} \\ &:= \sum_{i=1}^{n} p_{in}\delta_{(X_i)} \end{aligned}$$

where K_n is given by (5.1).

Then we have consistency of the predictor:

Corollary 5.2
If (a), (b) and (c) in Theorem 5.1 hold then

$$\mathrm{E}\left\| \sum_{i=1}^{n} p_{in}\delta_{(X_i)} - P_{X_{n+1}}^{X_n} \right\|_{H_G}^2 \underset{n\to\infty}{\longrightarrow} 0.$$

The i.i.d. case

In the i.i.d. case a consequence of the previous result is a Glivenko–Cantelli type theorem. We complete it with an exponential bound due to C. Guilbart.

Theorem 5.3 (*Glivenko–Cantelli theorem*)
Let $(X_t, t \in \mathbb{Z})$ be an i.i.d. E_0-valued sequence with common distribution μ. Then

$$\mathrm{E} \parallel \mu_n - \mu \parallel_{H_G}^2 = \mathcal{O}\left(\frac{1}{n}\right) \tag{5.5}$$

where $\mu_n = \left(\sum_{i=1}^{n} \delta_{(X_i)}\right)/n$.

Moreover, if $G(x, y) = \sum_{i=1}^{\infty} |\alpha_i| f_i \otimes f_i$, where $\alpha = \sum_i |\alpha_i| < \infty$ then, for all $\eta > 0$,

$$P(\parallel \mu_n - \mu \parallel_{H_G} \geq \eta) \leq 2n_0(\eta) \exp\left(-\frac{n\eta^2}{4\alpha}\right), \quad n \geq 1 \tag{5.6}$$

where

$$n_0(\eta) = \min\left\{ p : \sum_{i>p} |\alpha_i| < \frac{\eta^2}{8} \right\}.$$

PROOF:
Here $P_{X_{n+1}}^{X_n} = \mu$, thus $r = f_\mu(\cdot) = \int_{E_0} G(x, \cdot)\, \mathrm{d}\mu(x)$. Now if $K_n \equiv 1$ we clearly have $K_n r = r$. Applying Corollary 5.1 with $k_n = 0$ one obtains (5.5).

Concerning (5.6) we refer to Guilbart (1979). ■

Example 5.1
If $G(x, y) = \mathrm{e}^{xy} = \sum_{i=0}^{\infty} x^i y^i / i!$, $0 \leq x, y \leq 1$, we can choose $n_0(\eta)$ as the smallest integer such that $(n_0(\eta) + 1)! \geq [8\mathrm{e}/\eta^2] + 1$. ◇

5.4 Predicting the conditional distribution function

The general results presented before are not very easy to handle in practice. We now consider a more concrete situation: consider real data $Y_0, \ldots, Y_n$ with a common continuous and strictly increasing distribution function, say F. If F is known one may construct new data $X_i = F(Y_i)$, $0 \leq i \leq n$ with uniform distribution μ on $[0, 1]$.

Thus, we now suppose that $(E_0, \mathcal{B}_{E_0}) = ([0, 1], \mathcal{B}_{[0,1]})$ and choose

$$g(x, \cdot) := g(x)(\cdot) = \mathbb{1}_{[x,1[}(\cdot), \quad 0 \leq x \leq 1,$$

then

$$r(x, \cdot) := r(x)(\cdot) = P(X_{n+1} \leq \cdot | X_n = x), \quad 0 \leq x \leq 1,$$

is the conditional distribution function of X_{n+1} with respect to X_n.

A simple predictor of $g(X_n)$ is based on the kernel

$$K_{n,0} = k_n \sum_{j=1}^{k_n} \mathbb{1}_{\mathbf{I}_{jn}} \otimes \mathbb{1}_{\mathbf{I}_{jn}}$$

where

$$\mathrm{I}_{jn} = \left[\frac{j-1}{k_n}, \frac{j}{k_n}\right[, \quad 1 \leq j \leq k_n.,$$

hence

$$r_{n,0}(X_n)(\cdot) = \frac{k_n}{n} \sum_{i=1}^{n} \left(\sum_{j=1}^{k_n} \mathbb{1}_{\mathrm{I}_{jn}}(X_{i-1}) \mathbb{1}_{\mathrm{I}_{jn}}(X_n) \right) \mathbb{1}_{[X_i,1[}(\cdot).$$

In order to apply Theorem 5.1 we take $H = L^2(\mu)$, then if conditions in Theorem 5.1 are satisfied with $q = \infty$, one has

$$\mathrm{E} \parallel r_{n,0}(x) - r(x) \parallel^2_{L^2_{L^2(\mu)}(\mu)} \rightarrow 0$$

provided $k_n \rightarrow \infty$ and $k_n^3/n \rightarrow 0$.

Now, our aim is to construct a prediction interval. For this purpose we need a sharper result; we suppose that the density $f_{(X_0,X_1)}$ of (X_0, X_1) does exist, is continuous and satisfies the Lipschitz condition

$$(\mathbf{L})\ \ |f_{(X_0,X_1)}(x,y) - f_{(X_0,X_1)}(x',y)| \leq \ \ell|x'-x|,$$

$0 \leq x, x' \leq 1$, $0 \leq y \leq 1$, where ℓ is a constant.

Then, one may state a uniform result.

Theorem 5.4

Let $X = (X_t,\ t \in \mathbb{Z})$ be a strictly stationary geometrically α-mixing, such that $P_{X_0} \approx \mathcal{U}_{[0,1]}$ and (L) holds, then

$$\sup_{0\leq y\leq 1} |r_{n,0}(X_n, y) - r(X_n, y) = \mathcal{O}(n^{-\delta}) \quad a.s. \tag{5.7}$$

for all $\delta \in]0, 1/6[$.

PROOF:
Set

$$\overline{r}_{n,0}(X_n, y) = \sum_{j=1}^{k_n} b_{jn}(y) e_{jn}(X_n),$$

where

$$b_{jn}(y) = \mathrm{E}\widehat{b}_{jn}(y) = \mathrm{E}\left(\frac{1}{n}\sum_{i=1}^{n} e_j(X_{i-1})g(X_i, y)\right)$$
$$= \int_0^1 r(x,y)e_{jn}(x)\mathrm{d}x$$

and

$$e_{jn} = k_n^{1/2}\mathbb{1}_{\mathrm{I}_{jn}}, \quad 1 \leq j \leq k_n.$$

From (L) it follows that

$$|r(x',y) - r(x,y)| \leq \ell|x'-x|,\ 0 \leq x, x', y \leq 1,$$

where $r(x, y) = P(X_{n+1} \leq y|X = x)$.
Now, let $\hat{j}_n$ be the unique j such that $X_n \in \mathrm{I}_{jn}$; one has

$$\overline{r}_{n,0}(X_n, y) = k_n^{1/2} b_{\hat{j}_n,n}(y) = k_n \int_{\mathrm{I}_{\hat{j}_n,n}} r(x,y)dx$$
$$= r(\xi_n, y)$$

where $\xi_n \in I_{\hat{j}_n,n}$. Therefore

$$R_n := \sup_{0 \le y \le 1} |r(X_n, y) - \overline{r}_{n,0}(X_n, y)| \le \ell(X_n - \xi_n) \le \ell k_n^{-1}. \tag{5.8}$$

On the other hand

$$\begin{aligned}\Delta_n &:= \sup_{0 \le y \le 1} |r_{n,0}(X_n, y)| - \overline{r}_{n,0}(X_n, y)| \\ &\le k_n^{1/2} \sum_{j=1}^{k_n} \| \widehat{b}_{jn} - b_{jn} \|_\infty .\end{aligned}$$

Then, in order to study Δ_n, we first specify variations of $\widehat{b}_{jn}(y)$ and $b_{jn}(y)$: consider the intervals $J_{mn} = [(m-1)/\nu_n, m/\nu_n[, m = 1, \ldots, \nu_n - 1$ and $J_{\nu_n m} = [(\nu_n - 1)/\nu_n, 1]$, where ν_n is a positive integer; then, if $y, y' \in J_{mn}$ (with, for example, $y < y'$), we have

$$\begin{aligned}|\widehat{b}_{jn}(y') - \widehat{b}_{jn}(y)| &\le \frac{k_n^{1/2}}{n} \sum_{i=1}^{n} \mathbb{1}_{[y,y'[}(X_{i-1}) \\ &\le \frac{k_n^{1/2}}{n} \sum_{i=1}^{n} \mathbb{1}_{J_{mn}}(X_{i-1}) := k_n^{1/2} \overline{Z}_{mn},\end{aligned}$$

similarly

$$\begin{aligned}|b_{jn}(y') - b_{jn}(y)| &= |\mathrm{E}(e_j(X_0) \mathbb{1}_{[y,y'[}(X_1))| \\ &\le k_n^{1/2} P(X_1 \in J_{mn}) = k_n^{1/2} \nu_n^{-1},\end{aligned}$$

hence

$$\begin{aligned}k_n^{1/2} \sup_{y \in J_{m,n}} |\widehat{b}_{jn}(y) - b_{jn}(y)| &\le k_n^{1/2} (\widehat{b}_{jn} - b_{jn})((m-1)\nu_n^{-1}) \\ &\quad + k_n \nu_n^{-1} + k_n \overline{Z}_{mn},\end{aligned}$$

therefore

$$e_{jn} := k_n^{1/2} \| \widehat{b}_{jn} - b_j \|_\infty \le k_n^{1/2} B_n + k_n \nu_n^{-1} + k_n Z_n$$

where

$$B_n = \max_{1 \le m \le \nu_n} |(\widehat{b}_{jn} - b_{jn})((m-1)\nu_n^{-1})|$$

and

$$Z_n = \max_{1 \le m \le \nu_n} \overline{Z}_{mn}.$$

Now, for all $\delta_n > 0$,

$$P(\Delta_n > \delta_n) \leq \sum_{j=1}^{k_n} P(c_{jn} > \delta_n k_n^{-1}),$$

and we have

$$\{c_{jn} > \delta_n k_n^{-1}\} \Rightarrow \left\{B_n > \frac{\delta_n}{2k_n^{3/2}}\right\} \cup \left\{Z_n > \frac{\delta_n}{2k_n^2} - \frac{1}{\nu_n}\right\}.$$

Now, applying ν_n times the exponential inequality (3.25) (see Lemma 3.8) one obtains

$$P\left(B_n > \frac{\delta_n}{2k_n^{3/2}}\right) \leq \nu_n\left[2\exp\left(-\frac{\delta_n^2}{32k_n^4}q\right) + 11\left(\left(1+\frac{8k_n^2}{\delta_n}\right)^{1/2} q\alpha\left(\left[\frac{n}{q}\right]\right)\right)\right]. \quad (5.9)$$

Similarly, noting that

$$\mathrm{E}(\overline{Z}_{mn}) = P(X_0 \in \mathrm{J}_{mn}) = \nu_n^{-1}$$

one may use ν_n times the same inequality, provided $\gamma_n = \delta_n/(2k_n^2) - 2/\nu_n > 0$ for obtaining

$$P(Z_n > \gamma_n) \leq \nu_n\left[2\exp\left(-\frac{\gamma_n^2}{8}q\right) + 11\left(1+\frac{4}{\gamma_n}\right)^{1/2} q\alpha\left(\left[\frac{n}{q}\right]\right)\right]. \quad (5.10)$$

Combining (5.9) and (5.10) yields

$$\begin{aligned} P(\Delta_n > \delta_n) \leq k_n\nu_n \Bigg[& 2\exp\left(-\frac{\delta_n^2}{32k_n^4}q) + 11(1+\frac{8k_n^2}{\delta_n}\right)^{1/2} q\, a_0 \mathrm{e}^{-b_0\left[\frac{n}{q}\right]} \\ & +2\exp\left(-\frac{\gamma_n^2}{8}q\right) + 11\left(1+\frac{4}{\gamma_n}\right)^{1/2} q a_0 \mathrm{e}^{-b_0\left[\frac{n}{q}\right]}\Bigg] \end{aligned} \quad (5.11)$$

where $a_0 > 0$ and $b_0 > 0$ are such that $\alpha(k) \leq a_0\mathrm{e}^{-b_0 k}$, $k \geq 1$.

Now, choose $q = q_n \simeq n^{\gamma}$, $k_n \simeq n^{\beta}$, $\nu_n \simeq n^{\nu}$, $\delta_n \simeq n^{-\delta}$ with $\nu > 2\beta + \delta$, $1 > \alpha > 4\beta + 2\delta$, $\beta > \delta$, then (5.11) implies

$$\sum_n P(\Delta_n > n^{-\delta}) < \infty$$

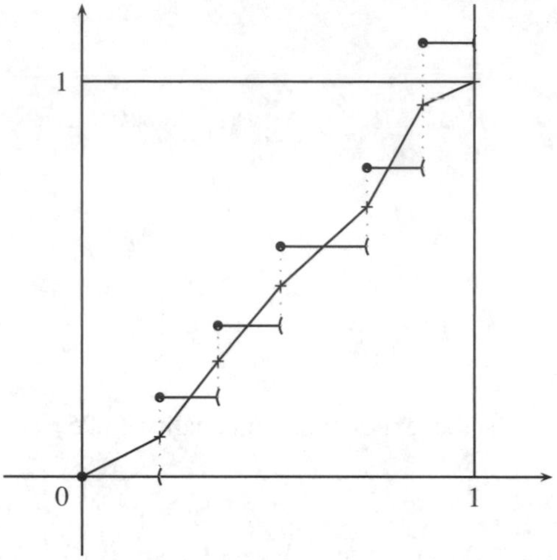

Figure 5.1 Linear interpolation.

and, (5.8) gives

$$R_n = \mathcal{O}(n^{-\beta}),$$

hence (5.7) since α may be taken arbitrarily close to 1 and δ arbitrarily close to β. ■

Prediction interval

$r_{n,0}(X_n, \cdot)$ is a nondecreasing step function such that $r_{n,0}(X_{n,0}) = 0$. By linear interpolation (see Figure 5.1) one may slightly modify $r_{n,0}$ in order to obtain a continuous distribution function $G_n(X_n, \cdot)$ on $[0, 1]$.

Now let $\alpha \in]0, 1[$ and y_n and z_n be such that

$$G_n(y_n) = \frac{\alpha}{2}$$

and

$$G_n(z_n) = 1 - \frac{\alpha}{2}$$

then using Theorem 5.4 it can be proved that

$$P(X_{n+1} \in [y_n, z_n] | X_n) \to 1 - \alpha, \quad 0 < \alpha < 1$$

thus $[y_n, z_n]$ is a prediction interval for X_{n+1} of asymptotic level $1 - \alpha$.

Notes

As indicated in the text, Theorems 5.1 and 5.2 are slight improvements of similar results in Bosq (1983b). Guilbart spaces were introduced in Guilbart's Thesis and developed in Guilbart (1979). See also Berlinet and Thomas-Agnan (2004). Prediction of the conditional distribution function is a slight improvement of Bosq (2005b). A partial extension to continuous time is considered in Bosq and Delecroix (1985).

Part III

Inference by Kernels

6

Kernel method in discrete time

6.1 Presentation of the method

Let $X_1, \dots, X_n$ be independent and identically distributed (i.i.d.) observed real random variables with density f, and consider the basic problem of estimating f from these data. Histograms are quite popular nonparametric density estimators, indeed their implementation and their interpretation are simple (even for nonstatisticians). On the other hand, their various drawbacks lead to the development of more sophisticated estimators, like the kernel ones. Let us briefly recall that histograms are obtained when the sample space is divided into disjoint bins and one considers the proportion of observations falling in each bin (see section 4.1 for its mathematical definition). As a result, these estimators are discontinuous and strongly depend on the choice of the end points of the bins. To remove such dependence, it is more convenient to consider bins which are centred at the point of estimation x, which leads to the naive kernel estimator:

$$f_n^0(x) = \frac{1}{nh_n} \sum_{i=1}^{n} \mathbb{1}_{\left[x-\frac{h_n}{2}, x+\frac{h_n}{2}\right]}(X_i) = \frac{1}{nh_n} \sum_{i=1}^{n} \mathbb{1}_{\left[-\frac{1}{2}, \frac{1}{2}\right]}\left(\frac{x - X_i}{h_n}\right), \qquad x \in \mathbb{R}.$$

Here (h_n) is a given positive sequence called 'the bandwidth' (equivalent to the histogram's binwidth). The accuracy of f_n^0 depends strongly on the choice of h_n that must reconcile the smallness of $\left[x - \frac{h_n}{2}, x + \frac{h_n}{2}\right]$ with a large number of observations falling in this interval. Since this expected number has the same order as $nh_n f(x)$ (provided $f(x) > 0$ with x a point of continuity), we obtain the natural conditions: $h_n \to 0^+$ and $nh_n \to +\infty$ as $n \to +\infty$.

Inference and Prediction in Large Dimensions D. Bosq and D. Blanke

Now, to get a smoother density estimate, the indicator function is replaced by a kernel, that is, a smooth nonnegative function (usually a symmetric probability density):

$$f_n(x) = \frac{1}{nh_n} \sum_{i=1}^{n} K\left(\frac{x - X_i}{h_n}\right), \qquad x \in \mathbb{R},$$

leading to the general form of univariate *kernel density estimators*. In this way, one may assign a greater weight to observations which are close to x, whereas remote ones may receive little or no weight.

Finally, the kernel method may be interpreted as a regularization of the *empirical measure*:

$$\mu_n = \frac{1}{n} \sum_{i=1}^{n} \delta_{(X_i)}$$

which is not absolutely continuous with respect to Lebesgue measure. Namely, the kernel estimator consists in the convolution of μ_n with the kernel K:

$$f_n = \mu_n * \frac{1}{h_n} K\left(\frac{\cdot}{h_n}\right).$$

Now for the purpose of performing nonparametric prediction, we have to consider the closely related problem of nonparametric regression estimation. More precisely, let $(X_1, Y_1), \ldots, (X_n, Y_n)$ be i.i.d. bi-dimensional random variables, such that a specified version r of the regression of Y_i on X_i does exist: $r(x) = \mathrm{E}(Y_i | X_i = x)$, $x \in \mathbb{R}$. In contrast to the traditional nonlinear regression (where r is explicit but with finitely many unknown parameters), no functional form (except some kind of smoothness conditions) is required or imposed on r. In this way, one avoids the serious drawback of the choice of an inadequate regression model.

Like the histogram, the regressogram is easy to construct and to interpret. The sample space of the X_i's is partitioned into bins and the estimator is obtained by averaging Y_i's with the corresponding X_i's in the considered bin. A smoother estimate is the kernel one, obtained with a distance-weighted average of Y_i where weights depend on the distance between X_i and the point of evaluation:

$$r_n(x) = \frac{\sum_{i=1}^{n} Y_i K\left(\frac{x - X_i}{h_n}\right)}{\sum_{i=1}^{n} K\left(\frac{x - X_i}{h_n}\right)} := \sum_{i=1}^{n} \mathcal{W}_{i,n}(x) Y_i.$$

Here the $\mathcal{W}_{i,n}(x)$'s represent the randomized weights defined by

$$\mathcal{W}_{i,n}(x) = K\left(\frac{x - X_i}{h_n}\right) \left(\sum_{i=1}^{n} K\left(\frac{x - X_i}{h_n}\right)\right)^{-1}.$$

This estimator was conjointly introduced by Nadaraja (1964) and Watson (1964). Note that in the case where $\sum_{i=1}^{n} K((x - X_i)/h_n) = 0$, commonly used settings for the weights are given by $\mathcal{W}_{i,n}(x) \equiv n^{-1}$ or $\mathcal{W}_{i,n}(x) \equiv 0$.

In this chapter, first we recall some basic properties of these estimators in the i.i.d. case and next, in our viewpoint of prediction, we focus on the properties of convergence of kernel functional estimators (density, regression) in the case of correlated sequences of random variables.

6.2 Kernel estimation in the i.i.d. case

For the whole chapter, we consider a d-dimensional kernel satisfying the following conditions:

$$\mathcal{K}: \begin{cases} K: \quad \mathbb{R}^d \to \mathbb{R} \\ K \text{ is a bounded symmetric density such that, if } u = (u_1, \ldots, u_d) \\ \lim_{\|u\| \to \infty} \| u \|^d K(u) = 0 \\ \text{and } \int_{\mathbb{R}^d} |v_i| \, |v_j| \, K(v) \mathrm{d}v < \infty, \quad i,j = 1, \ldots, d \end{cases} \tag{6.1}$$

where $\| \cdot \|$ denotes the euclidian norm over $\mathbb{R}^d$. Typical examples of multivariate kernels satisfying conditions $\mathcal{K}$ are: the naive kernel $K(u) = \mathbb{1}_{[-1/2,1/2]^d}(u)$, the normal kernel $K(u) = (2\pi\sigma^2)^{-d/2} \exp(- \| u \|^2 /(2\sigma^2))$, the Epanechnikov kernel $K(u) = \left(\frac{3}{4}\sqrt{5}\right)^d \prod_{i=1}^{d} (1 - u_i^2/5) \mathbb{1}_{[-\sqrt{5},\sqrt{5}]}(u_i)$.

Suppose that $X_1, \ldots, X_n$ are i.i.d. $\mathbb{R}^d$-valued random variables with density f, the associated multivariate kernel estimator is then defined as:

$$f_n(x) = \frac{1}{nh_n^d} \sum_{i=1}^{n} K\left(\frac{x - X_i}{h_n}\right) := \frac{1}{n} \sum_{i=1}^{n} K_h(x - X_i), \qquad x \in \mathbb{R}^d, \tag{6.2}$$

where $K_h(u) := h_n^{-d} K(u/h_n)$, $u = (u_1, \ldots, u_d) \in \mathbb{R}^d$.

Note that more general kernels are considered in the literature. Here, we restrict our study to positive ones in order to obtain estimators that are probability densities. In the following, we need an approximation of identity:

Lemma 6.1 *(Bochner lemma)*
If $H \in L^1(\mathbb{R}^d) \cap L^\infty(\mathbb{R}^d)$ is such that $\int H = 1$ and

$$\| x \|^d \, |H(x)| \underset{\|x\| \to \infty}{\longrightarrow} 0,$$

if $g \in L^1(\mathbb{R}^d)$, then for $H_\varepsilon(\cdot) := \varepsilon^{-d} H(\cdot/\varepsilon)$, one has $g * H_\varepsilon \underset{\varepsilon \to 0}{\longrightarrow} g$ at every continuity point of g.

PROOF:
The condition $H \in L^\infty(\mathbb{R}^d)$ ensures that $g * H_\varepsilon$ is defined everywhere and we set $\Delta(x) = (g * H_\varepsilon)(x) - g(x)$. Since $\int H = \int H_\varepsilon = 1$, one may write $\Delta(x) = \int [g(x-y) - g(x)] H_\varepsilon(y) \mathrm{d}y$. First note that if g is a bounded function (not necessarily in $L^1(\mathbb{R}^d)$!), the Bochner lemma easily follows from the dominated convergence theorem. In the general case, if g is continuous at x, then given $\eta > 0$, we choose $\delta (= \delta_x) > 0$ such that for $\| y \| < \delta$, one has $|g(x-y) - g(x)| < \eta$. Next,

$$
\begin{aligned}
|\Delta(x)| &\leq \eta \int_{\|y\|<\delta} |H_\varepsilon(y)| \mathrm{d}y + \int_{\|y\|\geq\delta} |g(x-y) - g(x)| |H_\varepsilon(y)| \mathrm{d}y \\
&\leq \eta \| H \|_1 + \int_{\|y\|\geq\delta} |g(x-y)| |H_\varepsilon(y)| \mathrm{d}y + |g(x)| \int_{\|y\|\geq\delta} |H_\varepsilon(y)| \mathrm{d}y.
\end{aligned}
$$

Now since $H \in L^1$ and $\delta/\varepsilon \to \infty$ as $\varepsilon \to 0$, we have

$$
\int_{\|y\|\geq\delta} |H_\varepsilon(y)| \mathrm{d}y = \int_{\|t\|>\delta/\varepsilon} |H(t)| \mathrm{d}t \underset{\varepsilon \to 0}{\longrightarrow} 0.
$$

Moreover $g \in L^1(\mathbb{R}^d)$ and $\| t \|^d |H(t)| \to 0$ as $\| t \| \to \infty$ yield:

$$
\int_{\|y\|\geq\delta} |g(x-y)| |H_\varepsilon(y)| \mathrm{d}y \leq \delta^{-d} \sup_{\|t\|>\frac{\delta}{\varepsilon}} \| t \|^d |H(t)| \int_{\|y\|>\delta} |g(x-y)| \mathrm{d}y \to 0,
$$

as $\varepsilon \to 0$ concluding the proof. ■

Coming back to density estimation, we begin with a necessary and sufficient condition for the L^2-consistency of the estimator (6.2).

Theorem 6.1
The following assertions are equivalent:

(i) $h_n \to 0$, $nh_n^d \to \infty$.

(ii) $\forall x \in \mathbb{R}^d$, *for every density* f *continuous at* x, $\mathrm{E}(f_n(x) - f(x))^2 \underset{n\to\infty}{\longrightarrow} 0$.

PROOF: *(Outline)*

(1) (i) ⇒ (ii)

One may always write: $\mathrm{E}(f_n(x) - f(x))^2 = \mathrm{Var}\, f_n(x) + (\mathrm{E}\, f_n(x) - f(x))^2$. Now the Bochner lemma and $h_n \to 0$ imply that $\mathrm{E}\, f_n(x) \to f(x)$ and $nh_n^d \mathrm{Var}\, f_n(x) \to f(x) \| K \|_2^2$ as $n \to \infty$ at every continuity point of f. This relation, together with $nh_n^d \to \infty$, yields $\mathrm{Var}\, f_n(x) \to 0$.

(2) Let us assume (ii), we first show that necessarily $h_n \to 0$. First suppose that $h_n \to h > 0$, then for a bounded continuous f, one has $Ef_n(x) = \int K(y)f(x - h_n y)\mathrm{d}y \to \int K(y)f(x - hy)\mathrm{d}y$ and (ii) implies in particular that $\mathrm{E}f_n(x) \to f(x)$. So one has $\int h^{-d}K((x-z)/h)f(z)\mathrm{d}z = f(x)$ which is impossible since K and f are densities. Next if $h_n \to \infty$, one has

$$\mathrm{E}f_n(x) = h_n^{-d} \int K\left(\frac{x-z}{h_n}\right) f(z)\mathrm{d}z \leq \| K \|_\infty h_n^{-d} \to 0$$

which is contradictive with $\mathrm{E}f_n(x) \to f(x)$ as soon as $f(x) \neq 0$. Other cases (h_n not bounded, h_n bounded but $h_n \not\to 0$) may be handled similarly (see Parzen 1962). So one obtains $h_n \to 0$, implying in turn that $nh_n^d \mathrm{Var} f_n(x) \to f(x) \| K \|^2$ with the Bochner lemma. Since Var $f_n(x) \to 0$, one may conclude that $nh_n^d \to \infty$. ■

It is not difficult to get exact rates of convergence for the mean-square error of kernel estimators: as seen in the previous proof, if x is a continuity point for f, $nh_n^d \mathrm{Var}\ f_n(x) \to f(x) \| K \|_2^2$. The term of bias can be handled with some additional conditions on the smoothness of f. The exact behaviour of this L^2 error will appear in the next section dealing with the more general case of dependent variables. Even if the quadratic error is a useful measure of the accuracy of a density estimate, it is not completely satisfactory since it does note provide information concerning the shape of the graph of f whereas the similarity between the graph of f_n and the one of f is crucial for the user (especially in the viewpoint of prediction). A good measure of this similarity should be the uniform distance between f_n and f. We first recall a surprising result obtained by J. Geffroy (personal communication, 1973) and then generalized by Bertrand–Retali (1974).

Theorem 6.2
Let us denote by $\mathcal{F}$ the space of uniformly continuous densities and $\| f_n - f \|_\infty = \sup_{x\in\mathbb{R}} |f_n(x) - f(x)|$, $d = 1$, one has

(i) ($f \in \mathcal{F}$), $d(f_n, f) \xrightarrow{p} 0$ implies $nh_n/\ln n \to +\infty$.

(ii) If K has bounded variation and $nh_n/\ln n \to +\infty$ then $\| f_n - f \|_\infty \xrightarrow{a.s.} 0$.

In other words, for common kernels, convergence in probability yields almost sure convergence! Note that this result was originally established in dimension d, with a technical condition on kernels K (satisfied by kernels with bounded variations in the univariate case). Concerning rates of convergence, one has the following result of Stute (1984): if f is twice continuously differentiable with bounded partial derivatives of order two, then for all $A > 0$, $f > 0$ and h_n such that $nh_n^{4+d}/\ln h_n^{-d} \to 0$ as $n \to \infty$,

$$\sqrt{\frac{nh_n^d}{\ln h_n^{-d}}} \sup_{\|x\| \leq A} \frac{|f_n(x) - f(x)|}{\sqrt{f(x)}} \underset{n\to\infty}{\longrightarrow} \sqrt{2} \| K \|_2, \qquad (6.3)$$

with probability one. Note that uniformity over $\mathbb{R}^d$ has been recently obtained by Deheuvels (2000, $d = 1$) and Giné and Guillou (2002, $d \geq 1$).

Concerning regression estimation, there also exists an extensive literature for the Nadaraya–Watson estimator. Our main purpose being prediction, we do not present here the i.i.d. results, about rates of convergence and their optimality, choices of bandwidths and kernels. The interested reader may refer to the monographs (with references therein) of e.g. Devroye (1987); Efromovich (1999); Györfi *et al.* (2002); Härdle (1990); Silverman (1986). We now turn to the dependent case.

6.3 Density estimation in the dependent case

6.3.1 Mean-square error and asymptotic normality

Let $\xi = (\xi_t,\ t \in \mathbb{Z})$ be a $\mathbb{R}^{d_0}$-valued stochastic process. The Kolmogorov theorem yields that P_ξ is completely specified by the finite-dimensional distributions $P_{\xi_1,\dots,\xi_k}$, $k \geq 1$, see Ash and Gardner (1975). Thus, it is natural to estimate the associated densities if they do exist.

If $\xi_1,\dots,\xi_N$ are observed, with $N \geq k$, one may set $X_1 = (\xi_1,\dots,\xi_k)$, $X_2 = (\xi_2,\dots,\xi_{k+1}),\dots,X_n = (\xi_{N-k+1},\dots,\xi_N)$ where $n = N - k + 1$. This construction yields $\mathbb{R}^d$-valued random variables $X_1,\dots,X_n$ with $d = kd_0$. In all the following, we will assume that they admit the common density f.

Let us now evaluate the asymptotic quadratic error of f_n, see (6.1)–(6.2). We will show that, under the following mild conditions, this error turns out to be the same as in the i.i.d. case.

Assumptions 6.1 (A6.1)

(i) $f \in C_d^2(b)$ for some positive b where $C_d^2(b)$ denotes the space of twice continuously differentiable real-valued functions h, defined on $\mathbb{R}^d$ and such that $\| h^{(2)} \|_\infty \leq b$ with $h^{(2)}$ standing for any partial derivative of order two for h.

(ii) For each couple (s,t), $s \neq t$ the random vector (X_s, X_t) has density $f_{(X_s,X_t)}$ such that $\sup_{|s-t|\geq 1}, \| g_{s,t} \|_\infty < +\infty$ where $g_{s,t} = f_{(X_s,X_t)} - f \otimes f$.

(iii) (X_t) is assumed to be 2-α-*mixing* with $\alpha^{(2)}(k) \leq \gamma k^{-\beta}$, $k \geq 1$ for some positive constants γ and $\beta > 2$.

Now we state the result.

Theorem 6.3

(1) If condition A6.1(i) is fulfilled,

$$h_n^{-4}(\mathrm{E}f_n(x) - f(x))^2 \to b_2^2(x) := \frac{1}{4}\left(\sum_{1\leq i,j\leq d} \frac{\partial^2 f}{\partial x_i \partial x_j}(x) \int_{\mathbb{R}^d} v_i v_j K(v)\mathrm{d}v\right)^2. \quad (6.4)$$

(2) If conditions A6.1(ii)–(iii) are fulfilled with f continuous at x, then

$$nh_n^d \mathrm{Var}\, f_n(x) \to f(x) \parallel K \parallel_2^2 . \tag{6.5}$$

(3) If Assumptions A6.1 are fulfilled with $f(x) > 0$, the choice $h_n = c_n n^{-1/(d+4)}$, $c_n \to c > 0$, leads to

$$n^{4/(d+4)} \mathrm{E}(f_n(x) - f(x))^2 \to c^4 b_2^2(x) + \frac{f(x)}{c^d} \parallel K \parallel_2^2 .$$

PROOF:

(1) One has

$$\mathrm{E}f_n(x) - f(x) = \int_{\mathbb{R}^d} K(v)(f(x - h_n v) - f(x))\mathrm{d}v.$$

By using the Taylor formula and the symmetry of K we get

$$\mathrm{E}\, f_n(x) - f(x) = \frac{h_n^2}{2} \int_{\mathbb{R}^d} K(v) \sum_{1 \le i,j \le d} v_i v_j \frac{\partial^2 f}{\partial x_i \partial x_j}(x - \theta h_n v)\mathrm{d}v$$

where $0 < \theta < 1$. Thus a simple application of the Lebesgue dominated convergence theorem gives the result.

(2) We have

$$nh_n^d \mathrm{Var}\, f_n(x) = h_n^d \mathrm{Var}\, K_h(x - X_1) + \frac{h_n^d}{(n-1)} \sum_{t=1}^{n} \sum_{\substack{t'=1 \\ t' \neq t}}^{n} \mathrm{Cov}(K_h(x - X_t), K_h(x - X_{t'})).$$

The Bochner lemma implies that $h_n^d \mathrm{Var}\, K_h(x - X_1) \to f(x) \parallel K \parallel_2^2$ whereas covariances may be handled as follows:

$$\begin{aligned} &\mathrm{Cov}(K_h(x - X_t), K_h(x - X_{t'})) \\ &\quad = \iint K_h(x-u)K_h(x-v) g_{t,t'}(u,v)\mathrm{d}u\mathrm{d}v \\ &\quad \le \min\left(\sup_{|t-t'| \ge 1} \parallel g_{t,t'} \parallel_\infty \parallel K \parallel_1^2,\ 4h_n^{-2d} \parallel K \parallel_\infty^2 \alpha^{(2)}(|t'-t|) \right) \end{aligned}$$

by condition A6.1(ii) and the *Billingsley inequality*. Next, splitting the sum of covariances into $\{|t-t'| \leq [h_n^{-2d/\beta}]\}$ and $\{|t-t'| > [h_n^{-2d/\beta}]\}$, one arrives at

$$\frac{h_n^d}{(n-1)} \sum_{1 \leq |t'-t| \leq n-1} \mathrm{Cov}(K_h(x-X_t), K_h(x-X_{t'})) = \mathcal{O}\left(h_n^{d(1-\frac{2}{\beta})} \right) \to 0,$$

with the help of the mixing condition since $\beta > 2$.

(3) The mean-square error follows immediately from its classical decomposition into $\text{Bias}^2 + \text{Variance}$. ■

Note that the i.i.d. case is a special case of Theorem 6.3.

Now, we give the limit in distribution of f_n in the univariate case ($d = 1$), which is useful for constructing confidence sets for $(f(x_1), \ldots, f(x_n))$.

Theorem 6.4

Suppose that $X = (X_k, k \in \mathbb{Z})$ is a real strictly stationary α-mixing sequence satisfying conditions A6.1(i)–(ii); in addition if $\sum_{k \geq 1} \alpha_k < \infty$, f is bounded, $h_n = c \cdot \ln \ln n \cdot n^{-1/5}$, $c > 0$ and K has a compact support, then

$$\sqrt{nh_n}\left(\frac{f_n(x_i) - f(x_i)}{f_n(x_i)^{1/2} \parallel K \parallel_2}, \quad 1 \leq i \leq m \right) \xrightarrow[n\to\infty]{\mathcal{D}} N^{(m)}$$

for any finite collection $(x_i,\ 1 \leq i \leq m)$ of distinct real numbers such that $f(x_i) > 0$ and where $N^{(m)}$ denotes a random vector with standard normal distribution in $\mathbb{R}^m$.

PROOF:
See Bosq, Merlevède and Peligrad (1999) ■

6.3.2 Almost sure convergence

We now turn to almost sure convergence of the kernel estimator. To this end, we need the following exponential type inequality.

Proposition 6.1

Let $(Z_t, t \in \mathbb{Z})$ be a zero-mean real valued process such that, for a positive constant M, $\sup_{1 \leq t \leq n} \parallel Z_t \parallel_\infty \leq M$. Then for each integer $q \in [1, n/2]$ and each positive κ, ε,

$$P\left(|\sum_{i=1}^n Z_i| > n\varepsilon \right) \leq \frac{8M}{\varepsilon\kappa}(1+\kappa)\alpha_Z\left(\left[\frac{n}{2q} \right] \right) + 4\exp\left(-\frac{n^2\varepsilon^2/q}{8(1+\kappa)^2\sigma^2(q) + \frac{2M}{3}(1+\kappa)n^2q^{-2}\varepsilon} \right) \quad (6.6)$$

where for $p = n/(2q)$,

$$\sigma^2(q) = \max_{j=0,\dots,2q-1} \left(\sum_{\ell=1}^{[p]+2} \operatorname{Var} Z_{[jp]+\ell} + \sum_{\ell=1}^{[p]+2} \sum_{\substack{k=1 \\ k\neq\ell}}^{[p]+2} |\operatorname{Cov}(Z_{[jp]+\ell}, Z_{[jp]+k})| \right). \tag{6.7}$$

Note that in the stationary case, one has

$$\sigma^2(q) \leq ([p]+2) \left(\operatorname{Var} Z_1 + 2 \sum_{\ell=1}^{[p]+1} |\operatorname{Cov}(Z_0, Z_\ell)| \right).$$

PROOF:
We consider the auxiliary continuous time process $Y_t = Z_{[t+1]}$, $t \in \mathbb{R}$. Clearly $S_n := \sum_{i=1}^n Z_i = \int_0^n Y_u du$. Now define blocks of variables $V_n(j)$, $j = 1, \dots, 2q$ by setting $V_n(j) = \int_{(j-1)p}^{jp} Y_u du$ with $p = n/2q$ for a given integer q.
Clearly, for any positive ε,

$$P(|S_n| > n\varepsilon) \leq P\left(\left| \sum_{j=1}^q V_n(2j-1) \right| > \frac{n\varepsilon}{2} \right) + P\left(\left| \sum_{j=1}^q V_n(2j) \right| > \frac{n\varepsilon}{2} \right).$$

The two terms may be handled similarly. Consider e.g. the first one: we use recursively the Rio (1995) coupling lemma to approximate $V_n(1), \dots, V_n(2q-1)$ by independent random variables $V_n^*(1), \dots, V_n^*(2q-1)$ such that $P_{V_n(2j-1)} = P_{V_n^*(2j-1)}$, $j = 1, \dots, q$ and

$$\mathrm{E}|V_n^*(2j-1) - V_n(2j-1)| \leq 4 \parallel V_n(2j-1) \parallel_\infty \alpha_Z([p]). \tag{6.8}$$

Now for any positive κ, ε, one may write

$$P\left(\left| \sum_{j=1}^q V_n(2j-1) \right| > \frac{n\varepsilon}{2} \right) \leq P\left(\left| \sum_{j=1}^q V_n^*(2j-1) \right| > \frac{n\varepsilon}{2(1+\kappa)} \right) + P\left(\left| \sum_{j=1}^q V_n(2j-1) - V_n^*(2j-1) \right| > \frac{n\varepsilon\kappa}{2(1+\kappa)} \right).$$

Since the $V_n^*(2j-1)$ are independent, the *Bernstein inequality* implies

$$P\left(\left| \sum_{j=1}^q V_n^*(2j-1) \right| > \frac{n\varepsilon}{2(1+\kappa)} \right)$$

$$\leq 2\exp\left(-\frac{n^2\varepsilon^2}{8(1+\kappa)^2\sum_{j=1}^{q}\operatorname{Var}V_n^*(2j-1)+\frac{4}{3}(1+\kappa)\parallel V_n(2j-1)\parallel_\infty n\varepsilon}\right)$$

$$\leq 2\exp\left(-\frac{n^2\varepsilon^2/q}{8(1+\kappa)^2\sigma^2(q)+\frac{2M}{3}(1+\kappa)(n/q)^2\varepsilon}\right)$$

since $\operatorname{Var} V_n^*(2j-1) = \operatorname{Var} V_n(2j-1) = \mathrm{E}\left(\int_{(2j-2)p}^{(2j-1)p} Z_{[u+1]}\mathrm{d}u\right)^2$ and

$$\mathrm{E}\left(\int_{(2j-2)p}^{(2j-1)p} Z_{[u+1]}\mathrm{d}u\right)^2$$

$$\leq \max_{j=0,\ldots,2q-1} \mathrm{E}(([jp]+1-jp)Z_{[jp]+1}+Z_{[jp]+2}+\cdots+Z_{[(j+1)p]}$$

$$+((j+1)p-[(j+1)p])Z_{[(j+1)p]+1})^2$$

$$\leq \max_{j=0,\ldots,2q-1}\left(\sum_{\ell=1}^{[p]+2}\operatorname{Var} Z_{[jp]+\ell}+\sum_{\ell=1}^{[p]+2}\sum_{\substack{k=1\\k\neq\ell}}^{[p]+2}|\operatorname{Cov}(Z_{[jp]+\ell},Z_{[jp]+k})\right)$$

$$=: \sigma^2(q).$$

Next from the Markov inequality and (6.8), one obtains

$$P\left(\left|\sum_{j=1}^{q} V_n(2j-1)-V_n^*(2j-1)\right|>\frac{n\varepsilon\kappa}{2(1+\kappa)}\right)$$

$$\leq\frac{2(1+\kappa)}{n\varepsilon\kappa}\mathrm{E}\left|\sum_{j=1}^{q} V_n(2j-1)-V_n^*(2j-1)\right|\leq\frac{4M}{\varepsilon\kappa}(1+\kappa)\alpha_Z\left(\frac{n}{2q}\right).$$

Since the same bound holds for $V_n(2j)$, inequality (6.6) is proved. ■

Our following result deals with pointwise almost sure convergence of f_n for GSM processes with

$$\alpha(u)\leq\beta_0\mathrm{e}^{-\beta_1 u},\quad u\geq 1\quad(\beta_0>0,\beta_1>0),$$

which clearly implies condition A6.1(iii).

Theorem 6.5
Let $(X_t, t \in \mathbb{Z})$ be a GSM $\mathbb{R}^d$-valued process,

(i) *if f is continuous at x and h_n such that $(\ln n)^\rho/(nh_n^d) \underset{n\to\infty}{\longrightarrow} 0$ with $\rho > 2$, then*

$$\left(\frac{nh_n^d}{(\ln n)^\rho}\right)^{1/2} |f_n(x) - \mathrm{E}f_n(x)| \underset{n\to\infty}{\longrightarrow} 0,$$

almost completely,

(ii) *if Assumptions A6.1 hold with $f(x) > 0$, the choice $h_n = c_n(\ln n/n)^{1/(d+4)}$ where $c_n \to c > 0$ implies that, for all $x \in \mathbb{R}^d$,*

$$\overline{\lim_{n\to\infty}} \left(\frac{n}{\ln n}\right)^{\frac{2}{4+d}} |f_n(x) - f(x)| \leq 2c^{-d/2} \parallel K \parallel_2 \sqrt{f(x)} + c^2|b_2(x)| \qquad (6.9)$$

almost surely with $b_2(x)$ given by (6.4).

These results may be easily interpreted: for geometrically strong mixing processes, one obtains a quasi-optimal rate of pointwise convergence for the stochastic term even if joint densities are not defined. Such a result is particularly useful for estimation of the finite-dimensional densities of a process (if they do exist), as well as for the prediction problem (see Section 6.5). Moreover, under the general Assumptions A6.1, the result is improved with a rate and an asymptotic constant similar to the one of the i.i.d. case, see (6.3).

PROOF:
We apply inequality (6.6) to the random variables

$$Z_i = K_h(x - X_i) - \mathrm{E}K_h(x - X_i)$$

with the choice: $q_n = [n/(2p_0 \ln n)]$ ($p_0 > 0$). Note that, for a positive kernel K, one has $M = \parallel K \parallel_\infty h_n^{-d}$.

(1) For $\sigma^2(q_n)$ defined by (6.7), the condition $(\ln n)^\rho/(nh_n^d) \to 0$, together with the Cauchy–Schwarz inequality and Bochner lemma, 6.1, imply that $\sigma^2(q_n) = \mathcal{O}(p_n^2 h_n^{-d})$. Now, the result follows from $\varepsilon = \eta(\ln n)^{\rho/2}/(nh_n^d)^{1/2}$ ($\rho > 2$), $p_0 > 5/(2\beta_1)$ and Borel–Cantelli lemma.

(2) We have to refine the bound for $\sigma^2(q_n)$. Actually, it is easy to infer that condition A6.1(ii) yields

$$\begin{aligned}\sigma^2(q_n) &\leq (p_n + 2)\mathrm{Var}\, K_h(x - X_1) + (p_n + 2)^2 \sup_{|t-t'|\geq 1} \parallel g_{t,t'} \parallel_\infty \\ &\leq p_n h_n^{-d} \parallel K \parallel_2^2 f(x)(1 + o(1))\end{aligned}$$

since $p_n = n/(2q_n)$ has a logarithmic order and $h_n = c_n((\ln n)/n)^{1/(4+d)}$, $c_n \to c$. In this way,

$$P\left(\frac{1}{n}\left|\sum_{i=1}^{n} Z_{i,n} - \mathrm{E}Z_{i,n}\right| > \eta\sqrt{\frac{\ln n}{nh_n^d}}\right) \leq \frac{8\beta_0 c^{\frac{d}{2}}}{\eta\kappa}(1+\kappa) \parallel K \parallel_\infty \frac{n^{\frac{2+d}{4+d}-\beta_1 p_0}}{(\ln n)^{\frac{2+d}{4+d}}}$$
$$+4\exp\left(-\frac{\eta^2 \ln n}{4(1+\kappa)^2 \parallel K \parallel_2^2 f(x)(1+o(1))}\right).$$

Therefore $\sum_n P((n/(\ln n))^{2/(4+d)}|f_n(x) - \mathrm{E}f_n(x)| > \eta c_n^{-d/2}) < \infty$ for all $\eta > 2(1+\kappa) \parallel K \parallel_2 \sqrt{f(x)}$ and $p_0 > 2/\beta_1$, so that the Borel–Cantelli lemma implies:

$$\overline{\lim_{n\to\infty}} \left(\frac{n}{\ln n}\right)^{\frac{2}{4+d}} |f_n(x) - \mathrm{E}\, f_n(x)| \leq 2c^{-d/2}(1+\kappa) \parallel K \parallel_2 \sqrt{f(x)} \quad \text{a.s.}$$

for all positive κ, hence (6.9) from (6.4). ■

Now we may state uniform results on a (possibly) increasing sequence of compact sets of $\mathbb{R}^d$ and then over the entire space.

Theorem 6.6

Suppose that X is a GSM process with a bounded density f and that K satisfies a Lipschitz condition, also let ℓ and $c_{(d)}$ be some constants ($\ell \geq 0$, $c_{(d)} > 0$).

(1) *If condition A6.1(i) is satisfied then, the choice $h_n = c_n((\ln n)^2/n)^{1/(d+4)}$, with $c_n \to c > 0$ yields*

$$\sup_{\|x\| \leq c_{(d)} n^\ell} |f_n(x) - f(x)| = \mathcal{O}\left(\frac{(\ln n)^2}{n}\right)^{\frac{2}{d+4}},$$

almost surely.

(2) *If Assumptions A6.1 are fulfilled, the choice $h_n = c_n((\ln n)/n)^{1/(d+4)}$ with $c_n \to c > 0$ yields*

$$\sup_{\|x\| \leq c_{(d)} n^\ell} |f_n(x) - f(x)| = \mathcal{O}\left(\left(\frac{\ln n}{n}\right)^{\frac{2}{d+4}}\right)$$

almost surely.

(3) *If in addition K has a compact support, f is ultimately decreasing, and $\mathrm{E} \parallel X_0 \parallel < \infty$, then, almost surely,*

$$\sup_{x\in\mathbb{R}^d} |f_n(x) - f(x)| = \mathcal{O}\left(\left(\frac{\ln n}{n}\right)^{\frac{2}{d+4}}\right).$$

PROOF:
First, the Taylor formula gives the following uniform bound for the bias:

$$\overline{\lim_{h_n \to 0}} h_n^{-2} \sup_{x \in \mathbb{R}^d} |\mathrm{E} f_n(x) - f(x)| \leq \frac{b}{2} \sum_{i,j=1}^{d} \int |u_i u_j| K(u) \mathrm{d}u.$$

For convenience, we take $\|\cdot\|$ as the sup norm on $\mathbb{R}^d$, defined by $\| (x_1, \ldots, x_d) \| = \sup_{1 \leq i \leq d} |x_i|$. Let $D_n := \{x : \| x \| \leq c_{(d)} n^\ell\}$; one may cover this compact set by M_n^d hypercubes, $D_{k,n}$, centred in $x_{k,n}$ and defined by:

$$D_{k,n} = \{x : \| x - x_{k,n} \| \leq c_{(d)} n^\ell / M_n\} \quad \text{for} \quad 1 \leq k \leq M_n^d$$

with $D_{k,n}^\circ \cap D_{k',n}^\circ = \varnothing$, $1 \leq k \neq k' \leq M_n^d$. One has,

$$\begin{aligned} &\sup_{\|x\| \leq c_{(d)} n^\ell} |f_n(x) - \mathrm{E} f_n(x)| \\ &\quad \leq \max_{1 \leq k \leq M_n^d} \sup_{x \in D_{k,n}} |f_n(x) - f_n(x_{k,n})| + \max_{1 \leq k \leq M_n^d} |f_n(x_{k,n}) - \mathrm{E} f_n(x_{k,n})| \\ &\quad\quad + \max_{1 \leq k \leq M_n^d} \sup_{x \in D_{k,n}} |\mathrm{E} f_n(x_{k,n}) - \mathrm{E} f_n(x)| \\ &\quad =: A_1 + A_2 + A_3. \end{aligned}$$

Since K satisfies a Lipschitz condition, there exists $L > 0$ such that

$$|f_n(x) - f_n(x_{k,n})| \leq L \frac{\| x - x_{k,n} \|}{h_n^{d+1}}, \quad x \in D_{k,n}, \quad k = 1, \ldots, M_n.$$

If ψ_n denotes the rate of convergence, it follows that

$$\psi_n(A_1 + A_3) = \mathcal{O}\left(\frac{n^\ell \psi_n}{M_n h_n^{d+1}} \right).$$

Next, we choose

$$M_n = (\log_m n) n^\ell \psi_n h_n^{-(d+1)}, \qquad m \geq 1, \tag{6.10}$$

which implies $\psi_n(A_1 + A_3) = o(1)$. For the term A_2, we proceed as follows:

$$P\left(\max_{1 \leq k \leq M_n^d} |f_n(x_{k,n}) - \mathrm{E} f_n(x_{k,n})| > \varepsilon \right) \leq \sum_{k=1}^{M_n^d} P |f_n(x_{k,n}) - \mathrm{E} f_n(x_{k,n})| > \varepsilon).$$

(1) Similar to the proof of Theorem 6.5(1), the choice $\varepsilon = \eta(\ln n)/(nh_n^d)^{1/2}$ leads to an upper bound not depending on $x_{k,n}$, and of order

$$\mathcal{O}(M_n^d \exp(-c_1\eta^2(\ln n))) + \mathcal{O}\left(M_n^d n^{\frac{2+d}{4+d}-\beta_1 p_0}(\ln n)^{c_2}\right)$$

where $c_1 > 0$, c_2 are given constants. The result follows from (6.10), with $\psi_n = (n(\ln n)^{-2})^{2/(d+4)}$, and the Borel–Cantelli lemma as soon as η is chosen large enough and $p_0 > (\ell d + d + 3)/\beta_1$.

(2) We set $\varepsilon = \eta(\ln n)^{1/2}(nh_n^d)^{-1/2}$, $\eta > 0$, and we follow the steps of the proof of Theorem 6.5(2). Using the uniform bound $h_n^{-d} \parallel f \parallel_\infty \parallel K \parallel_2^2$ for $\mathrm{Var}(K_h(x - X_1))$, we get:

$$P(\psi_n|f_n(x_{k,n}) - \mathrm{E}f_n(x_{k,n})| > \eta) \leq \frac{8\beta_0}{\eta\kappa}(1+\kappa) \parallel K \parallel_\infty \left(\frac{n}{\ln n}\right)^{\frac{2+d}{4+d}} n^{-\beta_1 p_0}$$
$$+ 4\exp\left(-\frac{\eta^2 \ln n}{4(1+\kappa)^2 \parallel K \parallel_2^2 \parallel f \parallel_\infty (1+o(1))}\right)$$

where $o(1)$ is uniform with respect to $x_{k,n}$. For $\eta > 2(1+\kappa) \parallel K \parallel_2 \parallel f \parallel_\infty^{1/2} (\ell d + d + 1)^{1/2}$ and $p_0 > (\ell d + d + 3)\beta_1^{-1}$, we may deduce, with the choice of M_n given in (6.10) and $\psi_n = (n/\ln n)^{2/(d+4)}$, that $\sum_{k=1}^{M_n^d} P(\psi_n|f_n(x_{k,n}) - \mathrm{E}f_n(x_{k,n})| > \eta) \leq \lambda_1 n^{-\lambda_2}$ with $\lambda_1 > 0$ and $\lambda_2 > 1$ yielding the result.

(3) Finally we turn to uniform behaviour of f_n over the entire space. One may write

$$\sup_{\|x\| \in \mathbb{R}^d} \psi_n|f_n(x) - f(x)| \leq \sup_{\|x\| \leq c_{(d)}n^\ell} \psi_n|f_n(x) - f(x)|$$
$$+ \sup_{\|x\| > c_{(d)}n^\ell} \psi_n|f_n(x)| + \sup_{\|x\| > c_{(d)}n^\ell} \psi_n f(x).$$

The first term has been studied just above. Now we have

$$\left\{\sup_{1\leq j\leq n} \parallel X_j \parallel \leq \frac{c_{(d)}n^\ell}{2}, \parallel x \parallel > c_{(d)}n^\ell\right\} \subset \left\{\parallel \frac{x - X_j}{h_n} \parallel \geq \frac{c_{(d)}n^\ell}{2h_n}\right\}.$$

Since K has a compact support, say $[-c_K, c_K]$, $K((x - X_j)/h_n) = 0$ as soon as $\parallel \frac{x-X_j}{h_n} \parallel > c_K$. Now let n_0 be a positive real number such that for all $n \geq n_0$, $c_{(d)}n^\ell/(2h_n) \geq c_K$, we get for all $\parallel x \parallel > c_{(d)}n^\ell$,

$$\left\{\sup_{1\leq j\leq n} \parallel X_j \parallel \leq \frac{c_{(d)}}{2}n^\ell\right\} \subset \left\{\sup_{\|x\| > c_{(d)}n^\ell} \psi_n|f_n(x)| = 0\right\},$$

then for all $\eta > 0$, $n \geq n_0$,

$$P\left(\sup_{\|x\|>c_{(d)}n^{\ell}} \psi_n|f_n(x)| > \eta\right) \leq P\left(\sup_{1\leq j\leq n} \| X_j \|> \frac{c_{(d)}}{2}n^{\ell}\right) = O\left(\frac{1}{n^{\ell-1}}\right)$$

from Markov inequality.

Now the Borel–Cantelli lemma implies that for all $\ell > 2$, $\sup_{\|x\|>c_{(d)}n^{\ell}} \psi_n|f_n(x)| \to 0$ almost surely.

On the other hand,

$$\psi_n \sup_{\|x\|>c_{(d)}n^{\ell}} f(x) \leq \frac{\psi_n}{c_{(d)}n^{\ell}} \sup_{\|x\|>c_{(d)}n^{\ell}} \| x \| f(x)$$

where for all $\ell > 2$, $\psi_n n^{-\ell} \to 0$ as $n \to \infty$. Finally, since f is ultimately decreasing (and integrable!), one obtains $\sup_{\|x\|>c_{(d)}n^{\ell}} \| x \| f(x) \to 0$ as $n \to \infty$. ■

To conclude this section, we consider the case of compactly supported densities not fulfilling condition A6.1(i). The following result illustrates the bad behaviour of bias near the edge of the support.

Theorm 6.7

Suppose that X is a GSM process. If f is bounded and $f(x) = \tilde{f}(x)\prod_{i=1}^{d} \mathbb{1}_{[a_i,b_i]}(x_i)$ *where* $\tilde{f} \in C_d^2(b)$ *for some positive b, one obtains for* $h_n = c_n((\ln n)^2/n)^{1/(d+4)}$, $c_n \to c > 0$, *that*

$$\sup_{x\in\prod_{i=1}^{d}[a_i+\varepsilon_n, b_i-\varepsilon_n]} |f_n(x) - f(x)| = \mathcal{O}\left(\left(\frac{(\ln n)^2}{n}\right)^{\frac{2}{d+4}}\right)$$

almost surely and as soon as, $K = \otimes_1^d \tilde{K}$ for a real normal kernel $\tilde{K}$ with variance $\sigma^2 (\sigma > 0)$ and $\varepsilon_n \geq c' h_n (\ln n)^{1/2}$ with $c' \geq (2\sigma)/\sqrt{5}$.

PROOF:

We have:

$$\sup_{x\in\prod_{i=1}^{d}[a_i+\varepsilon_n, b_i-\varepsilon_n]} |f_n(x) - f(x)| \leq \sup_{x\in\prod_{i=1}^{d}[a_i, b_i]} |f_n(x) - \mathrm{E}f_n(x)| + \sup_{x\in\prod_{i=1}^{d}[a_i+\varepsilon_n, b_i-\varepsilon_n]} |\mathrm{E}f_n(x) - f(x)|.$$

The stochastic term should be handled as in the proof of Theorem 6.6(1) (with $\ell = 0$). Concerning the bias we obtain, for $x \in \prod_{i=1}^{d}[a_i, b_i]$, that

$$\mathrm{E}f_n(x) - f(x) = \int_{\prod_{i=1}^{d}\left[\frac{x_i-b_i}{h_n}, \frac{x-a_i}{h_n}\right]} K(z)(\tilde{f}(x - h_n z) - \tilde{f}(x))\mathrm{d}z$$
$$- \tilde{f}(x) \int_{\mathbb{R}^d \setminus \prod_{i=1}^{d}\left[\frac{x_i-b_i}{h_n}, \frac{x_i-a_i}{h_n}\right]} K(z)\mathrm{d}z.$$

A Taylor expansion leads to

$$\int_{\prod_{i=1}^{d}\left[\frac{x_i-b_i}{h_n}, \frac{x_i-a_i}{h_n}\right]} K(z)(\tilde{f}(x - h_n z) - \tilde{f}(x))\mathrm{d}z$$
$$= -h_n \sum_{j=1}^{d} \frac{\partial \tilde{f}}{\partial x_j}(x) \int_{\prod_{i=1}^{d}\left[\frac{x_i-b_i}{h_n}, \frac{x_i-a_i}{h_n}\right]} zK(z)\mathrm{d}z$$
$$+ \frac{h_n^2}{2} \sum_{k,\ell=1}^{d} \int_{\prod_{i=1}^{d}\left[\frac{x_i-b_i}{h_n}, \frac{x_i-a_i}{h_n}\right]} z_k z_\ell K(z) \frac{\partial^2 \tilde{f}}{\partial x_\ell \partial x_k}(x - \theta z)\mathrm{d}z.$$

First since K is a product of univariate normal kernels, the choice made for ε_n induces that

$$\sup_{x_i \in [a_i+\varepsilon_n, b_i-\varepsilon_n]} \left| \int_{\frac{x_i-b_i}{h_n}}^{\frac{x_i-a_i}{h_n}} z_i \tilde{K}(z_i)\, \mathrm{d}z_i \right| = \mathcal{O}\left(\mathrm{e}^{-\frac{\varepsilon_n^2}{2h_n^2\sigma^2}} \right) = o(h_n).$$

As usual, $\tilde{f} \in C_d^2(b)$ implies that the second term is $\mathcal{O}(h_n^2)$ and the result follows since

$$\sup_{x_i \in [a_i+\varepsilon_n, b_i-\varepsilon_n]} \int_{-\infty}^{\frac{x_i-b_i}{h_n}} \tilde{K}(t)\mathrm{d}t$$

and

$$\sup_{x_i \in [a_i+\varepsilon_n, b_i-\varepsilon_n]} \int_{\frac{x_i-a_i}{h_n}}^{+\infty} \tilde{K}(t)\mathrm{d}t$$

are of order $o(h_n^2)$. ■

6.4 Regression estimation in the dependent case

6.4.1 Framework and notations

Let $Z_t = (X_t, Y_t)$, $t \in \mathbb{Z}$, be an $\mathbb{R}^d \times \mathbb{R}^{d'}$-valued stochastic process defined on a probability space $(\Omega, \mathcal{A}, P)$ and m be a Borelian function of $\mathbb{R}^{d'}$ into $\mathbb{R}$ such that

$\omega \mapsto m^2(Y_t(\omega))$ is P-integrable for each $t \in \mathbb{Z}$. Assuming that the Z_t's have the same distribution with density $f_Z(x,y)$ we wish to estimate the regression parameter $r(\cdot) = \mathrm{E}(m(Y_0)|X_0 = \cdot)$, given the observations $Z_1, \ldots, Z_n$. One may define the functional parameters

$$f(x) = \int f_Z(x,y)\mathrm{d}y, \quad x \in \mathbb{R}^d \tag{6.11}$$

$$\varphi(x) = \int m(y) f_Z(x,y)\mathrm{d}y, \quad x \in \mathbb{R}^d \tag{6.12}$$

and

$$r(x) = \begin{cases} \varphi(x)/f(x) & \text{if } f(x) > 0, \\ \mathrm{E}m(Y_0) & \text{if } f(x) = 0. \end{cases} \tag{6.13}$$

Clearly, $r(\cdot)$ is a version of $\mathrm{E}(m(Y_0)|X_0 = \cdot)$. We call $r(\cdot)$ a *regression parameter*. Typical examples of regression parameters are

$$r(x) = P(Y \in B | X = x), \qquad (B \in \mathcal{B}_{\mathbb{R}^{d'}}),$$
$$r(x) = E(Y^k | X = x), \qquad (k \geq 1,\ d' = 1),$$

leading to the conditional variance of Y:

$$\mathrm{Var}(Y|X = x) = \mathrm{E}(Y^2|X = x) - (\mathrm{E}(Y|X = x))^2, \quad (d' = 1).$$

We consider a d-dimensional kernel estimator defined as:

$$r_n(x) = \begin{cases} \varphi_n(x)/f_n(x), & \text{if } f_n(x) > 0, \\ \frac{1}{n}\sum_{i=1}^n m(Y_i), & \text{if } f_n(x) = 0, \end{cases} \tag{6.14}$$

where f_n is given by (6.2) with K a positive kernel satisfying conditions $\mathcal{K}$ given by (6.1) (so that $f_n(x) > 0$) and

$$\varphi_n(x) = \frac{1}{nh_n^d}\sum_{i=1}^n m(Y_i) K\left(\frac{x - X_i}{h_n}\right), \quad x \in \mathbb{R}^d. \tag{6.15}$$

In some cases, we also use the following variant:

$$\tilde{r}_n(x) = \begin{cases} \tilde{\varphi}_n(x)/f_n(x), & \text{if } f_n(x) > 0, \\ \frac{1}{n}\sum_{i=1}^n m(Y_i), & \text{if } f_n(x) = 0, \end{cases} \tag{6.16}$$

where

$$\tilde{\varphi}_n(x) = \frac{1}{nh_n^d}\sum_{i=1}^n m(Y_i)\mathbb{1}_{\{|m(Y_i)| \leq b_n\}} K\left(\frac{x - X_i}{h_n}\right) \tag{6.17}$$

and (b_n) is a positive sequence such that $b_n \to \infty$. Note that $b_n \equiv \infty$ yields the classical regression estimator (6.14).

6.4.2 Pointwise convergence

In this context of regression estimation, we begin by an analogue of Theorem 6.5, i.e. a strong consistency result involving a sharp rate and a specified asymptotic constant. This constant involves the conditional variance parameter $V(x)$ defined by

$$V(x) = \text{Var}(m(Y_0)|X_0 = x) = \text{E}m^2(Y_0)|X_0 = x) - r^2(x). \tag{6.18}$$

We consider the following assumptions.

Assumptions 6.2 (A6.2)

(i) $\varphi \in C_d^2(b)$ for some positive b;

(ii) $\text{E}(m^2(Y_0)|X_0 = \cdot)f(\cdot)$ is both continuous and bounded away from zero at x;

(iii) either $\text{E}(|m(Y_0)|^s|X_0 = \cdot)f(\cdot)$ is continuous at x and integrable over $\mathbb{R}^d$ for some $s \geq \max(1 + 5/d, 2 + 1/d)$, or there exist $\lambda > 0$ and $\nu > 0$ such that $\text{E}(\exp(\lambda|m(Y_0)|^\nu)) < +\infty$;

(iv) for each $s \neq t$, (Z_s, Z_t) has a density $f_{(Z_s,Z_t)}$ such that, if $G_{s,t} := f_{(Z_s,Z_t)} - f_Z f_Z$, one has

$$\sup_{|s-t|\geq 1} \sup_{(x_1,x_2)\in\mathbb{R}^{2d}} \int_{\mathbb{R}^{2d'}} |G_{s,t}(x_1, y_1, x_2, y_2)| \text{d}y_1 \text{d}y_2 < \infty;$$

(v) (Z_t) is a GSM process with $\alpha_Z(u) \leq \beta_0 \text{e}^{-\beta_1 u}$, $u \geq 1$, $\beta_0 > 0$, $\beta_1 > 0$.

Note that, in condition A6.2(iii), the integrability is simply equivalent to the moment assumption $\text{E}|m(Y_0)|^s < \infty$ for some $s \geq \max(1 + 5/d, 2 + 1/d)$. The alternative condition involving the existence of an exponential moment for $m(Y_0)$ is then clearly more restrictive but does not require any continuity assumption. In particular it is fulfilled in the concrete situations where $m(\cdot)$ is a bounded function (e.g. $m = \mathbb{1}_B$, $B \in \mathcal{B}_{\mathbb{R}^{d'}}$) or m is any polynomial function and Y_0 is Gaussian. Finally, condition A6.2(iv) is easily satisfied as soon as f and $f_{(X_s,X_t)}$ are taken to be uniformly bounded ($s \neq t$).

Theorem 6.8

(1) If K satisfies $\int \| u \|^3 K(u)\text{d}u < \infty$, if

$$h_n = c_n \left(\frac{\ln n}{n}\right)^{1/(d+4)}, \quad b_n = c_n' n^{2d/((4+d)\max(5,d+1))}$$

$(c_n \to c > 0,\ c'_n \to c' > 0)$, then condition A6.1(i) and Assumptions A6.2 yield

$$\overline{\lim_{n\to\infty}} \left(\frac{n}{\ln n}\right)^{\frac{2}{4+d}} |\tilde{r}_n(x) - r(x)| \le C(x, c, K, \varphi, f), \quad a.s.$$

where $C(x, c, K, \varphi, f) = (2c^{-d/2} \parallel K \parallel_2 \sqrt{f(x)V(x)} + c^2 |b_{\varphi,f}(x)|)/f(x)$ *and*

$$b_{\varphi,f}(x) = \frac{1}{2}\sum_{i,j=1}^{d}\left(\frac{\partial^2 \varphi}{\partial x_i \partial x_j}(x) - r(x)\frac{\partial^2 f}{\partial x_i \partial x_j}(x)\right)\int_{\mathbb{R}^d} u_i u_j K(u)\mathrm{d}u.$$

(2) *If $m(Y_0)$ admits some exponential moment, the same result holds for the estimator r_n defined by* (6.14).

PROOF:
Let us consider the following decomposition (with x omitted):

$$\begin{aligned}\psi_n(r_n - r) &= \psi_n(\tilde{r}_n - r) + \psi_n(r_n - \tilde{r}_n) \\ &= \frac{\psi_n(\tilde{\varphi}_n - rf_n)}{f_n} + \frac{\psi_n(\varphi_n - \tilde{\varphi}_n)}{f_n}\end{aligned} \tag{6.19}$$

where $\tilde{\varphi}_n$ is given by (6.17) and $\psi_n = (n/\ln n)^{2/(4+d)}$. Note that, by Theorem 6.5(1), one has $f_n(x) \to f(x) \neq 0$, almost surely.

(1) We begin with the truncated estimator, while indicating, during the proof, the variants used in the second part of the theorem. One has:

$$\psi_n|\tilde{\varphi}_n - rf_n| \le \psi_n|\tilde{\varphi}_n - rf_n - \mathrm{E}(\tilde{\varphi}_n - rf_n)| + \psi_n|\mathrm{E}(\tilde{\varphi}_n - rf_n)| := A_n + B_n.$$

(a) Study of A_n

We consider (6.6) with the choices $q_n = [n/(2p_0 \ln n)] (p_0 > 2/\beta_1)$, for a positive η, $\varepsilon_n = \eta(\ln n/nh_n^d)^{1/2}$ and variables $Z_{i,n} = W_{i,n} - \mathrm{E}W_{i,n}$ where

$$W_{i,n} = K_h(x - X_i)[m(Y_i)\mathbb{1}_{\{|m(Y_i)| \le b_n\}} - r(x)].$$

First, one has $|Z_{i,n}| \le 2 \parallel K \parallel_\infty h_n^{-d} b_n(1 + o(1))$. Next concerning $\sigma^2(q_n)$, defined in (6.7), we obtain

$$\begin{aligned}h_n^d \mathrm{Var}\, Z_{1,n} \le\ & h_n^{-d}\mathrm{E}K^2\left(\frac{x - X_0}{h_n}\right)(m(Y_0) - r(x))^2 \\ &+ h_n^{-d}\mathrm{E}K^2\left(\frac{x - X_0}{h_n}\right) m(Y_0)(2r(x) - m(Y_0))\mathbb{1}_{\{|m(Y_0)| > b_n\}}.\end{aligned}$$

Note that this last term reduces to zero in the case where m is bounded and $b_n \equiv \| m \|_\infty$. On one hand, the Bochner lemma and condition A6.2(ii) yield

$$h_n^{-d}\mathrm{E}K^2\left(\frac{x-X_0}{h_n}\right)(m(Y_0)-r(x))^2 \to f(x)V(x) \| K \|_2^2 .$$

On the other hand, with condition A6.2(iii) and the Bochner lemma, one gets

$$\begin{aligned}\mathrm{E}K^2\left(\frac{x-X_0}{h_n}\right)&m^2(Y_0)\mathbb{1}_{\{|m(Y_0)|\geq b_n\}}\\ &=\int K^2\left(\frac{x-t}{h_n}\right)\mathrm{E}(m^2(Y_0)\mathbb{1}_{\{|m(Y_0)|\geq b_n\}}|X_0=t)f(t)\mathrm{d}t\\ &\leq \mathrm{E}(|m(Y_0)|^s|X_0=x)f(x) \| K \|_2^2\, b_n^{2-s}h_n^d(1+o(1))=o(h_n^d)\end{aligned}$$

as $s>2$. If an exponential moment exists, one applies Cauchy–Schwarz and Markov inequalities to get a bound of order $\mathcal{O}\big(\exp\big(-\frac{\lambda}{2}b_n^\nu\big)\big)=o(h_n^d)$ if $b_n=(\delta\ln n)^{1/\nu}$ with $\delta>2/\lambda$.
Finally with the help of conditions A6.2(ii), (iv), one arrives at

$$\sigma^2(q_n)\leq p_n h_n^{-d}f(x)V(x) \| K \|_2^2\,(1+o(1))+p_n^2 b_n^2(1+o(1)).$$

Therefore, since p_n has a logarithmic order and $b_n=c_n' n^{2d/((4+d)\max(5,d+1))}$, condition A6.2(v) and the Borel–Cantelli lemma entail

$$\overline{\lim_{n\to\infty}} A_n \leq 2\sqrt{V(x)f(x)} \| K \|_2, \quad \text{a.s.}$$

(b) Study of B_n

Since $r(x)=\int m(y)f_{(X_0,Y_0)}(x,y)\mathrm{d}y/f(x)$, one may write

$$\begin{aligned}\mathrm{E}\big(\tilde{\varphi}_n(x)-r(x)f_n(x)\big)&=\int K_h(x-t)f(t)(r(t)-r(x))\mathrm{d}t\\ &\quad-\mathrm{E}m(Y_0)\mathbb{1}_{\{|m(Y_0)|>b_n\}}K_h(x-X_0)\\ &=B_1-B_2.\end{aligned}$$

First, $f,\varphi\in C_d^2(b)$, $\int \| u \|^3 K(u)\mathrm{d}u<+\infty$, $\int u_iK(u)\mathrm{d}u=0$ and Taylor formula imply

$$\psi_n|B_1|\to\frac{1}{2}\left|\sum_{i,j=1}^d\left\{\frac{\partial^2\varphi}{\partial x_i\partial x_j}(x)-r(x)\frac{\partial^2 f}{\partial x_i\partial x_j}(x)\right\}\int u_iu_jK(u)\mathrm{d}u\right|$$

and with condition A6.2(iii), one gets also that $\psi_n|B_2|=\mathcal{O}(\psi_n b_n^{1-s})=o(1)$ since $s\geq\max(1+5/d,2+1/d)$. Finally this last bound turns out to be $\mathcal{O}(\psi_n h_n^{-d}\exp(-\lambda b_n^\nu/2))=o(1)$ in the case of an exponential moment as soon as $b_n=(\delta\ln n)^{1/\nu}$ with $\delta>2/\lambda$.

(2) If $\mathrm{E}e^{\lambda|m(Y_0)|^\nu} < \infty$ for some positive λ and ν (and $b_n = (\delta \ln n)^{1/\nu}$, $\delta > 2/\lambda$) or if m is bounded (and $b_n = \| m \|_\infty$), results concerning $\psi_n|\tilde{\varphi}_n - rf_n|$ hold true. In order to conclude, note that, for all positive ε,

$$\sum_n P(\psi_n|\varphi_n(x) - \tilde{\varphi}_n(x)| \geq \varepsilon) \leq \sum_n nP(|m(Y_0)| > b_n) < +\infty$$

using Markov inequality and previous choices of b_n. Details are omitted. ■

6.4.3 Uniform convergence

Similar to density estimation, uniform convergence of a regression estimator may be obtained over compact sets, but in general, not over the whole space, even if some information about the behaviour of $r(x)$ for large $\| x \|$ is available.

As a simple example, let us consider the case where the Z_t's are i.i.d. bivariate Gaussian variables with standard margins and $\mathrm{Cov}(X_t, Y_t) = \rho$. In this case the classical estimator of $r(x) = \rho x$ is $\rho_n x$ where $\rho_n = n^{-1}\sum_{i=1}^n X_i Y_i$ and clearly $\sup_{x\in\mathbb{R}} |\rho_n x - \rho x| = +\infty$ almost surely. In fact it is impossible to construct a regression estimator R_n such that $\sup_{x\in\mathbb{R}} |R_n(x) - \rho x| \to 0$ a.s. since such a property implies that, whatever $(u_n) \uparrow \infty$, $u_n(R_n(u_n)/u_n - \rho) \to 0$ a.s. and consequently it should be possible to obtain an estimator of ρ with an arbitrary sharp rate of convergence.

However it is possible to establish uniform convergence over fixed or suitable increasing sequences of compact sets. We need the following definition.

Definition 6.1
A sequence (D_n) of compact sets in $\mathbb{R}^d$ is said to be regular *(with respect to f) if there exists a sequence (β_n) of positive real numbers and nonnegative constants ℓ, ℓ' such that for each n*

$$\inf_{x\in D_n} f(x) \geq \beta_n \quad \textit{and} \ \mathrm{diam}(D_n) \leq c_{(d)}(\ln n)^{\ell'} n^{\ell} \tag{6.20}$$

where $c_{(d)} > 0$ and $\mathrm{diam}(D_n)$ *denotes the diameter of D_n.*
We first consider convergence on a fixed compact set D: as for density estimation, we obtain a quasi-optimal rate of convergence under the following mild conditions.

Assumptions 6.3 (A6.3)

(i) $f \in C_d^2(b)$, $\varphi \in C_d^2(b)$ for some positive b;

(ii) f and φ are bounded functions, $\inf_{x\in D} f(x) > 0$;

(iii) $\mathrm{E}(m^2(Y_0)|X_0 = \cdot)f(\cdot)$ is a bounded function over $\mathbb{R}^d$,

(iv) the strong mixing coefficient α_Z of (Z_t) satisfies $\alpha_Z(u) \leq \beta_0 e^{-\beta_1 u}$, $u \geq 1$, $\beta_0 > 0$, $\beta_1 > 0$.

Theorem 6.9

(1) Suppose that Assumptions A6.3 are satisfied. If $h_n = c_n((\ln n)^2/n)^{1/(d+4)}$ $(c_n \to c > 0)$*,* $b_n = c'_n h_n^{-2}(c'_n \to c' > 0)$ *and K satisfies a Lipschitz condition, one obtains that:*

$$\sup_{x\in D} |\tilde{r}_n(x) - r(x)| = \mathcal{O}\left(\left(\frac{(\ln n)^2}{n}\right)^{\frac{2}{4+d}}\right), \quad \text{a.s.}$$

(2) Under Assumptions A6.3(ii)–(iv), suppose moreover that $f(x) = \tilde{f}(x)\mathbb{1}_D(x)$*,* $\varphi(x) = \tilde{\varphi}(x)\mathbb{1}_D(x)$ *where* $D = \prod_{i=1}^d [a_i, b_i]$ *and* $\tilde{f}, \tilde{\varphi} \in C_d^2(b)$ *for some positive b. The previous choices for* h_n*,* b_n *yield*

$$\sup_{x\in D_\varepsilon} |\tilde{r}_n(x) - r(x)| = \mathcal{O}\left(\left(\frac{(\ln n)^2}{n}\right)^{\frac{2}{4+d}}\right),$$

almost surely, as soon as $K = \otimes_1^d \tilde{K}$ *for a real normal kernel* $\tilde{K}$ *with variance* σ^2 $(\sigma > 0)$ *and* $D_\varepsilon = \prod_{i=1}^d [a_i + \varepsilon_n, b_i - \varepsilon_n]$ *with* $\varepsilon_n \geq c' h_n (\ln n)^{1/2}$ *and* $c' \geq 2\sigma/\sqrt{5}$.

PROOF: *(Sketch)*

(1) We start from:

$$\sup_{x\in D} |\tilde{r}_n(x) - r(x)| \leq \frac{\sup_{x\in D} |\tilde{\varphi}_n(x) - \varphi(x)|}{\inf_{x\in D} f_n(x)} + \frac{\sup_{x\in D} |r(x)| \sup_{x\in D} |f_n(x) - f(x)|}{\inf_{x\in D} f_n(x)}.$$

Note that condition A6.3(ii) implies that $r(\cdot)$ is bounded on D.

From Theorem 6.6(1) (with $\ell = 0$), one may deduce that, with probability 1, $\sup_{x\in D} \psi_n |f_n(x) - f(x)| = \mathcal{O}(((\ln n)^2/n)^{2/(4+d)})$ and $\inf_{x\in D} f_n(x) \underset{n\to\infty}{\longrightarrow} \inf_{x\in D} f(x) > 0$. Concerning the first term, one has

$$|\tilde{\varphi}_n(x) - \varphi(x)| \leq |\mathrm{E}\tilde{\varphi}_n(x) - \varphi(x)| + |\tilde{\varphi}_n(x) - \mathrm{E}\tilde{\varphi}_n(x)|.$$

Conditions A6.3(i),(iii) allow us to write, uniformly in x,

$$\begin{aligned}\mathrm{E}\tilde{\varphi}_n(x) - \varphi(x) &= \int K_h(x-t)[\varphi(t) - \varphi(x)]\mathrm{d}t + \mathrm{E}m(Y_0)\mathbb{1}_{\{|m(Y_0)|>b_n\}} K_h(x - X_0) \\ &= \mathcal{O}(h_n^2) + \mathcal{O}(b_n^{-1}) = \mathcal{O}(h_n^2),\end{aligned}$$

since $b_n \simeq h_n^{-2}$. The stochastic term $|\tilde{\varphi}_n(x) - \mathrm{E}\tilde{\varphi}_n(x)|$ is handled similarly to the proof of Theorem 6.6(1) (with $\ell = 0$ but additional term b_n) by considering a suitable covering of D. Uniform bounds are obtained with the help of conditions A6.3(ii),(iii). Details are left to the reader.

(2) The result is obtained by proceeding similarly to Theorem 6.7 and by noting that, almost surely,

$$\inf_{x\in D_\varepsilon} f_n(x) \geq \inf_{x\in D} f_n(x) = \inf_{x\in D} f(x)(1+o(1)).$$

■

Results for the classical estimator r_n are straightforward if $m(Y_0)$ admits some exponential moment. This is the purpose of the following corollary.

Corollary 6.1

Results of Theorem 6.9 hold true with $\tilde{r}_n$ replaced by r_n under the additional condition: $\mathrm{E}e^{\lambda|m(Y_0)|^\nu} < +\infty$ *for some positive λ and ν.*

PROOF. *(Sketch)*

Let us start from (6.19). Uniform convergence of $\psi_n(\tilde{r}_n - r)$ is established in Theorem 6.9(1) with b_n of order h_n^{-2}. If $m(Y_0)$ admits some exponential moment (respectively m is bounded), the same result holds with $b_n = (\delta \ln n)^{1/\nu}$, $\delta > 2/\lambda$ (respectively $b_n = \| m \|_\infty$). The main change is in the bias, but

$$\mathrm{E}m(Y_0)\mathbb{1}_{\{|m(Y_0)|>b_n\}}K_h(x - X_0) = \mathcal{O}(h_n^{-d}\mathrm{e}^{-\frac{\lambda}{2}b_n^\nu}) = o(h_n^2),$$

by Markov inequality. Note that this last term disappears for bounded m if $b_n \equiv \| m \|_\infty$. Concerning $\sup_{x\in D} \psi_n|r_n - \tilde{r}_n|$, one has $\inf_{x\in D} f_n(x) \to \inf_{x\in D} f(x)$ almost surely and,

$$\sum_n P\left(\sup_{x\in D} \psi_n|\tilde{\varphi}_n(x) - \varphi_n(x)| > \varepsilon\right) \leq \sum_n nP(|m(Y_0)| > b_n) < +\infty \tag{6.21}$$

for all $\varepsilon > 0$, so that $\sup_{x\in D} \psi_n|r_n - \tilde{r}_n|$ is negligible. ■

Our last result considers the case of varying compact sets.

Theorem 6.10

(1) *Under the assumptions of Theorem 6.9(1), if D_n is a regular sequence of compact sets, the choices $h_n = c_n((\ln n)^2/n)^{1/(d+4)}$ $(c_n \to c > 0)$, $b_n = c_n' h_n^{-1}$ $(c_n' \to c' > 0)$ lead to*

$$\sup_{x\in D_n} |\tilde{r}_n(x) - r(x)| = \mathcal{O}\left(\left(\frac{(\ln n)^2}{n}\right)^{\frac{1}{d+4}} \cdot \beta_n^{-1}\right),$$

almost surely, with β_n defined in (6.20).

(2) *In addition, if* $\mathrm{E}e^{\lambda|m(Y_0)|^\nu} < +\infty$ *for some positive* λ *and* ν *(respectively if* m *is a bounded function),*

$$\sup_{x\in D_n} |r_n(x) - r(x)| = \mathcal{O}\left(\left(\frac{(\ln n)^2}{n}\right)^{\frac{2}{d+4}} \cdot \frac{b_n}{\beta_n}\right),$$

with $b_n = (\ln n)^{1/\nu}$ *(respectively* $b_n \equiv \| m \|_\infty$*).*

Proof: *(Sketch)*

(1) We omit x and write

$$\sup_{D_n} |\tilde{r}_n - r| \le \frac{\sup_{D_n} |\tilde{r}_n|}{\inf_{D_n} f} \sup_{D_n} |f_n - f| + \frac{1}{\inf_{D_n} f} \sup_{D_n} |\tilde{\varphi}_n - \varphi|.$$

First, by using (6.16)–(6.17) we easily get that $\sup_{x\in D_n} |\tilde{r}_n(x)| \le b_n$, next Theorem 6.6(1) implies that $\sup_{x\in D_n} |f_n(x) - f(x)| = \mathcal{O}(((\ln n)^2/n)^{2/(d+4)})$ almost surely. Next, following the proof of Theorem 6.9(1) one obtains almost surely that, for $b_n = c_n' h_n^{-1}$, $\sup_{x\in D_n} |\tilde{\varphi}_n(x) - \varphi(x)| = \mathcal{O}(b_n \cdot ((\ln n)^2/n)^{2/d+4})$.

(2) Note that,

$$\sup_{D_n} |r_n - r| \le \frac{1}{\inf_{D_n} f}\left\{\sup_{D_n} |r_n| \sup_{D_n} |f_n - f| + \sup_{D_n} |\tilde{\varphi}_n - \varphi| + \sup_{D_n} |\tilde{\varphi}_n - \varphi_n|\right\}.$$

The first term is handled with Theorem 6.6(1), noting that

$$P\left(\sup_{x\in D_n} |r_n(x)| \ge A\right) \le P\left(\sup_{i=1,\dots,n} |m(Y_i)| > A\right) \le nP(|m(Y_1)| \ge A) = \mathcal{O}(n^{-\gamma})$$

with $\gamma > 1$ if $A = (a_0 \ln n)^{1/\nu}$ for $a_0 > 2/\lambda$. The second term is derived similarly to Theorem 6.9(1), Corollary 6.1 while the third one follows from (6.21). ■

In conclusion, we give examples of expected rates of convergence in some specific cases.

Example 6.1 *(Case of* X_0 *with density* $f(x) \simeq \| x \|^{-p}$, $\| x \| \to \infty$, $p > d$*)*
Under the assumptions and choice of h_n proposed in Theorem 6.10, one obtains that,

$$\sup_{\|x\|\le n^\ell} |\bar{r}_n(x) - r(x)| = \mathcal{O}(n^{-a}(\ln n)^b), \quad \text{a.s.}$$

where a, b and $\bar{r}_n$ are respectively given by

(1) $a = 1/(d+4) - \ell p$ with $0 < \ell < 1/(p(d+4))$, $b = 2/(d+4)$, $\bar{r}_n \equiv \tilde{r}_n$,

(2) if moreover, $\mathrm{E}e^{\lambda|m(Y_0)|^\nu} < +\infty$ for some positive λ and ν, $a = 2/(d+4) - \ell p$ with $0 < \ell < 2/(p(d+4))$, $b = 4/(d+4) + 1/\nu$, $\bar{r}_n \equiv r_n$,

(3) if moreover m is a bounded function, $a = 2/(d+4) - \ell p$ with $0 < \ell < 2/(p(d+4))$, $b = 4/(d+4)$, $\bar{r}_n \equiv r_n$.

Example 6.2 *(Case of X_0 with density $f(x) \simeq \exp(-q \parallel x \parallel^p)$, $\parallel x \parallel \to \infty$, $q, p > 0$)*
Under the assumptions and choice of h_n proposed in Theorem 6.10, one obtains that, almost surely,

$$\sup_{\|x\| \leq (\varepsilon q^{-1} \ln n)^{1/p}} |\bar{r}_n(x) - r(x)| = \mathcal{O}(n^{-a}(\ln n)^b),$$

where a, b and $\bar{r}_n$ are respectively given by

(1) $a = 1/(d+4) - \varepsilon$ for each $0 < \varepsilon < 1/(d+4)$, $b = 2/(d+4)$, $\bar{r}_n \equiv \tilde{r}_n$,

(2) if moreover $\mathrm{E}e^{\lambda|m(Y_0)|^\nu} < +\infty$ for some positive λ and ν, $a = 2/(d+4) - \varepsilon$ for each $0 < \varepsilon < 2/(d+4)$, $b = 4/(d+4) + 1/\nu$, $\bar{r}_n \equiv r_n$,

(3) if moreover m is a bounded function, $a = 2/(d+4) - \varepsilon$ for each $0 < \varepsilon < 2/(d+4)$, $b = 4/(d+4)$, $\bar{r}_n \equiv r_n$. ◇

6.5 Nonparametric prediction by kernel

6.5.1 Prediction for a stationary Markov process of order k

Let $(\xi_t, t \in \mathbb{Z})$ be a strictly stationary $\mathbb{R}^{d_0}$-valued Markov process of order k, namely

$$\mathcal{L}(\xi_t | \xi_{t-s}, s \geq 1) = \mathcal{L}(\xi_t | \xi_{t-1}, \ldots, \xi_{t-k}), \qquad \text{a.s.}$$

or equivalently

$$\mathrm{E}(F(\xi_t) | \xi_{t-s}, s \geq 1) = \mathrm{E}(F(\xi_t) | \xi_{t-1}, \ldots, \xi_{t-k}) \qquad \text{a.s.}$$

for each Borelian real function F such that $\mathrm{E}(|F(\xi_0)|) < +\infty$.

Given the data $\xi_1, \ldots \xi_n$ we want to predict the nonobserved square integrable real random variable $m(\xi_{n+H})$ where $1 \leq H \leq n-k$. In the present case, note that $d = kd_0$ and $d' = d_0$.

For that purpose let us construct the associated process

$$Z_t = (X_t; Y_t) = ((\xi_t, \ldots, \xi_{t+k-1}); \xi_{t+k-1+H}), \quad t \in \mathbb{Z},$$

and consider the kernel regression estimator r_N (or $\tilde{r}_N$) based on the data $(Z_t,\ 1 \le t \le N)$ where $N = n - k + 1 - H$.

From r_N (or possibly $\tilde{r}_N$) we consider the predictor of $m(\xi_{n+H})$ defined by

$$\widehat{m|\xi_{n+H}|} = r_N(\xi_{n-k+1}, \ldots, \xi_n)$$

and we set

$$r(x) = \mathrm{E}(m(\xi_{k-1+H})|(\xi_0, \ldots, \xi_{k-1}) = x), \quad x \in \mathbb{R}^{kd_0}.$$

The empirical error

$$|\widehat{m|\xi_{n+H}|} - r(\xi_{n-k+1}, \ldots, \xi_n)| = |r_N(X_{n-k+1}) - r(X_{n-k+1})|$$

gives a good idea of the predictor's accuracy. Collecting results in the previous section, one may obtain various rough estimates for this error in the case $d' = d_0 = 1$ and $d = k$.

Example 6.3 *(Case of a fixed compact set D of $\mathbb{R}^k$)*
Under the assumptions and choices of h_N, b_N proposed in Theorem 6.9(1), one obtains that, almost surely,

$$|\tilde{r}_N(X_{n-k+1}) - r(X_{n-k+1})|\mathbb{1}_{X_{n-k+1} \in D} = \mathcal{O}\left(\left(\frac{(\ln n)^2}{n}\right)^{\frac{2}{4+k}}\right).$$

If the assumptions of Corollary 6.1 hold, one may replace $\tilde{r}_N(X_{n-k+1})$ by $r_N(X_{n-k+1})$. ◇

Example 6.4 *(Case where m is bounded)*
A typical example of such a situation is the case of *truncated prediction* where one is interested in predicting, for some fixed positive M, the variable $\xi_{n+H}\mathbb{1}_{|\xi_{n+H}| \le M}$ (so that, $m(x) = x\mathbb{1}_{|x| \le M}$).

(1) If X_0 is a.s. bounded with support $[a, b]^k$, one may derive that, under the conditions of Corollary 6.1,

$$|r_N(X_{n-k+1}) - r(X_{n-k+1})|\mathbb{1}_{X_{n-k+1} \in D_n} = \mathcal{O}\left(\left(\frac{(\ln n)^2}{n}\right)^{\frac{2}{4+k}}\right),$$

almost surely, and

$$D_n = [a + \varepsilon_n, b - \varepsilon_n]^k \quad \text{with} \quad \varepsilon_n = \frac{2\sigma}{\sqrt{5}} (\ln n)^{\frac{1}{2}+\frac{2}{4+k}} n^{-\frac{1}{4+k}}$$

whereas $P(X_{n-k+1} \notin D_n) = \mathcal{O}(\varepsilon_n)$.

(2) If X_0 has density $f(x) \simeq \exp(-q \parallel x \parallel^p)$, $\parallel x \parallel \to \infty$, $q, p > 0$, one may derive from Example 6.2(1) that, almost surely,

$$|r_N(X_{n-k+1}) - r(X_{n-k+1})| \mathbb{1}_{X_{n-k+1} \in D_n} = \mathcal{O}\Big(n^{-\left(\frac{2}{k+4}-\varepsilon\right)} (\ln n)^{\frac{4}{k+4}}\Big),$$

with $D_n = \{x : \parallel x \parallel \leq (\varepsilon q^{-1} \ln n)^{1/p}\}$ where $0 < \varepsilon < 2/(k+4)$.
Moreover, one has $P(X_{n-k+1} \notin D_n) = \mathcal{O}((\ln n)^\gamma n^{-\varepsilon})$ for some real number γ.

(3) If X_0 has density $f(x) \simeq \parallel x \parallel^{-p}$, $\parallel x \parallel \to \infty$, $p > k$, one may derive from Example 6.1(1) that, almost surely,

$$|r_N(X_{n-k+1}) - r(X_{n-k+1})| \mathbb{1}_{X_{n-k+1} \in D_n} = \mathcal{O}\Big(n^{-\left(\frac{2}{k+4}-\ell p\right)} (\ln n)^{\frac{4}{k+4}}\Big),$$

with $D_n = \{x : \parallel x \parallel \leq n^\ell\}$ where $0 < \ell < 2/(p(k+4))$.
In addition, one has $P(X_{n-k+1} \notin D_n) = \mathcal{O}(n^{-\ell(p-k)})$. ◇

Example 6.5 *(Case where $m(x) = x$)*

(1) If X_0 is a.s. bounded with support $[a, b]^k$, then $m(Y_0)$ admits an exponential moment. In this case, results obtained in the Example 6.4(1) hold true.

(2) If X_0 has density $f(x) \simeq \exp(-q \parallel x \parallel^p)$, $\parallel x \parallel \to \infty$, $q, p > 0$, one may derive from Example 6.2(2), that

$$|r_N(X_{n-k+1}) - r(X_{n-k+1})| \mathbb{1}_{X_{n-k+1} \in D_n} = \mathcal{O}\Big(n^{-\left(\frac{2}{k+4}-\varepsilon\right)} (\ln n)^{\frac{4}{k+4}+p^{-1}}\Big),$$

almost surely with $D_n = \{x : \parallel x \parallel \leq (\varepsilon q^{-1} \ln n)^{1/p}\}$ where $0 < \varepsilon < 2/(k+4)$.
Moreover, one has $P(X_{n-k+1} \notin D_n) = \mathcal{O}((\ln n)^\gamma n^{-\varepsilon})$ for some real number γ.

(3) If X_0 has density $f(x) \simeq \parallel x \parallel^{-p}$, $\parallel x \parallel \to \infty$, $p > k$, one may derive from Example 6.1(3) that, almost surely, for $b_n \simeq h_n^{-1}$,

$$|\tilde{r}_N(X_{n-k+1}) - r(X_{n-k+1})| \mathbb{1}_{X_{n-k+1} \in D_n} = \mathcal{O}\Big(n^{-\left(\frac{1}{k+4}-\ell p\right)} (\ln n)^{\frac{2}{k+4}}\Big),$$

with $D_n = \{x : \parallel x \parallel \leq n^\ell\}$ where $0 < \ell < 1/(p(k+4))$.
In addition, one has $P(X_{n-k+1} \notin D_n) = \mathcal{O}(n^{-\ell(p-k)})$. ◇

As indicated above, there is a loss of rate for prediction in the general case. This is a consequence of both phenomena: bad behaviour of the bias (if one considers fixed compact sets) and unpredictable behaviour of $r(x)$ for large values of $\| x \|$.

6.5.2 Prediction for general processes

The assumptions used in the above section allowed us to obtain some rates for the prediction error. However these assumptions may be too restrictive for applications. Actually, most of the stationary processes encountered in practice are not Markovian even if they can be approached by a kth order Markov process for a suitable k. In some cases the process is Markovian but k is unknown. Some methods for choosing k are available in the literature, particularly in the linear case: see Brockwell and Davis (1991), Gourieroux and Monfort (1983). Finally, in practice, k appears as a 'truncation parameter' which may depend on the number of observations.

In order to take that fact into account we are induced to consider associated processes of the form

$$Z_{t,n} = (X_{t,n}, Y_{t,n}) = ((\xi_t, \ldots, \xi_{t+k_n-1}), m(\xi_{t+k_n-1+H})), \quad t \in \mathbb{Z}, \quad n \geq 1,$$

where $\lim_{n\to\infty} k_n = \infty$ and $\lim_{n\to\infty} n - k_n = \infty$. Here the observed process (ξ_t) is $\mathbb{R}^{d_0}$-valued and strictly stationary.

The predictor of $m(\xi_{n+H})$ is defined as

$$r_N^*(X_{n-k_n+1}) = \frac{\sum_{t=1}^{N} Y_{t,n} K\left(\frac{X_{n-k_n+1,n} - X_{t,n}}{h_N}\right)}{\sum_{t=1}^{N} K\left(\frac{X_{n-k_n+1,n} - X_{t,n}}{h_N}\right)}$$

where $N = n - k_n + 1 - H$ and $K = K_0^{\otimes k_n}$ with K_0 a d_0-dimensional kernel.

Now some martingale considerations imply that $\mathrm{E}(m(\xi_{n+H})|\xi_n, \ldots, \xi_{n-k_n+1})$ is close to $\mathrm{E}(m(\xi_{n+H})|\xi_s,\ s \leq n)$ for large n. Then under regularity conditions comparable to those of Section 6.5.1, and using similar methods, it may be proved that

$$r_N^*(X_{n-k_n+1}) - \mathrm{E}(m(\xi_{n+H})|\xi_s,\ s \leq n) \xrightarrow{a.s.} 0$$

provided that $k_n = \mathcal{O}((\ln n)^{\delta})$ for some $\delta > 0$. There is clearly no hope of reaching a sharp rate in this general case. In fact, it can be proved that a $(\ln n)^{-\delta'}$ rate is possible. For precise results and details we refer to Rhomari (1994).

Notes

Kernel methods were first described in 1951, in an unpublished report by Fix and Hodges (see Silverman and Jones 1989). There were early studies by Rosenblatt

(1956) and Parzen (1962), who established Theorem 6.1. Concerning regression, the kernel estimator was simultaneously introduced by Nadaraja (1964) and Watson (1964).

Results of Section 6.3 and 6.4 are improvements of results that appeared in Bosq (1998). Concerning almost sure convergence, we have deliberately chosen a quite strong mixing condition (namely an exponential decrease of the coefficient) to present sharp results (that is with asymptotic constants similar to those of the i.i.d. case). For weakened dependence conditions, the interested reader may refer to e.g. the following works (and references therein): Liebscher (2001) for arithmetically strongly mixing processes; Doukhan and Louhichi (2001) (density); and Ango Nze *et al.* (2002) (regression) for a new concept of dependence including processes that are not mixing in general. For practical choices of h_n in the dependent case, we refer also to: Hall *et al.* (1995); Hart and Vieu (1990); Kim (1997).

Note that other topics are addressed in Bosq (1998, Chapter 3): especially the study of the mean-square error of prediction, the nonstationary case as well as related extensions (interpolation, chaos, regression with errors). Infinite-dimensional extensions appear in Dabo-Niang and Rhomari (2002) and Ferraty and Vieu (2006). Finally, a systematic comparison between prediction by ARMA models and nonparametric prediction appears in Carbon and Delecroix (1993).

7

Kernel method in continuous time

In this chapter we investigate the problem of nonparametric estimation of density when continuous data are available. This continuous time framework is especially of interest since many phenomena (in various domains such as physics, medicine, economics) lead to continuous time observations (possibly interpolated).

We shall see that the situation is somewhat different from Chapter 6. First, under conditions closely related to the discrete case, one obtains similar rates of convergence (depending both on dimension and regularity of the unknown density). Now in the continuous time framework, the crucial point is that regularity of sample paths may also act on rates of convergence: a quite irrelevant phenomenon in the discrete case! Actually, one should get a whole collection of optimal rates which are specific to the regularity of the underlying sample paths. As a by-product, these results will also be involved in Chapter 8 where only sampled data are at one's disposal and the question of an appropriate sampling scheme (well fitted to regularity of the underlying sample paths) naturally arises.

In Sections 7.1 and 7.2, we give rates of convergence for the kernel density estimator, while regression estimation is treated in Section 7.3 and applications to prediction are considered in Section 7.4.

7.1 Optimal and superoptimal rates for density estimation

Let $X = (X_t,\ t \in \mathbb{R})$ be an $\mathbb{R}^d$-valued continuous time process defined on a probability space $(\Omega, \mathcal{A}, P)$. In all the following, we assume that (X_t) is measurable (i.e. $(t, \omega) \mapsto X_t(\omega)$ is $\mathcal{B}_{\mathbb{R}} \otimes \mathcal{A}$-$\mathcal{B}_{\mathbb{R}^d}$ measurable).

Inference and Prediction in Large Dimensions D. Bosq and D. Blanke

Suppose that the X_t's have a common distribution μ. We wish to estimate μ from the data $(X_t,\ t \in [0, T])$. A primary estimator for μ is the *empirical measure* μ_T defined as

$$\mu_T(B) = \frac{1}{T}\int_0^T \mathbb{1}_B(X_t)\mathrm{d}t, \quad B \in \mathcal{B}_{\mathbb{R}^d}, \quad T > 0.$$

Now if μ has a density, say f, one may regularize μ_T by convolution, leading to the kernel density estimator:

$$f_T(x) = \frac{1}{T h_T^d}\int_0^T K\left(\frac{x - X_t}{h_T}\right)\mathrm{d}t, \quad x \in \mathbb{R}^d \tag{7.1}$$

where K is a *kernel* (see Chapter 6) satisfying conditions $\mathcal{K}$ given by (6.1) and the bandwidth $h_T \to 0^+$ as $T \to \infty$. We also set

$$K_h(u) = h_T^{-d} K\left(\frac{u}{h_T}\right), \quad u \in \mathbb{R}^d.$$

Since the bias has exactly the same expression as in the i.i.d. case, we will now focus on the variance of the kernel estimator. First we show that this variance has, at least, the same rate of convergence as in the discrete case: we will refer to this case as the 'optimal' one. Next, under more stringent conditions, we will establish that this rate can be improved up to a 'bandwidth-free' rate, that we will call the 'superoptimal' case or 'parametric' case.

7.1.1 The optimal framework

The following conditions are natural extensions of Assumptions A6.1 to the continuous time framework. As in Chapter 6, notice that neither strict nor large stationarity of X is required.

Assumptions 7.1 (A7.1)

(i) $f \in C_d^2(b)$ for some positive b where $C_d^2(b)$ is the space of twice continuously differentiable real-valued functions h, defined on $\mathbb{R}^d$, such that $\| h^{(2)} \|_\infty \le b$ and $h^{(2)}$ represents any partial derivative of order 2 for h.

(ii) There exists Γ, a Borelian set of $\mathbb{R}^2$ containing $D = \{(s, t) \in \mathbb{R}^2 : s = t\}$, such that for $(s, t) \notin \Gamma$, the random vector (X_s, X_t) has a density $f_{(X_s,X_t)}$ and $\sup_{(s,t)\notin\Gamma} \| g_{s,t} \|_\infty < \infty$ where $g_{s,t} = f_{(X_s,X_t)} - f \otimes f$. Moreover,

$$\overline{\lim_{T\to\infty}}\ \frac{1}{T}\int_{[0,T]^2\cap\Gamma} \mathrm{d}s\,\mathrm{d}t = \ell_\Gamma < +\infty.$$

(iii) (X_t) is supposed to be 2-α-*mixing* with $\alpha^{(2)}(|t - s|) \le \gamma|t - s|^{-\beta}$, $(s, t) \notin \Gamma$ for some positive constants γ and $\beta > 2$.

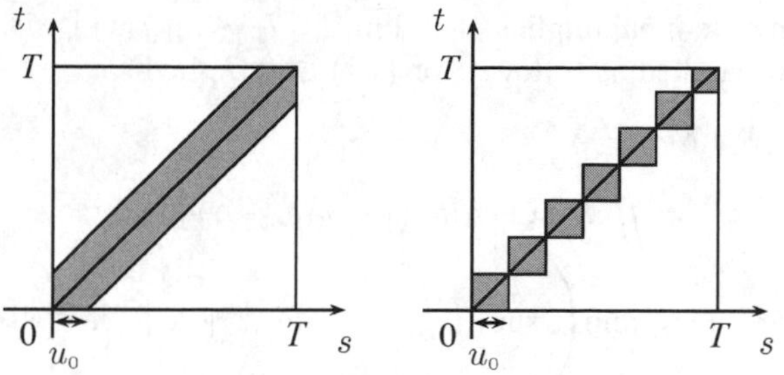

Figure 7.1
On the left: $\Gamma = \{(s,t) : |s-t| \leq u_0\}$, $\ell_\Gamma = 2u_0$.
On the right: $\Gamma = \bigcup_{n=0}^{[T/u_0]} \{(s,t) : [s/u_0] = [t/u_0] = n\}$, $\ell_\Gamma = u_0$.

Note that only condition A7.1(ii) differs from the discrete case: here the Borelian Γ is introduced to prevent the explosion of $f_{(X_s,X_t)}$ for small $|t-s|$, $s \neq t$.

Typical examples of Borelian Γ are furnished by Figure 7.1.

Theorem 7.1

(1) If conditions A7.1(ii)–(iii) are satisfied with f continuous at x, then $Th_T^d \to \infty$ yields

$$\overline{\lim_{T\to\infty}}\; Th_T^d \mathrm{Var} f_T(x) \leq \ell_\Gamma f(x) \parallel K \parallel_2^2 . \tag{7.2}$$

(2) If Assumptions A7.1 hold with $f(x) > 0$, the choice $h_T = c_T T^{-1/(d+4)}$, $c_T \to c > 0$, leads to

$$\overline{\lim_{T\to\infty}}\; T^{4/(d+4)} \mathrm{E}(f_T(x) - f(x))^2 \leq c^4 b_2^2(x) + \frac{f(x)}{c^d} \ell_\Gamma \parallel K \parallel_2^2$$

with constant $b_2(x)$ defined by (6.4).

PROOF:

(1) We consider the decomposition

$$\begin{aligned} Th_T^d \,\mathrm{Var}\, f_T(x) &= \frac{1}{Th_T^d} \int_{[0,T]^2 \cap \Gamma} \mathrm{Cov}\left(K\left(\frac{x - X_s}{h_T}\right), K\left(\frac{x - X_t}{h_T}\right)\right) \mathrm{d}s\,\mathrm{d}t \\ &\quad + \frac{1}{Th_T^d} \int_{[0,T]^2 \cap \Gamma^c} \mathrm{Cov}\left(K\left(\frac{x - X_s}{h_T}\right), K\left(\frac{x - X_t}{h_T}\right)\right) \mathrm{d}s\,\mathrm{d}t \\ &= I_T + J_T. \end{aligned} \tag{7.3}$$

The first integral may be bounded above by

$$I_T \leq h_T^{-d} \mathrm{E} K^2\left(\frac{x - X_0}{h_T}\right) \cdot \frac{1}{T} \int_{[0,T]^2 \cap \Gamma} \mathrm{d}s\,\mathrm{d}t \tag{7.4}$$

and the Bochner lemma implies that $\overline{\lim}_{T\to\infty} I_T \leq \ell_\Gamma f(x) \parallel K \parallel_2^2$. The second integral may be handled as follows: for $(s,t) \in [0,T]^2 \cap \Gamma^c$,

$$\begin{aligned}\text{Cov}\,(K_h(x-X_s),K_h(x-X_t)) &= \iint K_h(x-u)K_h(x-v)g_{s,t}(u,v)\mathrm{d}u\,\mathrm{d}v \\ &\leq \min\left(\sup_{(s,t)\notin\Gamma} \parallel g_{s,t} \parallel_\infty,\ 4h_T^{-2d} \parallel K \parallel_\infty^2 \alpha^{(2)}(|t-s|)\right) \qquad (7.5)\end{aligned}$$

by condition A7.1(ii) and the *Billingsley inequality*. Next, splitting the integrals into $\{(s,t) \in \Gamma^c : |s-t| \leq h_T^{-2d/\beta}\}$ and $\{(s,t) \in \Gamma^c : |s-t| > h_T^{-2d/\beta}\}$, one arrives at

$$J_T = \frac{h_T^d}{T}\int_{[0,T]^2\cap\Gamma^c} \text{Cov}(K_h(x-X_s),K_h(x-X_t)) = \mathcal{O}\left(h_T^{d(1-\frac{2}{\beta})}\right) = o(1),$$

since one has $\beta > 2$ in the mixing condition A7.1(iii).

(2) Since the bias has the same expression as in the discrete case, the mean-square error immediately follows. ■

In order to show that the above rate is achieved for some processes, let us consider the family $\mathcal{X}_0$ of processes $X = (X_t,\ t \in \mathbb{R})$ with common marginal f and satisfying Assumptions A7.1(ii)–(iii) uniformly, in the following sense: there exist positive constants L_0, δ_0, γ_0, β_0 such that, with obvious notation, for each $X \in \mathcal{X}_0$:

- $(1/T)\int_{[0,T]^2\cap\Gamma_X} \mathrm{d}s\,\mathrm{d}t \leq L_0(1+L_0/T)$,
- $\sup_{(s,t)\notin\Gamma_X} \parallel g_{s,t} \parallel_\infty \leq \delta_0$,
- $\gamma \leq \gamma_0$ and $\beta \geq \beta_0 > 2$.

Then we have:

Corollary 7.1
If f is continuous at x, the condition $Th_T^d \to \infty$ yields

$$\lim_{T\to\infty}\max_{X\in\mathcal{X}_0} Th_T^d \text{Var}_X f_T(x) = L_0 f(x) \parallel K \parallel_2^2$$

where Var_X *denotes variance if the underlying process is X.*

PROOF:
An easy consequence of (7.3)–(7.5) is

$$\overline{\lim_{T\to\infty}}\max_{X\in\mathcal{X}_0} Th_T^d \text{Var}_X f_T(x) \leq L_0 f(x) \parallel K \parallel_2^2 .$$

It remains to exhibit a process $\tilde{X}$ in $\mathcal{X}_0$ such that

$$Th_T^d \text{Var}_{\tilde{X}} f_T(x) \to L_0 f(x) \parallel K \parallel_2^2 .$$

To this aim let us consider a sequence $(Y_n,\ n \in \mathbb{Z})$ of i.i.d. $\mathbb{R}^d$-valued random variables with density f continuous at x and set $\tilde{X}_t = Y_{[t/L_0]},\ t \in \mathbb{R}$ (notice that the process $\tilde{X}$ is not stationary). One has $\tilde{X} \in \mathcal{X}_0$ with $\Gamma_{\tilde{X}} = \cup_{n\in\mathbb{Z}}\{(s,t) : [s/L_0] = [t/L_0] = n\}$, $\sup_{(s,t)\notin\Gamma_{\tilde{X}}} \parallel g_{s,t} \parallel_\infty = 0$, and $\alpha^{(2)}(|s-t|) = 0$ if $(s,t) \notin \Gamma_{\tilde{X}}$. Now for this particular process f_T takes a special form, namely

$$f_T(x) = \frac{[T/L_0]}{T/L_0} \hat{f}_T(x) + \frac{T - L_0[T/L_0]}{Th_T^d} K\left(\frac{x - Y_{[T/L_0]}}{h_T}\right)$$

where $\hat{f}_T$ is a kernel estimator of f associated with the i.i.d. sample $Y_0, \ldots, Y_{[T/L_0]-1}$.

Then, from the Bochner lemma, 6.1, it is easy to deduce that

$$Th_T^d \left(\frac{[T/L_0]}{T/L_0}\right)^2 \text{Var } \hat{f}_T(x) \to L_0 f(x) \parallel K \parallel_2^2$$

and

$$Th_T^d \left(\frac{T - L_0[T/L_0]}{Th_T^d}\right)^2 \text{Var } K\left(\frac{x - Y_{[T/L_0]}}{h_T}\right) \leq \frac{L_0^2}{Th_T^d} \parallel K \parallel_\infty^2 \to 0,$$

hence the result ■

Finally it can be shown (see Bosq 1998, Chapter 4) that the rate $T^{-2r/(2r+d)}$ is *minimax* over $\mathcal{X}_0$ for densities belonging to $C_{r,d}(\ell)$ $(r = k + \lambda, 0 < \lambda \leq 1, k \in \mathbb{N})$, where $C_{r,d}(\ell)$ denotes the space of k-times differentiable real-valued functions, defined on $\mathbb{R}^d$ and such that

$$\left| \frac{\partial f^{(k)}}{\partial x_1^{j_1} \cdots \partial x_d^{j_d}}(x') - \frac{\partial f^{(k)}}{\partial x_1^{j_1} \cdots \partial x_d^{j_d}}(x) \right| \leq \ell \parallel x' - x \parallel^\lambda;$$

$$x, x' \in \mathbb{R}^d,\ j_1 + \cdots + j_d = k.$$

Note that $C_d^2(b)$ is included in $C_{2,d}(b)$ with $k = \lambda = 1$.

7.1.2 The superoptimal case

Castellana and Leadbetter (1986) have pointed out that, in the stationary case, if the local dependence of X_0 and X_u, $u > 0$, is sufficiently restricted then it is possible to obtain a bandwidth-free rate for the variance of smoothed density estimators (including the kernel one). We present here their result as well as some extensions. For that purpose we suppose that $g_{s,t} = f_{(X_s,X_t)} - f \otimes f$ exists for $s \neq t$ and use the abbreviated notation g_u for $g_{0,u}$, $u > 0$.

Theorem 7.2
Let us assume that $g_{s,t} = g_{|t-s|}, t \neq s$.

(1) *If* $(y,z) \mapsto \int_{]0,+\infty[} |g_u(y,z)| \mathrm{d}u$ *is defined and bounded on* $\mathbb{R}^{2d}$, *then*

$$T \cdot \mathrm{Var} f_T(x) \leq 2 \sup_{(y,z)\in\mathbb{R}^{2d}} \int_0^{+\infty} |g_u(y,z)| \mathrm{d}u,$$

for any $T > 0$.

(2) *If* $u \mapsto \| g_u \|_\infty$ *is integrable on* $]0,+\infty[$ *(Castellana and Leadbetter's condition) and* g_u *is continuous at* (x,x) *for each* $u > 0$, *then*

$$\lim_{T\to+\infty} T \cdot \mathrm{Var} f_T(x) = 2 \int_0^\infty g_u(x,x) \mathrm{d}u.$$

In addition, if $f \in C_d^2(b)$, *the choice* $h_T = o(T^{-1/4})$ *implies*

$$\lim_{T\to+\infty} T \cdot \mathrm{E}(f_T(x) - f(x))^2 = 2 \int_0^\infty g_u(x,x) \mathrm{d}u.$$

PROOF:
Using the stationarity condition $g_{s,t} = g_{|t-s|}$, we get

$$\begin{aligned} T \cdot \mathrm{Var} f_T(x) &= \frac{1}{T} \int_0^T \int_0^T \mathrm{Cov}(K_h(x - X_t), K_h(x - X_s)) \mathrm{d}s \mathrm{d}t \\ &= 2 \int_0^T \left(1 - \frac{u}{T}\right) \iint_{\mathbb{R}^{2d}} K_h(x-y) K_h(x-z) g_u(y,z) \mathrm{d}y \, \mathrm{d}z \, \mathrm{d}u. \end{aligned} \quad (7.6)$$

Concerning the first part, one easily obtains that

$$\begin{aligned} T \cdot \mathrm{Var} f_T(x) &\leq 2 \iint_{\mathbb{R}^{2d}} K_h(x-y) K_h(x-z) \int_0^{+\infty} |g_u(y,z)| \, \mathrm{d}u \, \mathrm{d}y \, \mathrm{d}z \\ &\leq 2 \sup_{(y,z)\in\mathbb{R}^{2d}} \int_0^{+\infty} |g_u(y,z)| \, \mathrm{d}u \end{aligned}$$

from the Fubini theorem.

For the second part, note that in (7.6), the function

$$(u, v, w) \mapsto \left(1 - \frac{u}{T}\right) g_u(y, z) K_h(x - y) K_h(x - z) \mathbb{1}_{[0,T]}(u) \mathbb{1}_{\mathbb{R}^{2d}}(y, z)$$

is integrable since $\int_0^\infty \| g_u \|_\infty \, du < +\infty$. Using again Fubini, one gets

$$T \cdot \operatorname{Var} f_T(x) = 2 \iint_{\mathbb{R}^{2d}} \int_0^T \left(1 - \frac{u}{T}\right) K(v) K(w) g_u(x - h_T v, x - h_T w) du \, dy \, dz.$$

Next, one has

$$\left(1 - \frac{u}{T}\right) g_u(x - h_T v, x - h_T w) K(v) K(w) \mathbb{1}_{[0,T]}(u) \mathbb{1}_{\mathbb{R}^d \times \mathbb{R}^d}(v, w) \xrightarrow[T \to \infty]{} g_u(x, x) K(v) K(w) \mathbb{1}_{\mathbb{R}_+ \times \mathbb{R}^d \times \mathbb{R}^d}(u, v, w)$$

and

$$\left| \left(1 - \frac{u}{T}\right) g_u(x - h_T v, x - h_T w) K(v) K(w) \mathbb{1}_{[0,T]}(u) \mathbb{1}_{\mathbb{R}^d \times \mathbb{R}^d}(v, w) \right| \leq \| g_u \|_\infty K(v) K(w) \in L^1(\mathbb{R}_+ \times \mathbb{R}^d \times \mathbb{R}^d)$$

so the dominated convergence theorem yields

$$\lim_{T \to \infty} T \cdot \operatorname{Var} f_T(x) = 2 \int_0^{+\infty} g_u(x, x) du.$$

The mean-square error easily follows since the choice $h_T = o(T^{-1/4})$ implies that the squared bias is negligible with respect to the variance. ■

Note that this superoptimal mean-square rate of convergence remains valid for less regular densities: for example if f satisfies a Hölder condition of order λ $(0 < \lambda \leq 1)$, the choice $h_T = \mathcal{O}(e^{-T})$ yields the same result. However, from a practical point of view, this choice is somewhat unrealistic since it does not balance variance and squared bias for finite T. On the other hand, it is interesting to note that a choice of h_T which would give asymptotic efficiency for all continuous f is not possible since the bias depends on the modulus of continuity of f. An estimator which captures global efficiency will appear in Chapter 9.

We now give some examples of applicability of Theorem 7.2.

Example 7.1

Let $(X_t, t \in \mathbb{R})$ be a Gaussian real stationary process with zero mean and autocorrelation function $\rho(u) = 1 - a|u|^\theta + o(u^\theta)$ when $u \to 0$ for $0 < \theta < 2$. In this case, it is easy to verify that

$$|g_u(x, y)| \leq a|\rho(u)\mathbb{1}_{|u|>b} + (c + d|u|^{-\theta/2})\mathbb{1}_{0<|u|\leq b}$$

where a, b, c, d are suitable constants. Consequently, conditions in Theorem 7.2 are satisfied as soon as ρ is integrable on $]0, +\infty[$. ◇

Example 7.2
Let $(X_t,\ t \geq 0)$ be a real diffusion process defined by the stochastic differential equation

$$\mathrm{d}X_t = S(X_t)\mathrm{d}t + \sigma(X_t)\mathrm{d}W_t, \quad t \geq 0$$

where S and σ satisfy a Lipschitz condition and the condition

$$I = I(S) = \int_{\mathbb{R}} \sigma^{-2}(x) \exp\left(2 \int_0^x S(y)\sigma^{-2}(y)\mathrm{d}y\right)\mathrm{d}x < +\infty$$

and where $(W_t,\ t \geq 0)$ is a standard Wiener process.

It may be proved that such a process admits a stationary distribution with density given by

$$f(x) = I^{-1}\sigma^{-2}(x) \exp\left(2 \int_0^x S(y)\sigma^{-2}(y)\mathrm{d}y\right), \quad x \in \mathbb{R}.$$

Moreover, under some regularity assumptions on S and σ, the kernel estimator of f reaches the superoptimal rate T^{-1}. In particular if X_0 has the density f, conditions in Theorem 7.2 are satisfied, see Kutoyants (1997b); Leblanc (1997) and Veretennikov (1999). Moreover it can be proven that this estimator is also asymptotically efficient Kutoyants (1998, 2004). ◇

7.2 From optimal to superoptimal rates

7.2.1 Intermediate rates

One may naturally ask about the existence of intermediate rates of convergence (lying between the optimal and superoptimal cases). We begin by a result illustrating that conditions in Theorem 7.2 are, in some sense, necessary for obtaining the bandwidth-free rate $1/T$ for the variance.

Proposition 7.1
Let $(X_t,\ t \in \mathbb{R})$ be an $\mathbb{R}^d$-valued process such that

(a) $g_{s,t} = g_{|t-s|}$ exists for $s \neq t$ and for $u_0 > 0$,

$$\sup_{(y,z)\in\mathbb{R}^{2d}} \int_{u_0}^{\infty} |g_u(y,z)|\mathrm{d}u < \infty.$$

(b) f is continuous at x and $f_{(X_0,X_u)}$ is continuous at (x,x) for $u > 0$.

(c) $\int_0^{u_1} f_{(X_0,X_u)}(x,x)\mathrm{d}u = +\infty$ for $u_0 > u_1 > 0$.

Then, if K is a positive kernel,

$$\lim_{T\to\infty} T\cdot\operatorname{Var} f_T(x) = +\infty.$$

PROOF:
From (7.6), observe that

$$2\left|\int_{u_0}^{\infty}\left(1-\frac{u}{T}\right)\iint_{\mathbb{R}^{2d}} K_h(x-y)K_h(x-z)g_u(y,z)\mathrm{d}y\,\mathrm{d}z\,\mathrm{d}u\right|$$
$$\leq 2\sup_{(y,z)\in\mathbb{R}^{2d}}\int_{u_0}^{\infty}|g_u(y,z)|\mathrm{d}u = \mathcal{O}(1)$$

with condition (a). On the other hand, condition (b) implies

$$\lim_{h_T\to 0}\iint_{\mathbb{R}^{2d}} K_h(x-y)K_h(x-z)f(y)f(z)\mathrm{d}y\,\mathrm{d}z = f^2(x)$$

so that

$$T\cdot\operatorname{Var} f_T(x) =$$
$$2\int_0^{u_0}\left(1-\frac{u}{T}\right)\iint_{\mathbb{R}^{2d}} K_h(x-y)K_h(x-z)f_{(X_0,X_u)}(y,z)\mathrm{d}y\,\mathrm{d}z\,\mathrm{d}u + \mathcal{O}(1).$$

Now for $T \geq 2u_0$, using the affine transformation $(y,z)\mapsto(x-h_T v, x-h_T w)$ and the image measure theorem (see Rao 1984), one obtains that

$$T\cdot\operatorname{Var} f_T(x)$$
$$\geq 2\iint_{\mathbb{R}^{2d}}\int_0^{u_0} K(v)K(w)f_{(X_0,X_u)}(x-h_T v, x-h_T w)\mathrm{d}v\,\mathrm{d}w\,\mathrm{d}u + \mathcal{O}(1)$$

and the result follows from the Fatou lemma and conditions (b) and (c). ■

The next proposition illustrates, in the real Gaussian case, the link between sample path regularity and rate of convergence.

Proposition 7.2
Let $X = \{X_t,\ t\in\mathbb{R}\}$ be a real zero-mean and stationary Gaussian process. If X is mean-square continuous with autocorrelation function $\rho(\cdot)$ such that $|\rho(u)| < 1$ for $u > 0$ and $\int_{u_1}^{\infty}|\rho(u)|\mathrm{d}u < +\infty$ for some $u_1 > 0$, then if K is a positive kernel, one obtains for all real x

$$(a)\ \int_0^{u_1}\frac{1}{\sqrt{1-\rho^2(u)}}\mathrm{d}u = +\infty \ \Rightarrow\ T\cdot\operatorname{Var} f_T(x)\to+\infty,$$

(b) $\displaystyle\int_0^{u_1} \frac{1}{\sqrt{1-\rho^2(u)}}\,\mathrm{d}u < +\infty \quad \Rightarrow \quad T\cdot \mathrm{Var}\, f_T(x) \to \ell < +\infty.$

PROOF:
See Blanke and Bosq (2000). ■

From proposition 7.2 one gets the following alternative: if $1-\rho(u) \sim cu^{2\alpha}$ as $u \downarrow 0$ with $0<\alpha<1$, the bandwidth-free rate $1/T$ is reached and there exists an equivalent process with sample functions satisfying a Hölder condition of any order $a<\alpha$, with probability one (see Cramér and Leadbetter 1967, p.172). Such a condition is satisfied by the ⋆Ornstein–Uhlenbeck process⋆ (cf. Example 1.14). On the other hand, if the process is mean-square differentiable one has $1-\rho(u) \sim cu^2$ as $u \downarrow 0$ and Proposition 7.2(a) implies that the bandwidth-free rate is no longer possible. In the following, we will establish that the obtained rate is in fact $\ln(1/h_T)/T$.

7.2.2 Classes of processes and examples

Now we introduce our main assumptions on processes.

Assumptions 7.2 (A7.2)

(i) f is either bounded or continuous at x;

(ii) for $s \neq t$, the joint density $f_{(X_s,X_t)}$ of (X_s, X_t) does exist and for $t>s$, $f_{(X_s,X_t)} = f_{(X_0,X_{t-s})}$;

(iii) $\exists u_0 > 0 \;:\; \sup_{(y,z)\in\mathbb{R}^{2d}} \int_{u_0}^{+\infty} |g_u(y,z)|\mathrm{d}u < +\infty$, where g_u is defined by $g_u := f_{(X_0,X_u)} - f\otimes f,\ u>0$;

(iv) $\exists \gamma_0 > 0 \;:\; f_{(X_0,X_u)}(y,z) \le M(y,z)u^{-\gamma_0}$, for $(y,z,u) \in \mathbb{R}^{2d}\times]0,u_0[$, where $M(\cdot,\cdot)$ is either continuous at (x,x) and $\mathbb{R}^{2d}$-integrable or bounded.

Let us give some information about these assumptions. First A7.2(ii) is a weak stationarity condition whereas A7.2(iii) is a condition of asymptotic independence discussed in Vertennikov (1999) (for the special case of real diffusion processes) and Comte and Merlevède (2005). Condition A7.2(iv) is less usual but could be linked to the regularity of sample paths, in this way it is quite typical in our continuous time context. More precisely, let us define

$$Y_u := \left(\frac{X_u^{(1)} - X_0^{(1)}}{u^{\gamma_1}}, \dots, \frac{X_u^{(d)} - X_0^{(d)}}{u^{\gamma_d}} \right)$$

where $X_t^{(i)}$ is the ith component of $(X_t) = (X_t^{(1)}, \cdots, X_t^{(d)})$ and γ_i, $0<\gamma_i\le 1$, $i=1,\dots,d$ are the respective Hölderian coefficients; one may show that condition A7.2(iv) is equivalent to

$$f_{(X_0,Y_u)}\left(y, \frac{z-y}{u^{\gamma}}\right) \le M(y,z), \quad (y,z,u) \in \mathbb{R}^{2d}\times]0,u_0[\tag{7.7}$$

where $(z-y)/u^{\gamma} := ((z_1-y_1)/u^{\gamma_1}, \ldots, (z_d-y_d)/u^{\gamma_d})$ as soon as $\gamma_0 = \sum_{i=1}^{d} \gamma_i$. In this way γ_0 is linked to the 'total' sample path regularity and the condition: $\gamma_0 \in]0, d]$ naturally arises.

We now present typical processes satisfying Assumptions A7.2.

Example 7.3 (*Multidimensional homogeneous diffusions*: $\gamma_0 = d/2$)
Let $(X_t, t \geq 0)$ be an $\mathbb{R}^d$-valued diffusion process defined by the stochastic differential equation

$$\mathrm{d}X_t = S(X_t)\mathrm{d}t + \mathrm{d}W_t, \quad t \geq 0$$

where $(W_t, t \geq 0)$ is a standard d-dimensional Wiener process and $S : \mathbb{R}^d \to \mathbb{R}^d$ satisfies a Lipschitz condition, is bounded and such that for $M_0 \geq 0$, $r > 0$,

$$\left\langle S(x), \frac{x}{\| x \|} \right\rangle \leq -\frac{r}{\| x \|^p}, \quad 0 \leq p < 1, \| x \| \geq M_0. \tag{7.8}$$

Following the work of Qian *et al.* (2003), and Klokov and Veretennikov (2005), it may be shown that Assumptions A7.2 hold with $\gamma_0 = d/2$; (see also Bianchi 2007). ◇

Example 7.4 (*Case* $\gamma_0 = 1$)
This case may be associated with regular real processes satisfying condition (7.7), (see e.g. Blanke and Bosq 1997; Sköld and Hössjer 1999). Typical examples are: the Gaussian process defined in Example 7.1 with $\theta = 2$; bi-dimensional processes ($d = 2$) whose components are independent and with $\gamma_1 = \gamma_2 = 1/2$ (see Examples 7.1 and 7.2) and bi-dimensional diffusions processes in Example 7.3. ◇

Example 7.5 (*Case* $\gamma_0 > 1$)
This last case typically arises for dimensions $d > 1$, by combining the previous examples, see also Blanke and Bosq (2000); Sköld (2001). ◇

7.2.3 Mean-square convergence

We begin by giving upper bounds of convergence for the variance of the kernel estimator.

Theorem 7.3
Under Assumptions A7.2, one obtains

$$\overline{\lim_{T\to\infty}}\ \Psi(h_T, T, \gamma_0)\mathrm{Var} f_T(x) < +\infty;$$

where $\Psi(h_T, T, \gamma_0)$ *is given by*

$$\Psi(h_T, T, \gamma_0) = \begin{cases} T & \text{if} \quad \gamma_0 < 1, \\ \dfrac{T}{\ln(1/h_T)} & \text{if} \quad \gamma_0 = 1, \\ T h_T^{\frac{d}{\gamma_0}(\gamma_0 - 1)} & \text{if} \quad \gamma_0 > 1. \end{cases} \tag{7.9}$$

Note that Sköld (2001) obtains the exact behaviour of the variance under more restrictive conditions on the joint density $f_{(X_0,(X_u - X_0)/u^{\alpha})}$ for a given α. Furthermore in the case where no local information is available (in other words if one omits condition A7.2(iv)), one gets back the 'optimal case' by letting γ_0 tend to infinity. Finally, rates of Theorem 7.3 are preserved if condition A7.2(iii) is replaced by (X_t) GSM (in the case $\gamma_0 = 1$), or (X_t) 2-α-⋆mixing⋆ with coefficient $\beta \geq 2\gamma_0/(\gamma_0 - 1)$ (in the case $\gamma_0 > 1$).

PROOF: *(Elements)*
The stationarity condition A7.2(ii) allows us to decompose $T \operatorname{Var} f_T(x)$ into two terms involving the function g_u with $u \in]0, u_0[$ and $u \in]u_0, +\infty[$. The Fubini theorem and condition A7.2(iii) imply that the second term is $\mathcal{O}(1)$. Next, under condition A7.2(i), the main term, depending on γ_0, is given by

$$A = \frac{2}{T} \int_0^{u_0} (T - u) \iint_{\mathbb{R}^{2d}} K_h(x - y) K_h(x - z) f_{(X_0, X_u)}(y, z) \, \mathrm{d}y \, \mathrm{d}z \, \mathrm{d}u. \tag{7.10}$$

First if $\gamma_0 < 1$, condition A7.2(iv) implies, with the Bochner lemma 6.1, that $A = \mathcal{O}(1)$. Next for $\gamma_0 \geq 1$, one may bound A by

$$\begin{aligned} |A| \leq\ & 2 \int_0^{u_T} \iint_{\mathbb{R}^{2d}} |K(y)K(z)| f_{(X_0, X_u)}(x - h_T y, x - h_T z) \mathrm{d}y \, \mathrm{d}z \, \mathrm{d}u \\ & + 2 \int_{u_T}^{u_0} \iint_{\mathbb{R}^{2d}} |K(y)K(z)| f_{(X_0, X_u)}(x - h_T y, x - h_T z) \mathrm{d}y \, \mathrm{d}z \, \mathrm{d}u. \end{aligned}$$

where u_T is a positive sequence such that $u_T \to 0$ as $T \to \infty$. The Cauchy–Schwarz inequality implies that the first term is a $\mathcal{O}(u_T/h_T^d)$. Now, if $\gamma_0 = 1$, condition A7.2(iv) with the Fubini theorem implies that

$$\begin{aligned} & 2 \int_{u_T}^{u_0} \iint_{\mathbb{R}^{2d}} |K(y)K(z)| f_{(X_0, X_u)}(x - h_T y, x - h_T z) \mathrm{d}y \, \mathrm{d}z \, \mathrm{d}u \\ & \quad \leq 2(\ln u_0 - \ln u_T) \iint_{\mathbb{R}^{2d}} |K(y)K(z)| M(x - h_T y, x - h_T z) \mathrm{d}y \, \mathrm{d}z. \end{aligned}$$

Setting $u_T = h_T^d$, one obtains with either the Bochner lemma or dominated convergence theorem (if M is bounded) that $A = \mathcal{O}(\ln(1/h_T))$. For $\gamma_0 > 1$, the same methodology yields $A = \mathcal{O}(h_T^{d(1-\gamma_0)/\gamma_0})$ with the choice $u_T = h_T^{d/\gamma_0}$. Finally, note that for strongly mixing processes not satisfying condition A7.2(iii), a similar proof as for Theorem 7.1 yields the same rates of convergence as soon as (X_t) is GSM in the case $\gamma_0 = 1$, or (X_t) is 2-α-mixing with coefficient $\beta \geq 2\gamma_0/(\gamma_0 - 1)$ if $\gamma_0 > 1$. ■

Now, we turn to lower bounds of estimation. First, note that if one strengthens condition A7.2(iii) to $\int_{u_0}^{+\infty} \| g_u \|_\infty \, du < +\infty$ and if $M(\cdot,\cdot)$ is a bounded function in condition A7.2(iv), the case $\gamma_0 < 1$ is included in Theorem 7.2(2) and the variance has the bandwidth-free rate. If $\gamma_0 \geq 1$, we obtain the following result.

Theorem 7.4
Assume conditions A7.2(i)–(iii) and moreover that

$$f_{(X_0,X_u)}(y,z) \geq \frac{p_u(y,z)}{u^{\gamma_0}} \quad \text{with } \gamma_0 \geq 1,\ u \in \,]0, u_0[,\ (y,z) \in \mathbb{R}^{2d} \tag{7.11}$$

and

$$\varliminf_{T\to\infty} \inf_{u \in [\varepsilon_T^{d/\gamma_0}, u_0[} p_u(x - \varepsilon_T y, x - \varepsilon_T z) \geq p_0(x,y,z) > 0, \quad x \in \mathbb{R}^d$$

where $\varepsilon_T \to 0$ as $T \to \infty$. Then, for a positive kernel K,

(1) if $\gamma_0 = 1$, $\varliminf_{T\to\infty} (T/\ln h_T^{-1}) \operatorname{Var} f_T(x) > 0$;
(2) if $\gamma_0 > 1$, $\varliminf_{T\to\infty} T h_T^{d(\gamma_0-1)/\gamma_0} \operatorname{Var} f_T(x) > 0$.

The additional minorant condition is technical but it holds in particular for Gaussian processes (see Blanke and Bosq 2000) or d-dimensional homogeneous diffusion processes such that $f(x) > 0$ and with either bounded drift (see Bianchi 2007; Qian *et al.* 2003, Theorem 2], or drift with at most linear growth (see Qian and Zheng 2004, Theorem 3.1).

Proof: *(Elements)*
Again, the main proof concerns the term A, given by (7.10). For all $T \geq 2u_0$, one gets by the Fubini theorem

$$A \geq 2 \iint_{\mathbb{R}^{2d}} K_h(x-y) K_h(x-z) \int_0^{u_0} f_{(X_0,X_u)}(y,z) \mathrm{d}u \, \mathrm{d}y \, \mathrm{d}z,$$

where the integral is finite for λ^{2d}-almost all (y,z) (with λ^{2d} Lebesgue measure on $\mathbb{R}^{2d}$). Hence

$$A \geq \iint_{\mathbb{R}^{2d}} K(v) K(w) \int_{\epsilon_T}^{u_0} f_{(X_0,X_u)}(x - h_T v, x - h_T w) \mathrm{d}u \, \mathrm{d}v \, \mathrm{d}w$$

for all positive sequences $\epsilon_T \to 0$ as $T \to \infty$.

Now, we set $J = \int_{\epsilon_T}^{u_0} f_{(X_0,X_u)}(x - h_T v, x - h_T w)\,du$.

(1) If $\gamma_0 = 1$ and $\epsilon_T = h_T^d$, then

$$J \geq (\ln u_0 + \ln h_T^{-d}) \cdot \inf_{u \in [h_T^d, u_0[} p_u(x - h_T v, x - h_T w),$$

and the Fatou lemma gives

$$\underline{\lim}_{T\to\infty} \frac{T}{\ln h_T^{-d}} \operatorname{Var} f_T(x) \geq \iint_{\mathbb{R}^{2d}} p_0(x, v, w) K(v) K(w)\, dv\, dw > 0.$$

(2) In the case where $\gamma_0 > 1$, one may choose $\epsilon_T = h_T^{d/\gamma_0}$ and write:

$$J \geq \frac{h_T^{d/\gamma_0 - d} - u_0^{1-\gamma_0}}{\gamma_0 - 1} \inf_{u \in [h_T^{d/\gamma_0}, u_0[} p_u(x - h_T v, x - h_T w),$$

which yields, again by the Fatou lemma

$$\underline{\lim}_{T\to\infty} T h_T^{d - d/\gamma_0} \operatorname{Var} f_T(x) \geq \frac{1}{\gamma_0 - 1} \iint_{\mathbb{R}^{2d}} p_0(x, v, w) K(v) K(w)\, dv\, dw > 0.$$

■

With some additional assumptions on f, one may specify the mean-square rates of convergence.

Corollary 7.2
Under the assumptions of Theorems 7.3 and 7.4 and if $f \in C_d^2(b)$,

(1) for $\gamma_0 = 1$ and $h_T = c\,(\ln T/T)^{1/4}$ $(c > 0)$, one has

$$0 < \underline{\lim}_{T\to\infty} \frac{T}{\ln T} \operatorname{E}(f_T(x) - f(x))^2 \leq \overline{\lim}_{T\to\infty} \frac{T}{\ln T} \operatorname{E}(f_T(x) - f(x))^2 < +\infty;$$

(2) for $\gamma_0 > 1$ and $h_T = c\,T^{-\gamma_0/(4\gamma_0 + d(\gamma_0 - 1))}$ $(c > 0)$, one obtains

$$0 < \underline{\lim}_{T\to\infty} T^{\frac{4\gamma_0}{4\gamma_0 + d(\gamma_0 - 1)}} \operatorname{E}(f_T(x) - f(x))^2 \leq \overline{\lim}_{T\to\infty} T^{\frac{4\gamma_0}{4\gamma_0 + d(\gamma_0 - 1)}} \operatorname{E}(f_T(x) - f(x))^2 < +\infty.$$

As a by-product, we may also show that under the natural condition $\gamma_0 \in \,]0, d]$, these rates are minimax in the following specific sense: if one considers the set of processes with a given rate of convergence for the kernel estimator, there does not exist a better estimator over this set (see Blanke and Bosq 2000).

7.2.4 Almost sure convergence

We consider the following exponential type inequality extending the result of Proposition 6.1 to the continuous case.

Proposition 7.3
Let $(Z_t, t \in \mathbb{Z})$ be a zero-mean real valued process such that $\sup_{0 \le t \le T} \| Z_t \|_\infty \le M$, $\mathrm{E}(Z_t Z_s) = \mathrm{E}(Z_0 Z_{t-s}), t > s$ *and* $\int_0^\infty |\mathrm{E}Z_0 Z_u| \mathrm{d}u < +\infty$.
Then for each real number $p_T \in [1, T/2]$ *and each positive* κ, ε,

$$P\left(\left| \frac{1}{T} \int_0^T Z_t \mathrm{d}t \right| > \varepsilon \right) \le \frac{8M}{\varepsilon\kappa} (1+\kappa) \alpha_Z(p_T)$$

$$+ 4 \exp\left(- \frac{T\varepsilon^2}{8(1+\kappa)^2 \int_0^{p_T} |\mathrm{Cov}\,(Z_0, Z_u)| \mathrm{d}u + \frac{4M}{3}(1+\kappa) p_T \varepsilon} \right). \tag{7.12}$$

PROOF:
Similar to the proof of Proposition 6.1 (see also Bosq and Blanke 2004) and therefore omitted. ■

To obtain sharp almost sure results, we consider geometrically strongly mixing processes. Note that multidimensional diffusions processes of Example 7.3 are GSM as soon as $p = 0$ in (7.8) (see Veretennikov 1987).

Now for convenience, we suppose that the process is observed over intervals $[0, T_n]$ where $(T_n, n \ge 1)$ satisfies $T_{n+1} - T_n \ge \tau > 0$ and $T_n \uparrow \infty$: in other words sampled paths are displayed at increasing instants $T_1, T_2, \cdots$.

Theorem 7.5
Let $(X_t, t \in \mathbb{Z})$ be a GSM $\mathbb{R}^d$-valued process and suppose that Assumptions A7.2 hold.

(1) If h_{T_n} satisfies

$$\frac{T_n {h_{T_n}}^{2d}}{(\ln T_n)^3} \left(\mathbb{1}_{\{0<\gamma_0<1\}} + \ln(1/h_{T_n}) \mathbb{1}_{\{\gamma_0=1\}} + {h_{T_n}}^{d(\frac{1}{\gamma_0}-1)} \mathbb{1}_{\{\gamma_0>1\}} \right) \underset{T_n \uparrow \infty}{\longrightarrow} +\infty, \tag{7.13}$$

then for some constant $C_{\gamma_0}(x)$,

$$\overline{\lim_{T_n\uparrow\infty}}\sqrt{\frac{\Psi(h_{T_n},T_n,\gamma_0)}{\ln T_n}}|f_{T_n}(x)-\mathrm{E}f_{T_n}(x)|\leq C_{\gamma_0}(x)\ \ a.s.;$$

with $\Psi(h,T,\gamma_0)$ *given by (7.9);*

(2) *if* h_{T_n} *is such that* $T_nh^d{}_{T_n}/(\ln T_n)^2 \underset{T_n\uparrow\infty}{\longrightarrow} +\infty$ *but*

$$\frac{T_nh_{T_n}^{2d}}{(\ln T_n)^3}\left(\mathbb{1}_{\{0<\gamma_0<1\}}+\ln(1/h_{T_n})\mathbb{1}_{\{\gamma_0=1\}}+h_{T_n}{}^{d(\frac{1}{\gamma_0}-1)}\mathbb{1}_{\{\gamma_0>1\}}\right)\underset{T_n\uparrow\infty}{\longrightarrow}0,$$

then we have,

$$|f_{T_n}(x)-\mathrm{E}f_{T_n}(x)|=\mathcal{O}\left(\frac{(\ln T_n)^2}{T_nh_{T_n}^d}\right)\ \ a.s.$$

The proof of this result is similar to that of Theorem 6.5(2) with results of Proposition 7.3 and Theorem 7.3. The constant $C_{\gamma_0}(x)$ can be made explicit but it depends on unknown constants such as $M(x,x)$ and $\sup_{(y,z)\in\mathbb{R}^{2d}}\int_{u_0}^{\infty}|g_u(y,z)|\mathrm{d}u$.

Remark

Note that A7.2(iii) and the GSM property could be considered as redundant properties of asymptotic independence. Nevertheless, it may be shown that under (7.13), and for $\gamma_0\geq 1$, one has also

$$|f_{T_n}(x)-\mathrm{E}\,f_{T_n}(x)|=\mathcal{O}\Big((\ln T_n)^{1/2}\Psi^{-1/2}(h_{T_n},T_n,\gamma_0)\Big)$$

a.s. with condition A7.2(iii) replaced by '$f_{(X_0,X_u)}$ uniformly bounded for $u\geq u_0$' (whereas a logarithmic loss is observed for $\gamma_0<1$). This methodology will be used in the general case of regression estimation (see Section 7.3).
Next if $f\in C_d^2(b)$, and if $\max(1,\gamma_0)>d/2$, the choices

$$h_{T_n}(\gamma_0)=\begin{cases}\left(\dfrac{(\ln T_n)^4}{T_n}\right)^{1/2}, & \gamma_0<1\\ T_n^{-1/4}, & \gamma_0=1\\ \left(\dfrac{\ln T_n}{T_n}\right)^{\frac{\gamma_0}{4\gamma_0+d(\gamma_0-1)}}, & \gamma_0>1\end{cases}\tag{7.14}$$

give the almost sure rates:

$$\Psi_{T_n}^{-1}(\gamma_0)=\begin{cases}\left(\dfrac{\ln T_n}{T}\right)^{1/2}, & \gamma_0<1\\ \dfrac{\ln T_n}{T_n^{1/2}}, & \gamma_0=1\\ \left(\dfrac{\ln T_n}{T_n}\right)^{\frac{2\gamma_0}{4\gamma_0+d(\gamma_0-1)}}, & \gamma_0>1.\end{cases}\tag{7.15}$$

The technical condition $\max(1, \gamma_0) > d/2$ is a consequence of (7.13) and should be relaxed to $\max(1, \gamma_0) > d/r_o$ if, with obvious notation, $f \in C_d^{r_o}(b)$. If $\max(1, \gamma_0) \leq d/2$, one obtains the alternative result

$$|f_{T_n}(x) - f(x)| = \mathcal{O}\left(\left((\ln T_n)^2 T_n^{-1}\right)^{\frac{2}{2+d}}\right) \text{ a.s.} \tag{7.16}$$

for $h_{T_n} = c((\ln T_n)^2/T_n)^{1/(2+d)}$, $c > 0$.
As a by-product, for $\gamma_0 = d/2$ (e.g. for d-dimensional diffusions, $d \geq 2$), the choice $h_{T_n} = c((\ln T_n)^2/T_n)^{1/2(1+\gamma_0)}$ yields

$$|f_{T_n}(x) - f(x)| = \mathcal{O}\left(\left((\ln T_n)^2 T_n^{-1}\right)^{\frac{1}{1+\gamma_0}}\right) \text{ a.s.}$$

We now turn to uniform results over increasing compact sets.

Theorem 7.6
Suppose that Assumptions A7.2 hold with $f(\cdot)$ and $M(\cdot,\cdot)$ bounded functions; in addition if X is GSM with $f \in C_d^2(b)$ and K satisfies a Lipschitz condition, then

(1) for $\max(1, \gamma_0) > d/2$ *and* $h_{T_n}(\gamma_0)$, $\Psi_{T_n}(\gamma_0)$ *given by* (7.14)–(7.15),

$$\sup_{\|x\| \leq c_{(d)} T_n^{\ell}} |f_{T_n}(x) - f(x)| = \mathcal{O}(\Psi_{T_n}^{-1}(\gamma_0)) \text{ a.s.}$$

(2) for $\max(1, \gamma_0) \leq d/2$ *and* $h_{T_n} = c((\ln T_n)^2/T_n)^{1/(2+d)}$, $c > 0$,

$$\sup_{\|x\| \leq c_{(d)} T_n^{\ell}} |f_{T_n}(x) - f(x)| = \mathcal{O}\left(\left((\ln T_n)^2 T_n^{-1}\right)^{\frac{2}{2+d}}\right) \text{ a.s.}$$

Details and proofs of these results may be found in Blanke (2004). Under more stringent conditions and using the continuous-time version of the Borel–Cantelli lemma established in Bosq (1998), one should replace T_n by T. Moreover uniform convergence over $\mathbb{R}^d$ can also be obtained. Note also that the case $\gamma_0 < 1$ was first addressed by Bosq (1997). Finally for real ergodic diffusions, van Zanten (2000) obtains the rate $\mathcal{O}_P(1/\sqrt{T})$ and in a multidimensional context, Bianchi (2007) extends (7.16) and Theorems 7.5–7.6 to diffusions which have a subexponential mixing rate (i.e. for $u > 0$, $\alpha(u) \leq \beta_0 e^{-\beta_1 u^{\beta_2}}$, $\beta_0 > 0$, $\beta_1 > 0$, $0 < \beta_2 < 1$). Diffusions satisfying (7.8) with $0 < p < 1$ are subexponential mixing, (see Klokov and Veretennikov 2005). In this last case, similar almost sure rates of convergence (up to some additional logarithmic terms) are obtained.

7.2.5. An adaptive approach

As we have seen before, the choice of the optimal bandwidth depends on the parameter γ_0, which is not always known. In this section, we propose an algorithm providing an adaptive estimator, i.e. an estimator with automatic choice of the bandwidth. Here we suppose that the regularity of the density, say r_0, almost equals 2 and that adaptivity refers to γ_0 in the case $d \geq 2$.

Regarding (7.7), we use the natural condition $\gamma_0 \in]0, d]$. As before, we suppose that (T_n) is such that $T_{n+1} - T_n \geq \tau > 0$ and $T_n \uparrow \infty$. Now, we consider a grid Γ_{T_n} including tested values for γ_0:

$$\Gamma_{T_n} = \{\widetilde{\gamma}_0, 1, \gamma_{1,T_n}, \cdots, \gamma_{N_T,T_n}, d\},$$

where one sets

$$\begin{cases} 0 < \widetilde{\gamma}_0 < 1, \quad \gamma_{0,T_n} := 1, \\ \gamma_{j+1,T_n} - \gamma_{j,T_n} = \delta_{j,T_n} > 0, \quad j = 0, \ldots, N_T - 1. \end{cases}$$

Recall that for $\gamma_0 < 1$, the bandwidth choice does not depend on the exact value of γ_0, in this way $\widetilde{\gamma}_0$ symbolizes all values $\gamma_0 < 1$. Moreover, we suppose that (δ_{j,T_n}), (N_T) are such that

$$\begin{cases} 1 + \sum_{j=0}^{N_T-1} \delta_{j,T_n} \to d, & \sup\limits_{j=0,\ldots,N_T-1} \delta_{j,T_n} \to 0, \\ \delta_{j,T_n} \geq \delta_0 \dfrac{\ln \ln T_n}{\ln T_n}, & \delta_0 > \dfrac{4}{d}, \quad j = 0, \ldots, N_T - 1. \end{cases}$$

Such conditions discriminate the rates $\Psi_{T_n}(\gamma)$ for different γ in Γ_{T_n}. They hold clearly for

$$\delta_{j,T_n} \equiv \delta_{T_n} = \frac{d-1}{N_T} \quad \text{with} \quad N_T = O\left(\frac{\ln T_n}{(\ln \ln T_n)^2}\right).$$

The candidate γ_0^* is obtained by setting, for $x \in \mathbb{R}^d$,

$$\gamma_0^* = \min\{\gamma_1 \in \Gamma_{T_n} \ : \ \Psi_{T_n}(\gamma_2)|\hat{f}_{\gamma_1}(x) - \hat{f}_{\gamma_2}(x)| \leq \eta, \forall \gamma_2 \geq \gamma_1, \gamma_2 \in \Gamma_{T_n}\},$$

where

$$\hat{f}_\gamma(x) := \hat{f}_{T_n,\gamma}(x) = \frac{1}{T_n {h_{T_n}}^d(\gamma)} \int_0^{T_n} K\left(\frac{x - X_t}{h_{T_n}(\gamma)}\right) dt$$

with

$$h_{T_n}(\gamma) = \begin{cases} ((\ln T_n)^4 / T_n)^{1/2d} & \text{if} \quad \gamma < 1, \\ T_n^{-1/4} & \text{if} \quad \gamma = 1, \\ (\ln T_n / T_n)^{\frac{\gamma}{4\gamma + d(\gamma - 1)}} & \text{if} \quad \gamma > 1 \end{cases}$$

and

$$\Psi_{T_n}(\gamma) = \begin{cases} (T_n/\ln T_n)^{1/2} & \text{if} \quad \gamma < 1, \\ T_n^{1/2}/\ln T_n & \text{if} \quad \gamma = 1, \\ (T_n/\ln T_n)^{\frac{2\gamma}{4\gamma+d(\gamma-1)}} & \text{if} \quad \gamma > 1. \end{cases}$$

Finally the adaptive estimator of $f(x)$ is defined by

$$\hat{f}_{\gamma_0^*}(x) = \frac{1}{T_n h_{T_n}^d(\gamma_0^*)} \int_0^{T_n} K\left(\frac{x - X_t}{h_{T_n}(\gamma_0^*)}\right) \mathrm{d}t.$$

Theorem 7.7
Under the assumptions of Theorem 7.6, one obtains, for all $\gamma_0 \in]0, d]$ *with* $\max(1, \gamma_0) > d/2$ *and large enough* η,

$$|\hat{f}_{\gamma_0^*}(x) - f(x)| = \mathcal{O}\left(\Psi_{T_n}^{-1}(\gamma_0)\right) \ a.s.$$

Proof of this result is intricate and appears in Blanke (2004). We see, up to logarithmic terms, that rates of the previous section are not altered. Finally, in the special case $\max(1, \gamma_0) \leq d/2$, the same paper establishes that the adaptive estimator converges also at the alternative rate.

7.3 Regression estimation

Let $Z_t = (X_t, Y_t)$, $t \in \mathbb{R}$, be an $\mathbb{R}^d \times \mathbb{R}^{d'}$-valued measurable stochastic process defined on a probability space $(\Omega, \mathcal{A}, P)$. Let m be a Borelian function from $\mathbb{R}^{d'}$ to $\mathbb{R}$ such that $(w, t) \mapsto m^2(Y_t(w))$ is $P \otimes \lambda_T$-integrable for each positive T (λ_T stands for Lebesgue measure on $[0, T]$).

Assuming that the Z_t's have the same distribution with density $f_Z(x, y)$, we wish to estimate the regression parameter $\mathrm{E}(m(Y_0)|X_0 = \cdot)$ given the data $(Z_t, 0 \leq t \leq T)$. Functional parameters f, φ, r are unchanged from Chapter 6, see (6.11)–(6.13).

The kernel regression estimator is defined as

$$r_T(x) = \varphi_T(x)/f_T(x),$$

where

$$\varphi_T(x) = \frac{1}{T h_T^d} \int_0^T m(Y_t) K\left(\frac{x - X_t}{h_T}\right) \mathrm{d}t$$

and f_T is given by (7.1) with a positive kernel K, defined on $\mathbb{R}^d$, and satisfying conditions $\mathcal{K}$ given by (6.1).

In our viewpoint of prediction, we now focus on the almost sure convergence of this estimator. Note that results concerning mean-square error and limit in distribution have been obtained by Cheze-Payaud (1994) and Bosq (1998, Chapter 5) in the optimal and superoptimal cases.

7.3.1 Pointwise almost sure convergence

The pointwise study of $r_T(x)$ is considered under the following conditions.

Assumption 7.3 (A7.3)

(i) $f, \varphi \in C_d^2(b)$ for some positive b;

(ii) either (Z_t) is a strictly stationary process such that for $s \neq t$, (X_s, X_t) has density $f_{(X_s,X_t)}$; or for $s \neq t$, the joint density, $f_{(Z_s,\, Z_t)}$, of (Z_s, Z_t) exists with $f_{(Z_s,\, Z_t)} = f_{(Z_0,\, Z_{t-s})}$ for $t > s$;

(iii) there exists $u_0 > 0$ such that $f_{(X_0,X_u)}(\cdot,\cdot)$ is uniformly bounded for $u \geq u_0$;

(iv) there exists $\gamma_0 > 0$: $f_{(X_0,\, X_u)}(y,z) \leq M(y,z)u^{-\gamma_0}$, for $(y,z,u) \in \mathbb{R}^{2d} \times]0, u_0[$, with either $M(\cdot,\cdot)$ bounded or $M(\cdot,\cdot) \in L^1(\mathbb{R}^{2d})$ and continuous at (x,x);

(v) there exist $\lambda > 0$ and $\nu > 0$ such that $\mathrm{E}(\exp(\lambda |m(Y_0)|^{\nu})) < +\infty$;

(vi) the strong mixing coefficient α_Z of (Z_t) satisfies $\alpha_Z(u) \leq \beta_0 \mathrm{e}^{-\beta_1 u}$, $(u > 0$, $\beta_0 > 0$, $\beta_1 > 0)$.

As noted before, the GSM condition A7.3(vi) allows us to relax A7.2(iii) to A7.3(iii) (with only a logarithmic loss of rate in the case $\gamma_0 < 1$). Moreover, it should be noticed that conditions about the regularity of sample paths only involve the process (X_t) but not (Y_t). Finally, again we assume that sample paths are displayed over intervals $[0, T_n]$ with $T_{n+1} - T_n \geq \tau > 0$ and $T_n \uparrow \infty$.

Theorem 7.8
Suppose that Assumptions A7.3 hold with $f(x) > 0$. Then

(1) if $\max(1, \gamma_0) \geq d/2$ *and*

$$h_{T_n} = c\left(\left(\frac{(\ln T_n)^2}{T_n}\right)^{1/4} \mathbb{1}_{\gamma_0 \leq 1} + \left(\frac{(\ln T_n)^2}{T_n}\right)^{\frac{\gamma_0}{4\gamma_0 + d(\gamma_0 - 1)}} \mathbb{1}_{\gamma_0 > 1}\right), \quad (c > 0) \tag{7.17}$$

one obtains

$$|r_{T_n}(x) - r(x)| = \mathcal{O}\big(\Phi^{-1}(T_n, \gamma_0)\big) \quad a.s.$$

where $\Phi(T_n, \gamma_0) = T_n^a \cdot (\ln T_n)^{-b}$ *and*

$$\begin{cases} a = \frac{1}{2}, \; b = \frac{1}{\nu} + 1 & \text{if} \quad \gamma_0 \leq 1, \\ a = \frac{2\gamma_0}{4\gamma_0 + d(\gamma_0 - 1)}, \; b = \frac{1}{\nu} + \frac{4\gamma_0}{4\gamma_0 + d(\gamma_0 - 1)} & \text{if} \quad \gamma_0 > 1. \end{cases} \tag{7.18}$$

(2) If $\max(1, \gamma_0) < d/2$*, the choice* $h_{Tn} = c((\ln T_n)^2/T_n)^{1/(2+d)}$ *yields*

$$|r_{T_n}(x) - r(x)| = \mathcal{O}\left((\ln T_n)^{\frac{4}{2+d} + \frac{1}{\nu}} T_n^{-\frac{2}{2+d}} \right) \quad a.s.$$

Up to logarithmic terms, these rates have similar order as for density estimation (compare with (7.15)–(7.16)). Note that this logarithmic loss is overtaken if one strengthens Assumptions A7.3 with conditions that jointly involve the function m and the density of (Z_s, Z_t) (cf. Bosq 1998, Chapter 5).

PROOF: *(Elements)*
We start from decomposition (with omitted x):

$$|r_{Tn} - r| \leq |r_{Tn} - \widetilde{r}_{T_n}| + \frac{|\widetilde{r}_{T_n}|}{f} |f_{Tn} - f| + |\frac{|\widetilde{\varphi}_{T_n} - \mathrm{E}\widetilde{\varphi}_{T_n}|}{f} + \frac{|\mathrm{E}\widetilde{\varphi}_{T_n} - \varphi|}{f} \tag{7.19}$$

where $\widetilde{r}_{T_n} = \widetilde{\varphi}_{T_n}/f_{Tn}$, with $f_{Tn}(x) \to f(x) > 0$ a.s. (using (7.15) combined with the remark following Theorem 7.5), and moreover

$$\widetilde{\varphi}_{T_n}(x) = \frac{1}{T_n h_{T_n}^d} \int_0^{T_n} m(Y_t) \mathbb{1}_{|m(Y_t)| \leq b_{T_n}} K\left(\frac{x - X_t}{h_{Tn}} \right) \mathrm{d}t$$

with $b_{T_n} \to \infty$.

The choice $b_{T_n} = (\delta \ln T_n)^{1/\nu}$, with Markov inequality and condition A7.3(v), implies that $|\varphi_{T_n}(x) - \widetilde{\varphi}_{T_n}(x)| = o(\Phi^{-1}(T_n, \gamma_0))$ (uniformly in x), with probability one, as soon as $\delta > 4/\lambda$ where $\Phi^{-1}(T_n, \gamma_0)$ denotes any rate of convergence obtained in Theorem 7.8. On the other hand, since $|\widetilde{r}_{T_n}| \leq b_{T_n}$, the proposed choices of h_{Tn}, b_{T_n} and $f, \varphi \in C_d^2(b)$ respectively yield $(|\widetilde{r}_{T_n}|/f)|\mathrm{E}f_{Tn} - f| = \mathcal{O}(\Phi^{-1}(T_n, \gamma_0))$ and $|\mathrm{E}\widetilde{\varphi}_{T_n} - \varphi| = \mathcal{O}(h_{Tn}^2) + \mathcal{O}(h_{Tn}^{-d} \mathrm{e}^{-\lambda b_{T_n}/2}) = o(\Phi^{-1}(T_n, \gamma_0))$.

For $|\widetilde{\varphi}_{T_n}(x) - \mathrm{E}\widetilde{\varphi}_{T_n}(x)|$, we apply Proposition 7.3 to the zero-mean process (W_t) where

$$W_t = m(Y_t) \mathbb{1}_{|m(Y_t)| \leq b_{T_n}} K_h(x - X_t) - \mathrm{E}m(Y_0) \mathbb{1}_{|m(Y_0)| \leq b_{T_n}} K_h(x - X_0).$$

A7.3(ii) yields stationarity and $\sup_{t \in [0, T_n]} \parallel W_t \parallel_\infty \leq M_{T_n} := 2 \parallel K \parallel_\infty b_{T_n} h_{T_n}^{-d}$. Finally, $\mathrm{E}W_0^2 = \mathcal{O}(b_{T_n}^2 h_{Tn}^{-d})$ by the Bochner lemma.

The main task consists in the evaluation of $\int_0^{p_{T_n}} |\mathrm{E}W_0 W_u| \mathrm{d}u$. Note that one has

$$|\mathrm{E}W_0 W_u| \leq b_{T_n}^2 (\mathrm{E}K_{h_{T_n}}(x - X_0) K_{h_{T_n}}(x - X_u) + (\mathrm{E}K_h(x - X_0))^2). \tag{7.20}$$

We split this integral into three parts: the first one ($u \in [0, u_T]$) is controlled by the Cauchy–Schwarz inequality. On the other hand, the bound (7.20), Bochner lemma and condition A7.3(iv) are used for the second one ($u \in]u_T, u_0[$) while the third one ($u \in]u_0, p_T]$) follows from (7.20) and condition A7.3(iii). In this way,

$$\int_0^{p_{T_n}} |\mathrm{E}W_0 W_u| \mathrm{d}u = \left\{ \int_0^{u_{T_n}} + \int_{u_{T_n}}^{u_0} + \int_{u_0}^{p_{T_n}} \right\} |\mathrm{E}W_0 W_u| \mathrm{d}u$$

$$= b_{T_n}^2 \{ \mathcal{O}(u_{T_n} h_{T_n}^{-d}) + \mathcal{O}(\ln(u_{T_n}^{-1}) \mathbb{1}_{\gamma_0 \le 1} + u_{T_n}^{-\gamma_0+1} \mathbb{1}_{\gamma_0 > 1}) + \mathcal{O}(p_{T_n}) \}.$$

Finally the result follows from (7.12), the choices

$$u_{T_n} = h_{T_n}^{d} \mathbb{1}_{\gamma_0 \le 1} + h_{T_n}^{d/\gamma_0} \mathbb{1}_{\gamma_0 > 1}, \quad p_{T_n} \simeq p_0 \ln T_n$$

(p_0 a large enough positive constant) and for some positive η,

$$\varepsilon\ (= \varepsilon_{T_n}) = \eta \left(\frac{(\ln T_n)^{1+\frac{1}{\nu}}}{T_n^{1/2}} \mathbb{1}_{\gamma_0 \le 1, d \le 2} + \frac{(\ln T_n)^{1+\frac{1}{\nu}}}{T_n^{1/2} h_{T_n}^{\frac{d}{2\gamma_0}(\gamma_0 - 1)}} \mathbb{1}_{\gamma_0 > 1, d \le 2\gamma_0} \right.$$
$$\left. + \frac{(\ln T_n)^{2+\frac{1}{\nu}}}{T_n h_{T_n}^d} \mathbb{1}_{d > 2\max(1, \gamma_0)} \right).$$

The last term $|f_{T_n}(x) - f(x)|$ is handled similarly by setting, in the definition of W_t, $m(\cdot)\mathbb{1}_{|m(\cdot)| \le b_{T_n}} \equiv 1$. Details are left to the reader. ■

7.3.2 Uniform almost sure convergence

For the sake of simplicity, we suppose from now on that $\max(1, \gamma_0) \ge d/2$. Uniform convergence of a regression estimator will be established over fixed or suitable increasing sequences of compact sets, that are defined similarly to the discrete time case (see Definition 6.1).

Definition 7.1
A sequence (D_{T_n}) of compact sets in $\mathbb{R}^d$ is said to be regular (with respect to f) if there exists a sequence (β_{T_n}) of positive real numbers and nonnegative constants ℓ, ℓ' such that for each n

$$\inf_{x \in D_{T_n}} f(x) \ge \beta_{T_n} \text{ and } \operatorname{diam}(D_{T_n}) \le c_{(d)} (\ln T_n)^{\ell'} T_n^{\ell}$$

where $c_{(d)} > 0$, $\operatorname{diam}(D_{T_n})$ denotes the diameter of D_{T_n} and $T_{n+1} - T_n \ge \tau > 0$ with $T_n \uparrow \infty$.

Theorem 7.9
Suppose that Assumptions A7.3 hold with $\max(1, \gamma_0) \geq d/2$, $f(\cdot)$, $\varphi(\cdot)$ *and* $M(\cdot, \cdot)$ *bounded functions. In addition if K satisfies a Lipschitz condition, then*

$$\sup_{x \in D_{T_n}} |r_{T_n}(x) - r(x)| = \mathcal{O}\left(\Phi^{-1}(T_n, \gamma_0)\beta_{T_n}^{-1}\right) \quad a.s.$$

if D_{T_n} *is regular and such that* $\Phi^{-1}(T_n, \gamma_0)\beta_{T_n}^{-1} \to 0$, *with* $h_{T_n}(\gamma_0)$, $\Phi(T_n, \gamma_0)$ *given by* (7.17)–(7.18).

PROOF:
Based on decomposition (7.19), with a similar covering to Theorems 6.6 and 6.10, noting that obtained bounds are uniform in x under our assumptions. The main change occurs for the term $\sup_{x \in D_{T_n}} |r_{T_n}(x) - \widetilde{r}_{T_n}(x)| = o(\Phi^{-1}(T_n, \gamma_0)\beta_{T_n}^{-1})$ a.s. since $|\inf_{x \in D_{T_n}} f_{T_n}(x) - \inf_{x \in D_{T_n}} f(x)| = \mathcal{O}(\Phi^{-1}(T_n, \gamma_0)) = o(\beta_{T_n})$ almost surely.■

Explicit rates of convergence may be deduced from the obvious following corollaries.

Corollary 7.3
Under the assumptions of Theorem 7.9, if $\ell = \ell' = 0$ *and* $\beta_{T_n} \equiv \beta_0 > 0$, *then*

$$\sup_{x \in D} |r_{T_n}(x) - r(x)| = \mathcal{O}(\Phi^{-1}(T_n, \gamma_0))$$

a.s., with $h_{T_n}(\gamma_0)$, $\Phi(T_n, \gamma_0)$ *given by* (7.17)–(7.18).

Corollary 7.4
If the assumptions of Theorem 7.9 are satisfied and $f(x) \simeq \| x \|^{-p}$ *(with* $p > d$*) as* $\| x \| \to \infty$, *one obtains*

$$\sup_{\|x\| \leq c_{(d)} (\ln T_n)^{\ell'} T_n^{\ell}} |r_{T_n}(x) - r(x)| = \mathcal{O}(T_n^{-a} (\ln T_n)^b)$$

a.s. where b is some positive constant depending on ν, γ_0, ℓ *and* ℓ' *and, either*

$$a = \frac{1}{2} - \ell p \ \text{ if } \ 0 < \ell < (2p)^{-1} \ \text{ and } \ \gamma_0 \leq 1,$$

or

$$a = \frac{2\gamma_0}{4\gamma_0 + d(\gamma_0 - 1)} - \ell p \ \text{ if } \ 0 < \ell < \frac{2}{p(4+d)} \ \text{ and } \ \gamma_0 > 1.$$

Corollary 7.5
If the assumptions of Theorem 7.9 are satisfied and $f(x) \simeq \exp(-q \parallel x \parallel^p)$ (with $q, p > 0$) as $\|x\| \to \infty$, the choices $\ell = 0$ and $\ell' = p^{-1}$ imply

$$\sup_{\|x\| \leq c_{(d)}(\ln T_n)^{\ell'}} |r_{T_n}(x) - r(x)| = \mathcal{O}(T_n^{-a}(\ln T_n)^b)$$

a.s. where b is some positive constant depending on ν, γ_0, and, either

$$a = \frac{1}{2} - c_{(d)}^p q \text{ for each } 0 < c_{(d)} < (2q)^{-1/p} \text{ and for } \gamma_0 \leq 1,$$

or

$$a = \frac{2\gamma_0}{4\gamma_0 + d(\gamma_0 - 1)} - c_{(d)}^p q \text{ for each } 0 < c_{(d)} < \left(\frac{2}{q(4+d)}\right)^{1/p} \text{ and for } \gamma_0 > 1.$$

7.4 Nonparametric prediction by kernel

Let $(\xi_t, t \in \mathbb{Z})$ be a *strictly stationary* and GSM, $\mathbb{R}^{d_0}$-valued Markov process. Given the data $(\xi_t, t \in [0, T])$, we want to predict the real-valued square integrable random variable $\zeta_{T+H} = m(\xi_{T+H})$ where the horizon H satisfies $0 < H < T$. Particular interesting cases are obtained with $m(x) = x$ and $d_0 = 1$ (usual prediction) or $m(x) = \mathbb{1}_B(x)$ $(B \in \mathcal{B}_{\mathbb{R}^{d_0}})$, $d_0 \geq 1$ (prediction of zone alarms).

Now let us consider the associated process

$$Z_t = (X_t, Y_t) = (\xi_t, m(\xi_{t+H})), \quad t \in \mathbb{R}$$

and consider the kernel regression estimator based on the data $(Z_t, t \in [0, S])$ with $S = T - H$ (respectively $S_n = T_n - H$). The nonparametric predictor has the form

$$\widehat{\zeta}_{T+H} = \frac{\int_0^S m(\xi_{t+H}) K\left(\frac{\xi_{S+H} - \xi_t}{h_S}\right) \mathrm{d}t}{\int_0^S K\left(\frac{\xi_{S+H} - \xi_t}{h_S}\right) \mathrm{d}t}.$$

We now study the asymptotic behaviour of $\widehat{\zeta}_{T+H}$ as T tends to infinity, H remaining fixed. As usual $\widehat{\zeta}_{T+H}$ is an approximation of $r(\xi_T) = \mathrm{E}(\zeta_{T+H} | \xi_s, s \leq T) = \mathrm{E}(\zeta_{T+H} | \xi_T)$. Collecting results of the previous section, one may obtain sharp rates of convergence for this kernel predictor in the two specific cases:

Case I: $m(x) = x$, $d_0 = 1$, $\mathrm{E}(\exp(\lambda |\xi_0|^\nu)) < +\infty$ for some positive λ and ν. Moreover (ξ_t) satisfies conditions A7.3(i)–(iv) with $\gamma_0 \leq 1$.

Case II: $m(x) = \mathbb{1}_B(x)$ $(B \in \mathcal{B}_{\mathbb{R}^{d_0}})$ and (ξ_t) satisfies conditions A7.3(i)–(iv) with $\max(1, \gamma_0) \geq d_0/2$.

Example 7.6 *(Case I)*
Suppose that K satisfies a Lipschitz condition and $h_{S_n} = c((\ln S_n)^{1/2}/S_n^{1/4})$ $(c > 0)$.

(1) Let D be a fixed compact set such that $\inf_{x \in D} f(x) > 0$, then

$$|\widehat{\zeta}_{T_n+H} - r(\xi_{T_n})|\mathbb{1}_{\xi_{T_n} \in D} = \mathcal{O}\left(S_n^{-1/2}(\ln S_n)^{1+\frac{1}{\nu}}\right) \text{ a.s.}$$

(2) If $D_{S_n} = \{x : \| x \| \leq c_{(d)}(\ln S_n)^{1/\nu}\}$, then

$$|\widehat{\zeta}_{T_n+H} - r(\xi_{T_n})\mathbb{1}_{\xi_{T_n} \in D_{S_n}} = \mathcal{O}(S_n^{-a}(\ln S_n)^b)$$

a.s. where b is some positive constant depending on ν, γ_0, and,

$$a = \frac{1}{2} - c_{(d)}^{\nu}\lambda \text{ if } 0 < c_{(d)} < (2\lambda)^{-1/\nu}.$$

◇

Example 7.7 *(Case II)*
For D a fixed or an increasing compact set, we obtain similar rates, as in Corollary 7.3, 7.4 or 7.5, depending on γ_0 and ultimate behaviour of f. ◇

Notes

Some of the results presented in this chapter are new or extensions of those stated in Bosq (1998). We have quoted along the way some recent references on relevant topics. In addition, we mention that first results on density estimation for diffusion processes appeared in Banon (1978), followed by Banon and Nguyen (1978, 1981); Nguyen (1979); Nguyen and Pham (1980). The strong mixing case was studied by Delecroix (1980). Concerning intermediate rates, related results were also independently obtained by Sköld and Hössjer (1999), and Sköld (2001) under slightly different conditions.

We have focused our study on kernel estimators but, of course, other functional estimators share similar properties. In the optimal and superoptimal cases, we refer e.g. to results of Leblanc (1997) for wavelet estimators, Comte and Merlevède (2002) for adaptive projection estimators, Lejeune (2006) for frequency polygons and Labrador (2006) for a k_T-occupation time density estimator. Note that a last method, based on local time (see Bosq and Davydov 1999; Kutoyants 1997a) is extensively discussed in Chapter 9.

8

Kernel method from sampled data

Concerning functional estimation, a family of optimal rates of convergence is given in Chapter 7 when observations are delivered in continuous time. Now, we consider the practical case where the underlying process is in continuous time but data are collected in discrete time. In this context, several deterministic or random sampling schemes have been proposed and studied by various authors. By basing this chapter on results of the previous one, we especially investigate the case of high rate (deterministic) sampling, where observations are displayed at high frequency. The advantage of such modelling is, of course, reduction of the total length of observation but a drawback is that estimators could become inconsistent, due to the high local correlation between successive variables. Clearly, in the univariate case, a very irregular path should indicate that consecutive observed variables are not highly correlated, whereas more local dependance is involved if the sample path is smoother. In this chapter, we investigate the minimal spacing time which must be respected between two consecutive observations and we show that such choice depends both on dimension and regularity of the underlying continuous-time process. We illustrate these theoretical results by some numerical studies concerning the real Gaussian case for sample paths which are either differentiable or nondifferentiable. Analogous results stand for regression estimation and simulations are also performed in this framework.

Inference and Prediction in Large Dimensions D. Bosq and D. Blanke

8.1 Density estimation

Let $\{X_t,\ t \in \mathbb{R}\}$ be an $\mathbb{R}^d$-valued measurable process, defined on the probability space $(\Omega,\ \mathcal{A},\ P)$ and suppose that the X_t's have a common distribution μ admitting a density f w.r.t. Lebesgue measure λ over $\mathbb{R}^d$. We suppose that (X_t) is observed at equidistant times $t_{i,n}$, $i = 1, \ldots, n$ with $t_{i+1,n} - t_{i,n} = \delta_n$, so that $T_n = n\delta_n$ represents the total length of observation. We use the standard kernel density estimator defined as

$$f_n(x) = \frac{1}{nh_n^d} \sum_{i=1}^{n} K\left(\frac{x - X_{ti,n}}{h_n}\right) \tag{8.1}$$

where h_n is the bandwidth ($h_n \to 0$, $nh_n^d \to \infty$) and K, the kernel, satisfies conditions $\mathcal{K}$ given in (6.1).

8.1.1 High rate sampling

We begin with a result that illustrates that high rate sampling ($\delta_n \to 0$) does not necessarily provide a good approximation of the continuous time framework and, that estimators (8.1), computed from such a sampling, may have a quite erratic behaviour.

For that purpose, let us consider the sampling scheme where $T_n \equiv T$ and $\delta_n = T/n$. A special case is the dichotomic sampling with $n = 2^N, N = 1, 2, \ldots$ Parametric estimators based on this sampling are known to be consistent in various cases. A well known example is variance estimation of a Wiener process $(W_t,\ t \geq 0)$, where

$$\hat{\sigma}_n^2 = \frac{1}{T} \sum_{j=1}^{n-1} \left(W_{(j+1)T/n} - W_{jT/n}\right)^2$$

is clearly consistent in quadratic mean and almost surely.

Now, if $(X_t,\ t \in \mathbb{R})$ is a process with identically distributed margins, the associated kernel density estimator is written as

$$\hat{f}_n(x) = \frac{1}{nh_n^d} \sum_{j=1}^{n} K\left(\frac{x - X_{jT/n}}{h_n}\right), \quad x \in \mathbb{R}^d.$$

The next result shows that, in some cases, $\hat{f}_n$ can be totally inconsistent!

Proposition 8.1
Let $(X_t,\ t \in \mathbb{R})$ be a zero-mean real stationary Gaussian process with an auto-correlation function ρ satisfying

$$cu^\alpha \leq 1 - \rho^2(u) \leq c'u^\alpha, \quad 0 < u \leq T$$

where $0 < c < c' < 1$ *and* $0 < \alpha \leq 2/3$. *Then if* $h_n = n^{-\gamma}$ $(0 < \gamma < 1)$ *and if the kernel K is positive and satisfies* $\int u^4 K(u)\mathrm{d}u < +\infty$, *we have*

$$\varliminf_{n\to\infty} \operatorname{Var} \hat{f}_n(0) \geq \frac{4(c')^{-1/2}}{\pi(2-\alpha)(4-\alpha)} - \frac{1}{2\pi} > 0.$$

PROOF:
We may and do suppose that $T = 1$ and $\mathrm{E}X_0^2 = 1$. Now let us consider the decomposition

$$\operatorname{Var} \hat{f}_n(0) = V_n + C_{1n} + C_{2n} + C_{3n}$$

with $V_n = (1/nh_n^2)\operatorname{Var} K(X_0/h_n)$,

$$C_{1n} = \frac{2}{n}\sum_{j=1}^{n-1}\left(1 - \frac{j}{n}\right) f_{j/n}(0,0),$$

where $f_u := f_{(X_0, X_u)}$,

$$C_{2n} = \frac{2}{n}\sum_{j=1}^{n-1}\int\left(1 - \frac{j}{n}\right)(f_{j/n}(h_n y, h_n z) - f_{j/n}(0,0))K(y)K(z)\mathrm{d}y\,\mathrm{d}z$$

and

$$C_{3n} = -2n^{-1}\sum_{j=1}^{n-1}\left(1 - \frac{j}{n}\right)\left(\frac{1}{h_n}\int K\left(\frac{u}{h_n}\right)f(u)\mathrm{d}u\right)^2.$$

First, the Bochner lemma 6.1, and $nh_n \to \infty$ imply that $V_n \to 0$ and $C_{3n} \to -f^2(0)$. Since $f_{j/n}(0,0) = (2\pi)^{-1}(1 - \rho^2(j/n))^{-1/2}$ we have

$$C_{1n} \geq \frac{1}{\pi\sqrt{c'}}\frac{1}{n}\sum_{j=1}^{n-1}\left(1 - \frac{j}{n}\right)\left(\frac{j}{n}\right)^{-\alpha/2}$$

consequently

$$\varliminf_{n\to\infty} C_{1n} \geq \frac{4}{\pi\sqrt{c'}\,(2-\alpha)(4-\alpha)}.$$

Finally, using $1 - \rho^2(j/n) \geq c(j/n)^\alpha$ and the inequality $|\mathrm{e}^{-au} - 1| \leq au(1 + au/2)$ $(a > 0,\ u > 0)$, one obtains $C_{2n} \to 0$ as soon as $h_n \to 0$ and $0 < \alpha \leq 2/3$. ■

Notice that under slightly different hypotheses, it may be established that

$$\lim_{n\to\infty} \operatorname{Var} \hat{f}_n(0) = \frac{1}{\pi T}\int_0^T \frac{1-u}{(1-\rho^2(u))^{1/2}}\, du.$$

In conclusion it appears that the condition $h_n \to 0$ seems not to be appropriate in this context. Our next result gives another point of view, namely, considering $\hat{f}_n$ as an approximation of f_T, see (7.1), as soon as h_n tends to h_T. Note that an other kind of approximation will be derived in Chapter 9 (Theorem 9.9).

Theorem 8.1
If $(X_t,\ 0 \le t \le T)$ has cadlag sample paths, if K is uniformly continuous and $h_n \to h_T$, then

$$\hat{f}_n(x) \underset{n\to\infty}{\longrightarrow} f_T(x), \quad x \in \mathbb{R}^d.$$

PROOF:
We have

$$\hat{f}_n(x) = \int_{\mathbb{R}} K_{h_n}(x-u)\mathrm{d}\mu_n(u) \quad \text{and} \quad f_T(x) = \int_{\mathbb{R}} K_{h_T}(x-u)\mathrm{d}\mu_T(u)$$

where $\mu_n = \left(\sum_{i=1}^n \delta_{X_{jT/n}}\right)/n$ and μ_T are empirical measures.

Now let φ be a continuous real function defined on $\mathbb{R}$, then for all ω in Ω

$$\int_{\mathbb{R}} \varphi\, \mathrm{d}\mu_n = \frac{1}{T}\cdot\frac{T}{n}\sum_{j=1}^n \varphi(X_{jT/n}) \underset{n\to\infty}{\longrightarrow} \frac{1}{T}\int_0^T \varphi(X_t)\mathrm{d}t$$

since $t \mapsto \varphi \circ X_t(\omega)$ is Riemann integrable over $[0, T]$. In particular,

$$\int_{\mathbb{R}} K_{h_T}(x-u)\mathrm{d}\mu_n(u) \underset{n\to\infty}{\longrightarrow} \int_0^T K_{h_T}(x-u)\mathrm{d}\mu_T(u) = f_T(x).$$

On the other hand

$$\int_{\mathbb{R}} (K_{h_T}(x-u) - K_{h_n}(x-u))\mathrm{d}\mu_n(u) \underset{n\to\infty}{\longrightarrow} 0$$

since K is uniformly continuous. Hence the result. ■

Note that convergence in Theorem 8.1 is uniform with respect to x.

8.1.2 Adequate sampling schemes

In this part, we suppose again that $t_{i+1,n} - t_{i,n} = \delta_n \to 0$ as $n \to \infty$ but, now, with $T_n = n\delta_n \to \infty$. Our aim is to give minimal threshold δ_n (to get minimal time of observation T_n) such that i.i.d. rates are reached again by kernel estimators, while the underlying process is in continuous time. Such schemes will be referred in the following as 'adequate sampling schemes'. This search is performed with the results obtained in Chapter 7 in the ideal framework, where the whole sample path is observed. We reformulate Assumptions 7.2, since a strengthened condition of asymptotic independence is needed in this case of high rate sampling.

Assumptions 8.1 (A8.1)

(i) f is either bounded or continuous at x;

(ii) for $s \neq t$, the joint density $f_{(X_s,X_t)}$ of (X_s, X_t) does exist and for $t > s$, $f_{(X_s,X_t)} = f_{(X_0,X_{t-s})}$;

(iii) let $g_u := f_{(X_0,X_u)} - f \otimes f,\ u > 0$: $\exists u_0 > 0$: $\forall u \in [u_0, +\infty[, \| g_u \|_\infty \leq \pi(u)$ for a bounded and ultimately decreasing function π, integrable over $]u_0, +\infty[$;

(iv) $\exists \gamma_0 > 0$: $f_{(X_0,X_u)}(y,z) \leq M(y,z)u^{-\gamma_0}$, for $(y,z,u) \in \mathbb{R}^{2d} \times]0, u_0[$,

where $M(\cdot,\cdot)$ is either continuous at (x,x) and $\mathbb{R}^{2d}$-integrable or bounded.

Mean-square convergence

We focus on the asymptotic variance of the estimator (8.1) since, clearly, the bias does not depend on choice of the sampling scheme. The next result gives adequate sampling schemes, according to the value of γ_0.

Theorem 8.2

Suppose that Assumptions A8.1 are satisfied, and set

$$\begin{cases} \delta_n^*(\gamma_0) = h_n^d & \text{if} \quad \gamma_0 < 1, \\ \delta_n^*(\gamma_0) = h_n^d \ln(1/h_n) & \text{if} \quad \gamma_0 = 1, \\ \delta_n^*(\gamma_0) = h_n^{d/\gamma_0} & \text{if} \quad \gamma_0 > 1, \end{cases} \tag{8.2}$$

then for all h_n such that $nh_n^d \to \infty$, and δ_n such that $\delta_n \geq \kappa_n \delta_n^*(\gamma_0)$ $(\kappa_n \to \kappa > 0)$, one obtains

$$\operatorname{Var} f_n(x) = \mathcal{O}(n^{-1}h_n^{-d}), \tag{8.3}$$

more precisely if $\delta_n/\delta_n^*(\gamma_0) \to +\infty$, one has

$$\lim_{n\to\infty} nh_n^d \operatorname{Var} f_n(x) = f(x) \| K \|_2^2$$

at all continuity points x of f.

PROOF:
Recall that $\pi(u)$ is decreasing for u large enough, $u > u_1(> u_0)$ say. The stationarity condition A8.1(ii) allows us to decompose $nh_n^d \mathrm{Var} f_n(x)$ into $V_n + C_n$, with $V_n = h_n^{-d} \mathrm{Var} K((x - X_0)/h_n)$, and

$$C_n = 2h_n^{-d} \sum_{k=1}^{n-1} \left(1 - \frac{k}{n}\right) \mathrm{Cov}\left(K\left(\frac{x - X_0}{h_n}\right), K\left(\frac{x - X_{k\delta_n}}{h_n}\right)\right) \tag{8.4}$$
$$\leq C_{1n} + C_{2n} + C_{3n} + C_{4n}$$

with

$$C_{1n} = 2h_n^d \sum_{k=1}^{N_0} \iint K(y)K(z) f_{(X_0, X_{k\delta_n})}(x - h_n y, x - h_n z) \mathrm{d}y \, \mathrm{d}z,$$

$$C_{2n} = 2h_n^d N_0 \left(\int K(y) f(y) \mathrm{d}y \right)^2,$$

$$C_{3n} = 2h_n^d \sum_{k=N_0+1}^{N_1} \iint K(y)K(z) |g_{k\delta_n}(x - h_n y, x - h_n z)| \mathrm{d}y \, \mathrm{d}z,$$

$$C_{4n} = 2h_n^d \sum_{k=N_1+1}^{+\infty} \iint K(y)K(z) |g_{k\delta_n}(x - h_n y, x - h_n z)| \mathrm{d}y \, \mathrm{d}z$$

where one has set $N_0 = [u_0/\delta_n]$, $N_1 = \lceil u_1/\delta_n \rceil$, with $\lceil a \rceil$ denoting the minimal integer greater than or equal to a.

Under condition A8.1(i), V_n is either bounded or such that $V_n \to f(x) \parallel K \parallel_2^2$ by the Bochner lemma 6.1, at all continuity points of f. Similarly, one obtains $C_{2n} \to 0$. C_{1n} is handled with condition A8.1(iv),

$$C_{1n} \leq 2h_n^d \sum_{k=1}^{N_0} (k\delta_n)^{-\gamma_0} \iint K(y)K(z) M(x - h_n y, x - h_n z) \mathrm{d}y \, \mathrm{d}z.$$

If $\gamma_0 > 1$, one obtains that $C_{1n} = \mathcal{O}(h_n^d \delta_n^{-\gamma_0})$ which is either bounded if $\delta_n \geq \kappa_n h_n^{d/\gamma_0} =: \delta_n^*(\gamma_0)$ or negligible if $\delta_n/\delta_n^*(\gamma_0) \to +\infty$.

If $\gamma_0 = 1$, one has $C_{1n} = \mathcal{O}(h_n^d \delta_n^{-1} \ln(\delta_n^{-1}))$ which is bounded as soon as $\delta_n \geq \kappa_n h_n^d \ln(h_n^{-1}) =: \delta_n^*(\gamma_0)$ or negligible if $\delta_n/\delta_n^*(\gamma_0) \to +\infty$.

If $\gamma_0 < 1$, $C_{1n} = \mathcal{O}(h_n^d \delta_n^{-1})$ a bounded term if $\delta_n \geq \kappa_n h_n^d =: \delta_n^*(\gamma_0)$ or negligible if $\delta_n/\delta_n^*(\gamma_0) \to +\infty$.

Finally condition A8.1(iv) implies that terms C_{n3} and C_{n4} have the order $h_n^d \delta_n^{-1}$ which is either bounded or negligible under previous choices of δ_n. ■

As usual, mean-square convergence can be directly derived from the previous results with an additional regularity condition on f.

Corollary 8.1
If $f \in C_d^2(b)$ and $h_n = c_n n^{-1/(4+d)}$ $(c_n \to c > 0)$, then for $\delta_n^*(\gamma_0)$ given by (8.2) and all $\delta_n \geq \kappa_n \delta_n^*(\gamma_0)$ $(\kappa_n \to \kappa > 0)$, one obtains

$$\mathrm{E}(f_n(x) - f(x))^2 = \mathcal{O}\left(n^{-\frac{4}{4+d}}\right)$$

whereas if $\delta_n/\delta_n^*(\gamma_0) \to +\infty$, one gets that

$$n^{\frac{4}{4+d}}\mathrm{E}(f_n(x) - f(x))^2 \to c^4 b_2^2(x) + c^{-d} f(x) \parallel K \parallel_2^2$$

with $b_2(x)$ defined by (6.4).

Note that the fixed design ($\delta_n \equiv \delta > 0$) is self-contained in the second part of Theorem 8.2. In this framework, and if no local information on sample paths is available (see, for example, Assumptions A7.1), one can also establish that results in Chapter 6.3 remain valid with only slight modifications.

For high rate sampling ($\delta_n \to 0$), a natural consequence of our results is that the smaller γ_0, the more observations can be chosen near each other. In the real case ($d = 1$), the interpretation is immediate since irregular sample paths ($\gamma_0 < 1$) bring much more information to the statistician than more regular ones ($\gamma_0 = 1$), where the correlation between two successive variables is much stronger. Simulations that we have made (see Section 8.3) well confirm these theoretical results.

The previous sampling schemes are sharp, in the sense that kernel estimators should no longer reach the i.i.d. rate if one chooses $\delta_n = o(\delta_n^*(\gamma_0))$ with γ_0 the true regularity. Actually, if we denote by $T_n^* = n\delta_n^*(\gamma_0)$ the minimal time of observation, one may express the rate $n^{-4/(4+d)}$ in terms of T_n^*. Thereby, for $\gamma_0 < 1$, $\gamma_0 = 1$ and $\gamma_0 > 1$ the corresponding rates are respectively $(T_n^*)^{-1}$, $(\ln T_n^*)(T_n^*)^{-1}$ and $(T_n^*)^{-4\gamma_0/(4\gamma_0+d(\gamma_0-1))}$: we recognize the optimal rates established in Chapter 7.2.3 where the whole sample path is observed over $[0, T_n^*]$. Our next theorem refines the result (8.3) in the superoptimal case ($\gamma_0 < 1$): we establish that the variance error is exactly of order $(nh_n^d)^{-1}$ for $\delta_n = \delta_n^*(\gamma_0) = h_n^d$ (so that $T_n^* = nh_n^d$ and the obtained rate is superoptimal in terms of length of observation).

Assumptions (8.2) (A8.2)

(i) f is continuous at x;

(ii) if $g_{s,t} = f_{(X_s,X_t)} - f \otimes f$ for $s \neq t$, then $g_{s,t} = g_{0,|s-t|} =: g_{|s-t|}$,

(iii) for all $u > 0$, $g_u(\cdot,\cdot)$ is continuous at (x, x), $\parallel g_u \parallel_\infty \leq \pi(u)$ where $(1+u)\pi(u)$ is integrable over $]0, +\infty[$, and $u\pi(u)$ is bounded and ultimately decreasing;

(iv) $\sup_{(y,z)\in\mathbb{R}^{2d}} \left| \int_0^{+\infty} g_u(y,z)\mathrm{d}u - \sum_{k\geq 1} \delta_n g_{k\delta_n}(y,z) \right| \underset{\delta_n \to 0}{\longrightarrow} 0.$

Theorem 8.3
Suppose that Assumptions A8.2 are satisfied, then for $\delta_n = h_n^d \to 0$ *such that* $nh_n^d \to +\infty$, *one obtains*

$$\lim_{n\to\infty} nh_n^d \mathrm{Var}\, f_n(x) = f(x) \parallel K \parallel^2 + 2\int_0^{+\infty} g_u(x,x)\mathrm{d}u.$$

PROOF:
One has just to show that $C_n \to 2\int_0^\infty g_u(x,x)\mathrm{d}u$ with C_n given by (8.4) and $\delta_n = h_n^d$. For this purpose, write $C_n = 2\iint K(y)K(z)[C_{1n} + C_{2n} + C_{3n}]\mathrm{d}y\,\mathrm{d}z$ with

$$C_{1n}(y,z) = \sum_{k=1}^{n-1} \delta_n \left(1 - \frac{k}{n}\right) g_{k\delta_n}(x - h_n y, x - h_n z) - \sum_{k=1}^{\infty} \delta_n g_{k\delta_n}(x - h_n y, x - h_n z),$$

$$C_{2n}(y,z) = \sum_{k=1}^{\infty} \delta_n g_{k\delta_n}(x - h_n y, x - h_n z) - \int_0^\infty g_u(x - h_n y, x - h_n z)\mathrm{d}u,$$

$$C_{3n}(y,z) = \int_0^\infty g_u(x - h_n y, x - h_n z)\mathrm{d}u.$$

First, conditions A8.2(iii)–(iv) clearly imply that $\iint K(y)K(z)C_{2n}(y,z)\mathrm{d}y\,\mathrm{d}z \to 0$ and $\iint K(y)K(z)C_{3n}(y,z)\mathrm{d}y\,\mathrm{d}z \to \int_0^{+\infty} g_u(x,x)$ (by dominated convergence). Since $u\pi(u)$ is decreasing for u large enough, $u > u_1$ say, and $n\delta_n = nh_n^d \to \infty$, the term C_{1n} is uniformly bounded by

$$\sum_{k=n}^{\infty} \delta_n \pi(k\delta_n) + \frac{1}{n}\sum_{k=1}^{\lceil u_1/\delta_n \rceil} \delta_n k\pi(k\delta_n) + \frac{1}{nh_n^d}\sum_{k=\lceil u_1/\delta_n\rceil+1}^{n-1} \delta_n^2 k\pi(k\delta_n).$$

We have $\sum_{k=n}^{\infty} \delta_n\pi(k\delta_n) \le \int_{n\delta_n}^{\infty} \pi(u)\mathrm{d}u \to 0$ as $n \to \infty$. Since $u\pi(u)$ is bounded on $]0, u_1]$, one has also $n^{-1}\sum_{k=1}^{\lceil u_1/\delta_n\rceil} \delta_n k\pi(k\delta_n) \le (nh_n^d)^{-1} \sup_{u\in]0,u_1]} \pi(u) \to 0$. Finally the decreasing of $u\pi(u)$ over $[u_1, +\infty[$ implies that the last term is bounded by $(nh_n^d)^{-1}\int_{u_1}^{+\infty} u\pi(u)\mathrm{d}u \to 0$; this concludes the proof. ■

Almost sure convergence
For observations delivered at high rate sampling, we now study the pointwise behaviour of $|f_n(x) - f(x)|$.

Theorem 8.4
Let $X = (X_t,\ t \in \mathbb{R})$ *be a GSM process with* $f \in C_d^2(b)$.

(a) *Set $\delta_n \equiv \delta$ and suppose that $f_{(X_0,X_u)}$ is uniformly bounded for $u \geq \delta$. Under conditions A8.1(i)–(ii), and for $h_n = c_n(\ln n/n)^{1/4+d}$ $(c_n \to c > 0)$, one obtains*

$$\overline{\lim_{n\to+\infty}} \left(\frac{n}{\ln n}\right)^{\frac{2}{4+d}} |f_n(x) - f(x)| \leq 2c^{-d/2} \parallel K \parallel_2 \sqrt{f(x)} + c^2|b_2(x)| \ a.s.$$

with $b_2(x)$ given by (6.4).

(b) *Suppose now that conditions A8.1(i)–(ii),(iv) are satisfied with $f_{(X_0,X_u)}$ uniformly bounded for $u \geq u_0$. The previous result still holds for all $\delta_n \to 0$ such that $\delta_n/\delta_n^\star(\gamma_0) \to \infty$ with*

$$\begin{cases} \delta_n^\star(\gamma_0) = h_n^d \ln n & \text{if} \quad \gamma_0 \leq 1 \quad and \quad d \leq 2, \\ \delta_n^\star(\gamma_0) = h_n^{d/\gamma_0} \ln n & \text{if} \quad \gamma_0 > 1 \quad and \quad d \leq 2\gamma_0, \\ \delta_n^\star(\gamma_0) = (\ln n)^{3/2}(nh_n^d)^{-1/2} & \text{if} \quad d > 2\max(1,\gamma_0). \end{cases} \tag{8.5}$$

Similarly to Chapter 7, the GSM condition allows us to relax condition A8.1(iii) as soon as $f_{(X_0,X_u)}$ is bounded for $u \geq u_0$: one observes only a logarithmic loss in the case $\gamma_0 < 1$. Moreover, we get again the technical condition $\max(1,\gamma_0) \geq d/2$ that comes from the exponential inequality (6.6). As before, if one expresses the rate in terms of $T_n^\star = n\delta_n^\star(\gamma_0)$, we recognize the continuous time almost sure rates given by (7.15)–(7.16) (up to a logarithmic loss). Note that in the borderline case $\delta_n \simeq \delta_n^\star(\gamma_0)$, Theorem 8.4(b) remains true but with a much more complicated asymptotic constant.

PROOF: *(Sketch)*
The proof essentially involves the exponential inequality (6.6): we outline here only the main parts. We write $|f_n(x) - \mathrm{E}f_n(x)| = (1/n)\left|\sum_{i=1}^n Z_{n,i}\right|$ with $Z_{n,i} = K_h(x - X_{i\delta_n}) - \mathrm{E}K_h(x - X_{i\delta_n})$, so that for all i, $\mathrm{E}(Z_{ni}) = 0$ and $|Z_{ni}| \leq \parallel K \parallel_\infty h_n^{-d}$ (recall that K is a positive kernel).

(a) If $\delta_n \equiv \delta$, we apply (6.6) with the choice $q_n = [n/(2p_0 \ln n)]$ for some positive p_0. Next under the assumptions made, one obtains that

$$\sigma^2(q_n) \leq \frac{n}{2q_n} h_n^{-d} f(x) \parallel K \parallel_2^2 (1 + o(1)), \tag{8.6}$$

and the result follows with the Borel–Cantelli lemma under the choices $\varepsilon_n = \eta((\ln n)/(nh_n^d))^{1/2}$, p_0 large enough and $\eta > 2(1+\kappa) \parallel K \parallel_2 \sqrt{f(x)}$.

(b) For $\delta_n \to 0$, the condition A8.1(iv) allows us to keep the bound (8.6) for all δ_n such that $\delta_n/\delta_n^\star(\gamma_0) \to \infty$, under the choice $q_n = [n\delta_n/(2p_0 \ln n)]$. ■

We conclude this section by giving uniform results over increasing compact sets.

Theorem 8.5
Suppose that the assumptions of Theorem 8.4 hold with $f(\cdot)$ and $M(\cdot,\cdot)$ bounded functions, $f \in C_d^2(b)$ and K satisfies a Lipschitz condition.

(1) *For $\max(1,\gamma_0) \geq d/2$ and $\delta_n \geq \delta_0\, \delta_n^\star(\gamma_0)$, with $\delta_0 > 0$ and $\delta_n^\star(\gamma_0)$ given by (8.5), the choice $h_n = c_n((\ln n)/n)^{1/(d+4)}$ with $c_n \to c > 0$ yields*

$$\sup_{\|x\leq\|c_{(d)}n^\ell} |f_n(x) - f(x)| = \mathcal{O}\left(\left(\frac{\ln n}{n}\right)^{\frac{2}{d+4}}\right)$$

almost surely, for positive constants ℓ and $c_{(d)}$.

(2) *If $\max(1,\gamma_0) < d/2$, the same result holds for all $\delta_n \to 0$ such that $\delta_n \geq \delta_0\, n^{-2/(4+d)}(\ln n)^{(1+2/(4+d))} > 0$, $\delta_0 > 0$.*

The proof is similar to those of Theorems 6.6 and 8.4 and therefore omitted. Finally, since sampling schemes are governed by γ_0, that may be unknown, an adaptive estimator is proposed and studied in Blanke (2006) (even in the case where r_0, the regularity of f, is also unknown).

8.2 Regression estimation

Results obtained for density estimation with high rate sampled observations can naturally be applied to the nonparametric regression framework. As in section 7.3, $(Z_t,\, t \in \mathbb{R})$ denotes the $\mathbb{R}^d \times \mathbb{R}^{d'}$-valued measurable stochastic process (X_t, Y_t) (with common distribution admitting the density $f_Z(x,y)$), and we consider the same functional parameters m, φ, f, r.

Now, if (Z_t) is observed at times $t_{i,n} = i\delta_n$, $i = 1,\ldots,n$, the associated kernel estimator of r is given by:

$$r_n(x) = \varphi_n(x)/f_n(x),$$

where

$$\varphi_n(x) = \frac{1}{nh_n^d}\sum_{i=1}^n m(Y_{t_{i,n}})K\left(\frac{x - X_{t_{i,n}}}{h_n}\right)$$

and f_n is defined as in (8.1) with a positive kernel K (so that $r_n(x)$ is well defined).

As in the previous chapters, we focus on almost sure convergence of this estimator and consider the following set of conditions.

Assumptions 8.3 (A8.3)

(i) $f, \varphi \in C_d^2(b)$ for some positive b;

(ii) for $s \neq t$, the joint density $f_{(X_s,X_t)}$ of (X_s, X_t) does exist and for $t > s$, $f_{(X_s,X_t)} = f_{(X_0,X_{t-s})}$. In addition, there exists $u_0 > 0$ such that $f_{(X_0,X_u)}(\cdot,\cdot)$ is uniformly bounded for $u \geq u_0$;

(iii) there exists $\gamma_0 > 0$ such that $f_{(X_0,X_u)}(y,z) \leq M(y,z)u^{-\gamma_0}$, for $(y,z,u) \in \mathbb{R}^{2d} \times]0,u_0[$, with either $M(\cdot,\cdot)$ bounded or $M(\cdot,\cdot) \in L^1(\mathbb{R}^{2d})$ and continuous at (x,x);

(iv) there exist $\lambda > 0$ and $\nu > 0$ such that $\mathrm{E}(\exp(\lambda|m(Y_0)|^\nu)) < +\infty$;

(v) the strong mixing coefficient α_Z of (Z_t) satisfies $\alpha_Z(u) \leq \beta_0 \mathrm{e}^{-\beta_1 u}$, $(u > 0$, $\beta_0 > 0$, $\beta_1 > 0)$.

Theorem 8.6

(a) Set $\delta_n \equiv \delta$, under conditions A8.3(i)–(ii) (with $u_0 = \delta$), (iv)–(v) and for $h_n = c_n((\ln n)/n)^{1/(d+4)}$ ($c_n \to c > 0$), one obtains

$$|r_n(x) - r(x)| = \mathcal{O}\left(\left(\frac{\ln n}{n}\right)^{\frac{2}{d+4}} (\ln n)^{\frac{1}{\nu}}\right) \qquad a.s.$$

for all x such that $f(x) > 0$.

(b) Suppose now that Assumptions A8.3 are satisfied. The choice $\delta_n \to 0$ such that $\delta_n \geq \delta_0\, \delta_n^\star(\gamma_0)$, with $\delta_n^\star(\gamma_0)$ given by (8.5) and $\delta_0 > 0$, leads to the previous result with unchanged h_n.

PROOF: *(Elements)*
Since $f(x) > 0$, we start from the decomposition:

$$\begin{aligned} |r_n(x) - r(x)| &\leq \frac{|r_n(x)|}{f(x)}|f_n(x) - f(x)| + \frac{|\varphi_n(x) - \tilde{\varphi}_n(x)|}{f(x)} + \frac{|\tilde{\varphi}_n(x) - \varphi(x)|}{f(x)} \\ &=: A_1 + A_2 + A_3 \end{aligned} \tag{8.7}$$

where

$$\tilde{\varphi}_n(x) = n^{-1}\sum_{i=1}^{n} m(Y_{t_{i,n}})\mathbb{1}_{|m(Y_{t_{i,n}})| \leq b_n} K_h(x - X_{t_{i,n}})$$

with $b_n = (b_0 \ln n)^{1/\nu}$, for $b_0 > 2\lambda^{-1}$ (λ, ν defined in condition A8.3(iv)).

For A_1, one has for $\eta > 0$,

$$P(|r_n(x)| > b_n) \leq P\left(\sup_{i=1,\ldots,n} |m(Y_{t_{i,n}})| > (b_0 \ln n)^{1/\nu}\right) = \mathcal{O}(n^{1-\lambda b_0}),$$

so that a.s. $r_n(x) = \mathcal{O}(b_n)$ (uniformly in x) since $b_0 > 2\lambda^{-1}$.

Next if $\psi_n = (n/\ln n)^{2/(d+4)}$, one easily verifies that, under condition A8.3(i), $|\mathrm{E} f_n(x) - f(x)| = \mathcal{O}(\psi_n^{-1} b_n^{-1})$ as well as for $|f_n(x) - \mathrm{E} f_n(x)|$ by using the exponential inequality established in (6.6), the upper bound (8.6) and proposed choices of h_n, δ_n.

For A_2, one easily gets that

$$P(\psi_n b_n |\varphi_n(x) - \tilde{\varphi}_n(x)| > \eta) \leq nP(|m(Y_0)| \geq b_n) = \mathcal{O}(n^{-(\lambda b_0 - 1)}),$$

which implies $\psi_n b_n |\varphi_n(x) - \tilde{\varphi}_n(x)| \to 0$ a.s. as soon as $b_0 > 2\lambda^{-1}$.

For the last term A_3, one has

$$\begin{aligned} |\mathrm{E}\tilde{\varphi}_n(x) - \varphi(x)| &\leq |\mathrm{E}\tilde{\varphi}_n(x) - \mathrm{E}\varphi_n(x)| + |\mathrm{E}\varphi_n(x) - \varphi(x)| \\ &= \mathcal{O}\left(h_n^{-d} \mathrm{e}^{-\frac{\lambda}{2} b_n^{\nu}}\right) + \mathcal{O}(h_n^2), \end{aligned}$$

so that $\psi_n |\mathrm{E}\tilde{\varphi}_n(x) - \varphi(x)| \to 0$, under the proposed choices of h_n, b_n. On the other hand, the exponential inequality (6.6) applied to the variables

$$m(Y_{t_{i,n}}) \mathbb{1}_{|m(Y_{t_{i,n}})| \leq b_n} K_h(x - X_{t_{i,n}}) - \mathrm{E} m(Y_{t_{i,n}}) \mathbb{1}_{|m(Y_{t_{i,n}})| \leq b_n} K_h(x - X_{t_{i,n}})$$

yields the result $|\tilde{\varphi}_n(x) - \mathrm{E}\tilde{\varphi}_n(x)| = \mathcal{O}(\psi_n^{-1} b_n^{-1})$ with probability 1. ■

Concerning uniform convergence and following the remark of Section 6.4.3 we restrict our study to *regular* compact sets D_n, so that

$$\inf_{x \in D_n} f(x) \geq \beta_n \quad \text{and} \quad \mathrm{diam}(D_n) \leq c_{(d)} (\ln n)^{\ell'} n^{\ell},$$

see Definition 6.1.

Theorem 8.7

Suppose that Assumptions A8.3 hold with $f(\cdot)$, $\varphi(\cdot)$ and $M(\cdot, \cdot)$ bounded functions. In addition if K satisfies a Lipschitz condition, $h_n = c_n((\ln n)/n)^{1/(d+4)}$ $(c_n \to c > 0)$ and $\delta_n \geq \delta_0\, \delta_n^{\star}(\gamma_0)$, $(\delta_0 > 0)$ with $\delta_n^{\star}(\gamma_0)$ given by (8.5), one obtains

$$\sup_{x \in D_n} |r_n(x) - r(x)| = \mathcal{O}\left(\left(\frac{\ln n}{n}\right)^{\frac{2}{d+4}} \cdot \frac{(\ln n)^{1/\nu}}{\beta_n} \right) \quad a.s.$$

PROOF: *(Sketch)*
Based on the decomposition (8.7), using results of Theorem 8.6 and technical steps in the proof of Theorems 6.6 and 6.10. ■

Explicit rates of convergence may be deduced from the obvious following corollary.

Corollary 8.2
Suppose that the assumptions of Theorem 8.7 are satisfied, the previous choices of δ_n, h_n yield, almost surely,

(1) *if* $\ell = \ell' = 0$ *and* $\beta_n \equiv \beta_0 > 0$:

$$\sup_{x \in D} |r_n(x) - r(x)| = \mathcal{O}\left(\left(\frac{\ln n}{n} \right)^{\frac{2}{d+4}} (\ln n)^{1/\nu} \right);$$

(2) *if* $f(x) \simeq \| x \|^{-p}$ *(with* $p > d$*) as* $\| x \| \to \infty$,

$$\sup_{\|x\| \leq n^{\ell}} |r_n(x) - r(x)| = \mathcal{O}(n^{-a} (\ln n)^b)$$

where $a = 2/(d+4) - \ell p$ *with* $0 < \ell < 2/(p(d+4))$ *and* $b = 2/(d+4) + 1/\nu$;

(3) *if* $f(x) \simeq \exp(-q \| x \|^p)$, *(with* $q, p > 0$*) as* $\| x \| \to \infty$,

$$\sup_{\|x\| \leq (\varepsilon q^{-1} \ln n)^{1/p}} |r_n(x) - r(x)| = \mathcal{O}(n^{-a} (\ln n)^b),$$

where $a = 2/(d+4) - \varepsilon$ *for each* $0 < \varepsilon < 2/(d+4)$ *and with* $b = 2/(d+4) + 1/\nu$.

Using Corollary 8.2, results concerning prediction for strictly stationary GSM $\mathbb{R}^{d_0}$-valued Markov processes ($d_0 \geq 1$) are then straightforward (see Sections 6.5.1 and 7.4). In this way, given n observations, one obtains sharp rates of convergence for the nonparametric predictor (with horizon $H\delta_n$, $1 \leq H \leq n-1$), built with $N = n - H$ observations over $[0, N\delta_N^{\star}(\gamma_0)]$.

8.3 Numerical studies

We present some simulations performed with R, a software package actually developed by the R Development Core Team (2006). First, to illustrate our previous results on density estimation, we consider the real Gaussian case for two stationary processes: Ornstein–Uhlenbeck (nondifferentiable, $\gamma_0 = 0.5$) and CAR(2) (differentiable, $\gamma_0 = 1$). Figure 8.1 illustrates how much the two processes differ: typically estimation will be easier for the 'irregular' OU process than for the 'regular' CAR(2) process.

Ornstein–Uhlenbeck process
The ⋆Ornstein–Uhlenbeck (OU) process⋆ is the stationary solution of:

$$dX_t = -aX_t dt + b\, dW_t, \quad a > 0, b > 0.$$

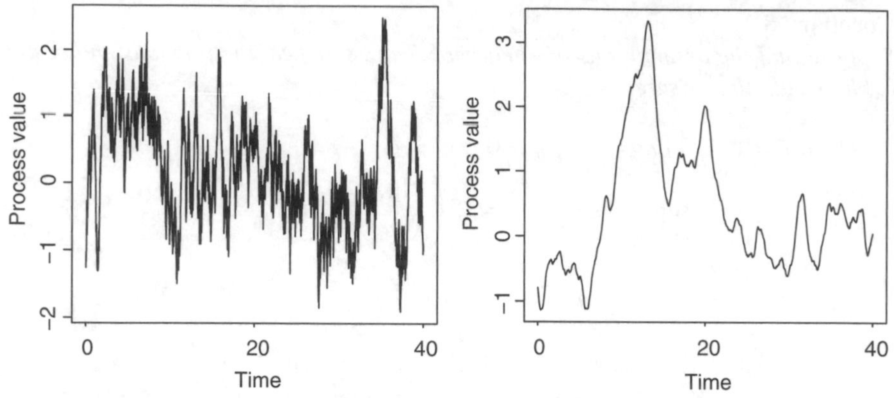

Figure 8.1 Simulated sample paths of (left) OU and (right) CAR(2) processes.

In the following, we choose $b = \sqrt{2}$ and $a = 1$ so that the X_t's have common density $\mathcal{N}(0, 1)$. We simulate this process using its (exact) autoregressive discrete form:

$$X_{(i+1)\tau_n} = \mathrm{e}^{-\tau_n} X_{i\tau_n} + Z_{(i+1)\tau_n}, \quad X_0 \sim \mathcal{N}(0, 1)$$

where $Z_{i\tau_n}$ are i.i.d. zero-mean Gaussian variables with variance $\sigma^2(1 - \mathrm{e}^{-2\tau_n})/2$.

From a simulated sample path (with $\tau_n = 10^{-2}$), we construct the estimator using the normal kernel $K(x) = (2\pi)^{-1/2}\exp(-x^2/2)$, $n = 1000$ observations and the bandwidth h_n equal to 0.3. We deliberately select a high sampling size to get a somewhat robust estimation, that allows us to show the effect of the sampling rate δ_n. We test the maximal sampling rate $\delta_n = 0.01$ ($T_n = 10$) and one close to the threshold $\delta_n^*(\gamma_0)$) with $\delta_n = 0.35$ ($T_n = 350$). Figure 8.2 shows that estimation is not

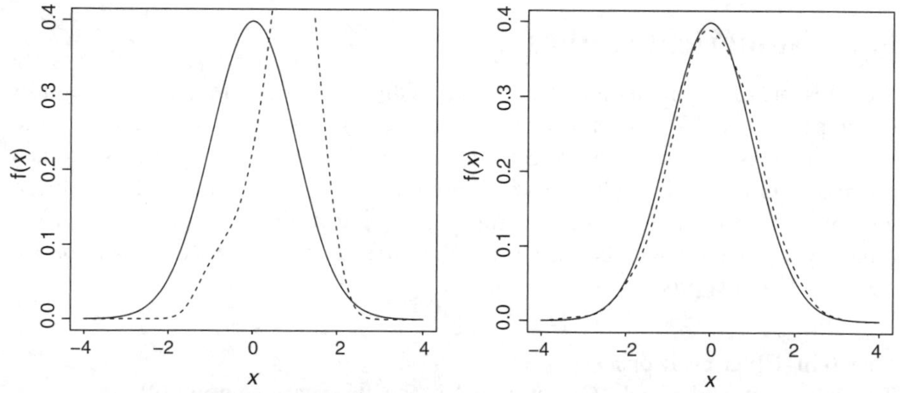

Figure 8.2 OU process: $\mathcal{N}(0, 1)$ density (solid) and estimated density (dash). On the left: $\delta_n = 0.01$, on the right: $\delta_n = 0.35$.

consistent in the first case, but that approximation turns out to be good for δ_n equal to 0.35.

CAR(2) process

A CAR(p) processes is a continuous-time AR process, see e.g. Brockwell (1993); Brockwell and Stramer (1995) and also the earlier reference of Ash and Gardner (1975). For $p = 2$, this process is the solution of the 2nd order differential equation:

$$\mathrm{d}X'_t = -(a_1 X_t + a_2 X'_t)\mathrm{d}t + b\mathrm{d}W_t$$

where X'_t represents the mean-square derivative of X_t. Note that the OU process is no more than a CAR(1) process (with $a_2 = 0$ and $\mathrm{d}X'_t$ replaced by $\mathrm{d}X_t$) and that CAR(k) processes are discussed in Example 10.10. The choices $a_1 = 1/2$, $a_2 = 2$ and $b = \sqrt{2}$ lead to a Gaussian process which is stationary, mean-square differentiable ($\gamma_0 = 1$) and with marginal distribution $\mathcal{N}(0, 1)$. Furthermore, sample paths can be simulated from the (exact) discrete form:

$$\begin{pmatrix} X_{(i+1)\tau_n} \\ X'_{(i+1)\tau_n} \end{pmatrix} = \mathrm{e}^{A\tau_n} \begin{pmatrix} X_{i\tau_n} \\ X'_{i\tau_n} \end{pmatrix} + \begin{pmatrix} Z^{(1)}_{(i+1)\tau_n} \\ Z^{(2)}_{(i+1)\tau_n} \end{pmatrix}, \quad X_0 \sim \mathcal{N}(0, 1), \quad X'_0 \sim \mathcal{N}\left(0, \frac{1}{2}\right),$$

where X_0 is independent from X'_0. Here A, a 2×2 matrix, and Σ the covariance matrix of $\begin{pmatrix} Z^{(1)}_{i\tau_n} \\ Z^{(2)}_{i\tau_n} \end{pmatrix}$ have rather complicated expressions that can be calculated from e.g. Tsai and Chan (2000).

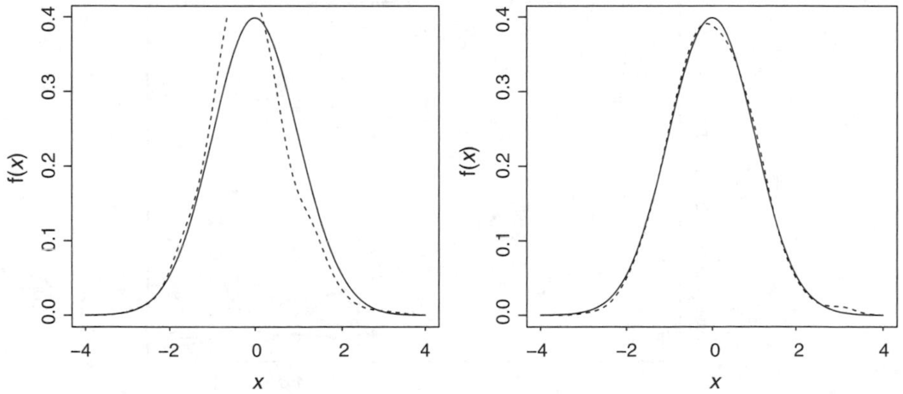

Figure 8.3 CAR(2): Estimated density (dash) and the $\mathcal{N}$(0,1) one (solid), $\delta_n = 0.35$ (left) and $\delta_n = 0.75$ (right).

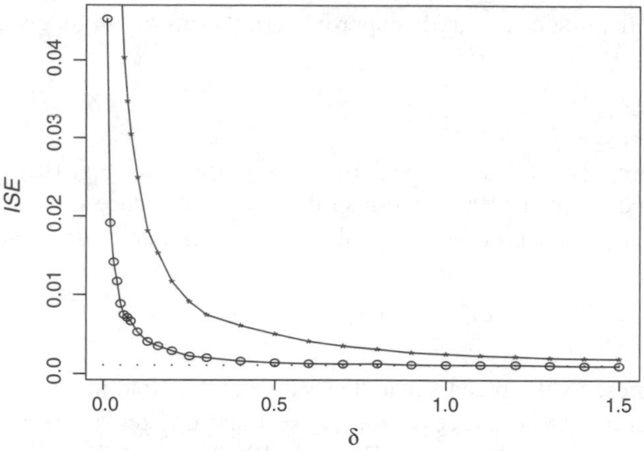

Figure 8.4 ISE(δ) for OU (circles) and CAR(2) (stars).

Using the same parameters as before ($n = 1000$, $h_n = 0.3$) we see, in Figure 8.3, estimations obtained respectively for $\delta_n = 0.35$ and $\delta_n = 0.75$ (on the same replication). It should be noted that $\delta_n = 0.35$ (corresponding to a good choice in the OU case) is not suitable for this smooth process whereas $\delta_n = 0.75$ ($T_n = 750$) furnishes a much better approximation of the $\mathcal{N}(0, 1)$ density.

Finally, we study the effect of various δ upon efficiency of estimation. We simulate $N = 100$ replications of OU and CAR(2) processes and compute an approximation of $ISE(\delta)$, a measure of accuracy defined by:

$$ISE(\delta) = \frac{1}{N}\sum_{j=1}^{N}\int (\hat{f}_{n,\delta,j}(x) - f(x))^2 \mathrm{d}x$$

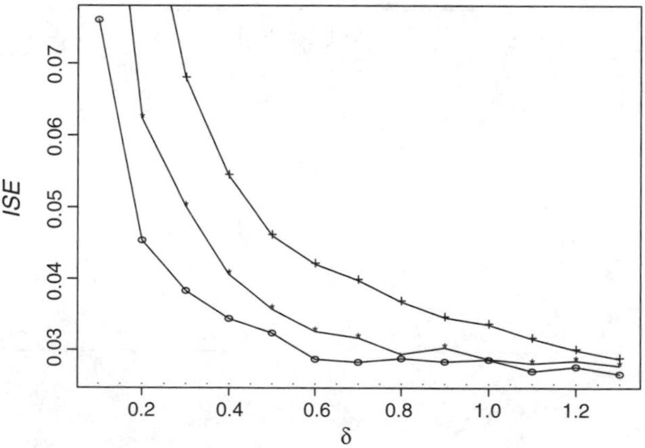

Figure 8.5 ISE(δ) for OU|OU (circles), OU|CAR(2) (stars) and CAR(2)|CAR(2) (+).

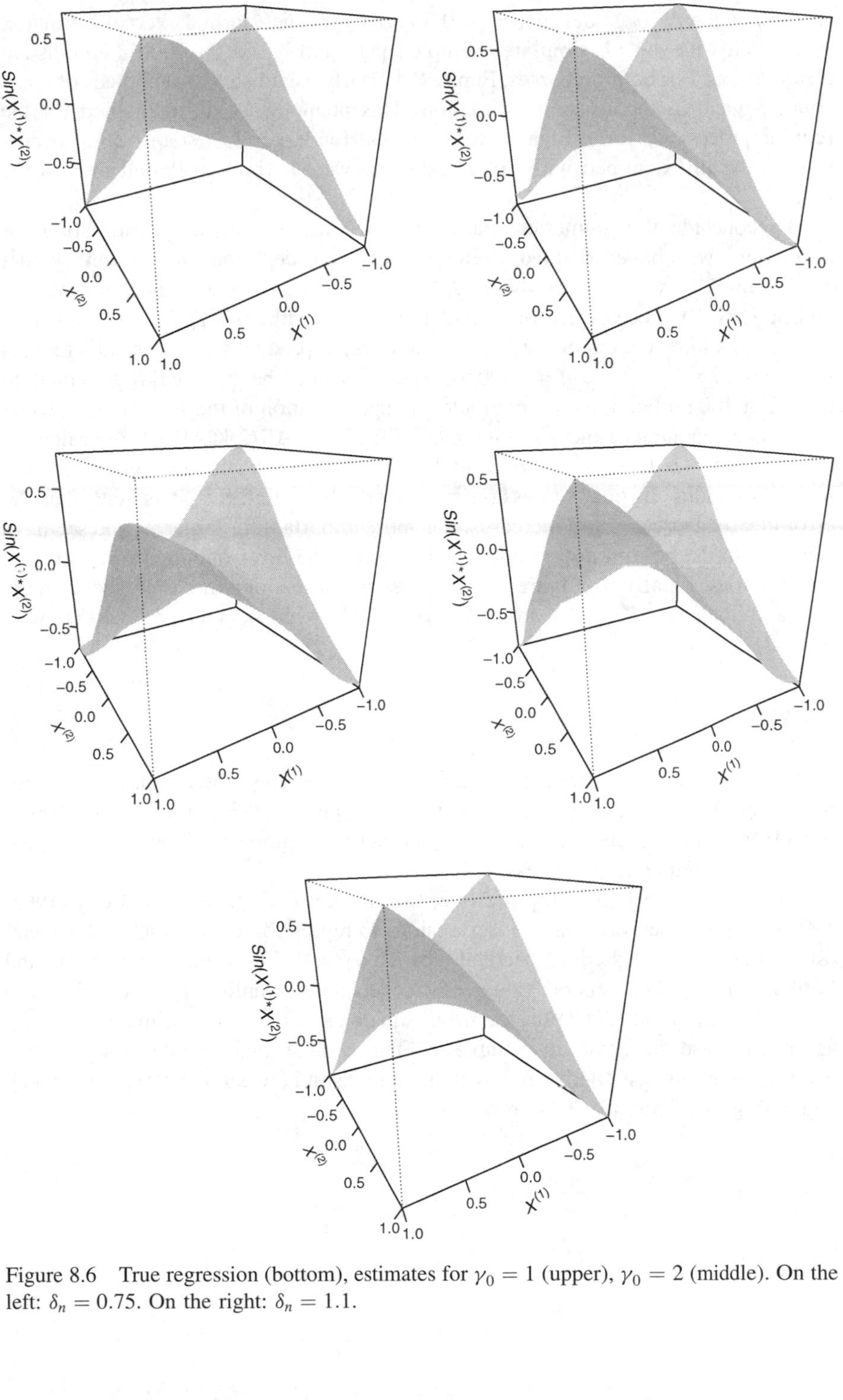

Figure 8.6 True regression (bottom), estimates for $\gamma_0 = 1$ (upper), $\gamma_0 = 2$ (middle). On the left: $\delta_n = 0.75$. On the right: $\delta_n = 1.1$.

where f is the $\mathcal{N}(0,1)$ density, $h_n = 0.3$ and $\hat{f}_{n,\delta,j}$ is the (normal) kernel estimator, associated with the jth simulated sampled path and based on 1000 δ-equidistant observations. For both processes, Figure 8.4 clearly establishes that choices of over-small δ yield to inconsistent estimation. This phenomenon is intensified for the regular process CAR(2). Finally, we note that for larger δ, a stabilization occurs around the ISE obtained with 1000 i.i.d. observations (bottom dotted line on the figure).

To conclude this numerical part, we now turn to regression estimation for processes we have defined before. We consider the regression model $Y_{t_{i,n}} = \sin(X_{t_{i,n}}^{(1)} \cdot X_{t_{i,n}}^{(2)}) + \varepsilon_{t_{i,n}}$ where $(X_t^{(1)})$, $(X_t^{(2)})$ and (ε_t) are independent OU and/or CAR(2) processes. From simulated sample paths of (Y_t, X_t), we construct the regression estimator using a tensorial product of normal kernels $K(x) = (\sqrt{2\pi})^{-1}\ \mathrm{e}^{-x^2/2}$, $n = 3000$ observations and the bandwidth h_n equal to 0.25. For 100 replications, we compute an approximation of the $ISE(\delta)$ error, when (X_t) has components either OU|OU, OU|CAR(2) or CAR(2)|CAR(2). We calculate this error on $[-1,1] \times [-1,1]$ to avoid the boundary effect, which is quite important in this multidimensional regression framework. In Figure 8.5, it is seen that the error clearly decreases as δ increases, it is more important for larger γ_0 and seems to stabilize to the bottom dotted line that represents the error obtained for 3000 i.i.d observations. Finally, in Figure 8.6, we present estimations in cases where (ε_t) is OU and for $i = 1, 2$, $(X_t^{(i)})$ are either both OU (with $\gamma_0 = 1$) or CAR(2) (with $\gamma_0 = 2$).

Notes

In the past, various deterministic or randomized sampling schemes have been proposed and studied, we refer e.g. to the works of Masry (1983); Prakasa Rao (1990); Wu (1997) and references therein. They do not take into account, the sample path properties of underlying processes.

Concerning the high sampling rate, the first results appeared in Bosq (1995, 1998) for processes with $\gamma_0 < 1$. Extensions to regression estimation, and wavelet estimators, were studied respectively by Bosq and Cheze-Payaud (1999); and Leblanc (1995). For general values of γ_0, adequate sampling schemes are given in Blanke and Pumo (2003) for the mean-square error of kernel estimators (as well as in the case of errors-in-variables). Finally, note that simulations are also performed in this last cited paper, with a comparison between Ornstein–Uhlenbeck and Wong (see Example 2.3) processes.

Part IV
Local Time

9

The empirical density

9.1 Introduction

Let $X_1, \ldots, X_n$ be observed real random variables with the same distribution μ. If μ possesses a density f with respect to Lebesgue measure λ, we have seen that an estimator of f may be obtained by regularizing the empirical measure $\mu_n = \left(\sum_{i=1}^{n} \delta_{(X_i)}\right)/n$ by projection (Chapter 3) or by convolution (Chapter 6). Such a regularization is necessary since μ_n is not *absolutely continuous*.

In continuous time the situation is somewhat different: under reasonable conditions the *empirical density* does exist!

In this chapter we study this estimator and show that the projection estimator and the kernel estimator are approximations of it.

The most amazing property of empirical density is unbiasedness, since it is well known that, in general, no unbiased estimator of density may exist. Its asymptotic properties are similar to these of the kernel estimator, in particular it reaches the so-called superoptimal rate.

9.2 Occupation density

Let $X = (X_t, t \in \mathbb{R})$ be a real measurable continuous time process, defined on the probability space $(\Omega, \mathcal{A}, P)$.

Consider the *occupation measure* on $[0, T]$ defined as

$$\nu_T(B) = \int_0^T \mathbb{1}_B(X_t)\mathrm{d}t, \quad B \in \mathcal{B}_{\mathbb{R}}.$$

$\nu_T(B)$ is the time spent by X in B in the time interval $[0, T]$.

Inference and Prediction in Large Dimensions D. Bosq and D. Blanke

If ν_T is absolutely continuous with respect to λ, (a version of) its density is denoted by $\ell_T(x,\omega)$, $x \in \mathbb{R}$, $\omega \in \Omega$ and is called *occupation density* on $[0,T]$. In the following we often write $\ell_T(x)$ instead of $\ell_T(x,\omega)$.

By definition we have

$$\int_0^T \mathbb{1}_B(X_t)\mathrm{d}t = \int_B \ell_T(x)\mathrm{d}x, \quad B \in \mathcal{B}_{\mathbb{R}}, \tag{9.1}$$

then, by using linearity and monotone convergence one obtains

$$\int_0^T \varphi(X_t)\mathrm{d}t = \int_{-\infty}^{+\infty} \varphi(x)\ell_T(x)\mathrm{d}x, \quad \varphi \in \mathcal{M}_+, \tag{9.2}$$

where $\mathcal{M}_+$ is the class of positive Borelian functions.

The following statements are classical criteria for existence of occupation density.

Theorem 9.1
Let $(X_t, 0 \le t \le T)$ be a measurable real process.

(1) If (X_t) has absolutely continuous sample paths, existence of ℓ_T is equivalent to

$$P(X'(t) = 0) = 0 \ \lambda \ a.e. \ on \ [0,T]. \tag{9.3}$$

Under this condition one has

$$\ell_T(x) = \sum_{\{t : X_t = x\}} \frac{1}{|X'(t)|}.$$

(2) (X_t) admits an occupation density $\ell_T \in L^2(\lambda \otimes P)$ if and only if

$$\liminf_{\eta \downarrow 0} \frac{1}{\eta} \int_{[0,T]^2} P(|X_t - X_s| \le \eta)\mathrm{d}s\,\mathrm{d}t < \infty. \tag{9.4}$$

PROOF:
See Geman and Horowitz (1973, 1980). ■

If (X_t) is a Gaussian process, (9.4) can be replaced by

$$\int_{[0,T]^2} [\mathrm{E}(X_t - X_s)^2]^{-1/2}\mathrm{d}s\,\mathrm{d}t < \infty,$$

if, in addition, it is stationary, a sufficient condition for existence of an occupation density is

$$\int_0^T [1-\rho(u)]^{-1/2}\mathrm{d}u < \infty$$

where ρ is the autocorrelation of (X_t).

Now, under more precise conditions, ℓ_T has some regularity properties. Consider the following assumptions:

Assumptions 9.1 (A9.1)

(i) The density $f_{s,t}(y,z)$ of (X_s, X_t) exists and is measurable over $(D^c \cap [0,T]^2) \times U$ where U is an open neighbourhood of $D = \{(x,x),\ x \in \mathbb{R}\}$,

(ii) The function $F_T(y,z) = \int_{[0,T]^2} f_{s,t}(y,z)\mathrm{d}s\,\mathrm{d}t$ is defined in a neighbourhood of D and is continuous at each point of D.

Let us set

$$\mathcal{K}_1 = \{K : \mathbb{R} \to \mathbb{R}, K \text{ is a bounded density, a.e. continuous, with compact support } S_K\},$$

and

$$Z_h^{(K)}(x) = \frac{1}{h}\int_0^T K\left(\frac{x - X_t}{h}\right)\mathrm{d}t, \quad x \in \mathbb{R}, h > 0, K \in \mathcal{K}_1;$$

then we have the following existence theorem for ℓ_T.

Theorem 9.2
If Assumptions A9.1 hold, then (X_t) has an occupation density ℓ_T such that

$$\sup_{a \le x \le b} \mathrm{E}\left|Z_h^{(K)}(x) - \ell_T(x)\right|^2 \underset{h \to 0}{\longrightarrow} 0 \tag{9.5}$$

$a, b \in \mathbb{R}, a \le b, K \in \mathcal{K}_1$.

Moreover $x \mapsto \ell_T(x)$ is continuous in mean square.

PROOF:
See Bosq and Davydov (1999). ■

Example 9.1
The process

$$X_t = U \sin t + V \cos t, \quad 0 \le t \le T,$$

where U and V are independent absolutely continuous random variables, admits an occupation density since

$$P(X'(t) = 0) = P(U \cos t - V \sin t = 0) = 0, \quad t \in [0, T].$$

◇

Example 9.2

The *Ornstein–Uhlenbeck process* has a square integrable occupation density, thanks to

$$\int_0^T [1 - \rho(u)]^{-1/2} \mathrm{d}u = \int_0^T (1 - \mathrm{e}^{-\theta u})^{-1/2} \mathrm{d}u < \infty,$$

moreover (9.5) holds since Assumptions A9.1 are satisfied. ◇

Example 9.3

Consider the process X defined by

$$X_t = \psi(Yt + Z), \quad t \in \mathbb{R}$$

where ψ is a measurable deterministic function, periodic with period δ ($\delta > 0$), Y and Z are independent random variables such that $P(Y \neq 0) = 1$ and Z has a uniform distribution over $[0, \delta]$. Then X is strictly stationary with δ/Y periodic sample paths. The case where $\psi(x) = \cos x$ is particularly interesting since a suitable choice of P_Y leads to a process X with arbitrary *autocovariance* (Wentzel 1975).

Now, if ψ is absolutely continuous, (9.3) holds iff $\psi' \neq 0$ almost everywhere. In particular, if $Y \equiv 1$ and $\delta = 1$ then ℓ_1 exists, does not depend on ω, and is nothing but the density of $\lambda\psi^{-1}$. Given a density q it is always possible to find ψ such that $\mathrm{d}\lambda\psi^{-1}/\mathrm{d}\lambda = q$. Thus it is possible to construct a process with $\ell_1 = q$. In particular, there exists X satisfying (9.3) and such that $\ell_1 \in L^2(\lambda \otimes P)$. ◇

9.3 The empirical density estimator

The occupation density generates a density estimator. The following lemma is simple but crucial.

Lemma 9.1

Let $(X_t, 0 \leq t \leq T)$ be a real measurable process such that $P_{X_t} = \mu$, $0 \leq t \leq T$, and which admits a measurable occupation density ℓ_T.

Then μ is absolutely continuous with density f and $\mathrm{E}(\ell_T/T)$ is a version of f.

PROOF:
From (9.1) it follows that

$$\frac{1}{T}\int_0^T \mathbb{1}_B(X_t)\mathrm{d}t = \int_B \frac{\ell_T(x)}{T}\mathrm{d}x, \quad B \in \mathcal{B}_{\mathbb{R}}. \tag{9.6}$$

Taking expectations in both sides of (9.6) and applying the Fubini theorem one obtains

$$\mu(B) = \int_B \mathrm{E}\frac{\ell_T(x)}{T}\mathrm{d}x, \quad B \in \mathcal{B}_{\mathbb{R}},$$

thus, μ is absolutely continuous with density $\mathrm{E}(\ell_T/T)$. ■

Since the *empirical measure* associated with the data $(X_t, 0 \le t \le T)$ is defined as

$$\mu_T(B) = \frac{1}{T}\int_0^T \mathbb{1}_B(X_t)\mathrm{d}t, \quad B \in \mathcal{B}_{\mathbb{R}}$$

it is natural to say that $f_{T,0} = \ell_T/T$ is the *empirical density* (ED). Lemma 9.1 shows that $f_{T,0}$ is an *unbiased estimator* of f.

Note that, if Assumptions A9.1 hold, Theorem 9.2 implies that $x \mapsto \mathrm{E}(\ell_T(x)/T)$ is the continuous version of f.

Now, other interesting properties of this empirical density estimator (EDE) are recursivity and invariance.

9.3.1 Recursivity

The occupation density is clearly an *additive functional*, that is

$$\ell_T(x) = \sum_{i=1}^m \ell_{[t_{i-1},t_i]}(x) \text{ (a.s.)} \tag{9.7}$$

where $0 = t_0 < t_1 < \cdots < t_m = T$ and $\ell_{[t_{i-1},t_i]}$ is density occupation on $[t_{i-1}, t_i]$.

From (9.7) one deduces the recursion formula

$$f_{n+1,0}(x) = \frac{n}{n+1}f_{n,0}(x) + \frac{1}{n}\ell_{[n,n+1]}(x), \quad n \in \mathbb{N}^*\text{(a.s.)},$$

that may be useful for computations.

9.3.2 Invariance

Definition 9.1
Let g_T^X be an estimator of the marginal density based on the data $(X_t, 0 \le t \le T)$. Let φ be a measurable real function. g_T is said to be φ-invariant if

$Y_t = \varphi(X_t), 0 \leq t \leq T$ implies

$$g_T^Y = \frac{\mathrm{d}(\gamma_T^X \varphi^{-1})}{\mathrm{d}\lambda} \tag{9.8}$$

where γ_T^X is the measure of density g_T^X.

The next statement gives invariance for ℓ_T.

Theorem 9.3

Let $(X_t, 0 \leq t \leq T)$ be a real measurable process with $P_{X_t} = \mu$, $0 \leq t \leq T$ and admitting the occupation density ℓ_T. Then, if φ is a strictly increasing derivable function, $f_{T,0}$ is φ-invariant.

PROOF:

Let μ_T^X and μ_T^Y be the empirical measures associated with (X_t) and (Y_t). Then

$$\mu_T^Y(B) = \frac{1}{T}\int_0^T \mathbb{1}_B(Y_t)\mathrm{d}t = \frac{1}{T}\int_0^T \mathbb{1}_{\varphi^{-1}(B)}(X_t)\mathrm{d}t = \mu_T^X[\varphi^{-1}(B)],$$

$B \in \mathcal{B}_{\mathbb{R}}$, thus $\mu_T^Y = \mu_T^X \varphi^{-1}$ which is equivalent to (9.8) with $g_T = f_{T,0}$. ■

An interesting consequence of Theorem 9.3 is invariance of $f_{T,0}$ with respect to translations and dilatations of time.

Of course, unbiasedness, recursivity and invariance are not satisfied by kernel and projection estimators.

9.4 Empirical density estimator consistency

Under mild conditions $f_{T,0}$ is a consistent estimator. The main reason is additivity: if $T = n$ (an integer) then

$$f_{n,0} = \frac{1}{n}\sum_{i=1}^{n} \ell_{[i-1,i]}, \tag{9.9}$$

thus, if X is a strictly stationary process, $f_{n,0}$ is nothing but the empirical mean of the strictly stationary discrete time process $(\ell_{[i-1,i]}, i \in \mathbb{Z})$. It follows that asymptotic properties of $f_{T,0}$ are consequences of various limit theorems for stationary processes.

Note also that, contrary to the cases of kernel or projection estimators, no smoothing or truncating parameter is needed for constructing the EDE.

Theorem 9.4

Let $X = (X_t, t \in \mathbb{R})$ be a strictly stationary real ergodic process with occupation density ℓ_T. Then, as $T \to \infty$,

(1) for almost all x

$$f_{T,0}(x) \to f(x) \text{ a.s.,} \tag{9.10}$$

(2) with probability 1

$$\| f_{T,0} - f \|_{L^1(\lambda)} \to 0, \tag{9.11}$$

hence

$$\mathrm{E} \| f_{T,0} - f \|_{L^1(\lambda)} \to 0, \tag{9.12}$$

and

$$\sup_{B \in \mathcal{B}_{\mathbb{R}}} |\mu_T(B) - \mu(B)| \to 0 \text{ a.s.,} \tag{9.13}$$

(3) if $\ell_T \in L^2(\lambda \otimes P)$, then, for almost all x,

$$\mathrm{E}(f_{T,0}(x) - f(x))^2 \to 0. \tag{9.14}$$

PROOF:
Without loss of generality we may suppose that

$$X_t = U_t X_0, \quad t \in \mathbb{R}$$

where $U_t \xi(\omega) = \xi T_t(\omega)$, $\omega \in \Omega$ and $(T_t, t \in \mathbb{R})$ is a measurable group of transformations on $(\Omega, \mathcal{A}, P)$ which preserves P (see Doob 1953).

On the other hand, since ℓ_T is increasing with T, we may and do suppose that T is an integer $(T = n)$.

Now, let us set $g_k(x) = U_k \ell_1(x), k \in \mathbb{Z}, x \in \mathbb{R}$ where ℓ_1 is a fixed version of the occupation density over $[0, 1]$. For all x, the sequence $(g_k(x), k \in \mathbb{Z})$ is stationary, and

$$f_{n,0}(x) = \frac{1}{n} \sum_{k=0}^{n-1} g_k(x),$$

then, the Birkhoff–Khinchine ergodic theorem (see Billingsley 1968) yields

$$f_{n,0}(x) \to g(x) \text{ a.s.,} \quad x \in \mathbb{R} \tag{9.15}$$

where $g(x) = \mathrm{E}^{\mathcal{I}} g_0(x)$ and $\mathcal{I}$ is the σ-algebra of invariant sets with respect to T_1.

Since ergodicity of X does not mean triviality of $\mathcal{I}$ we have to show that

$$g(x) = f(x), \quad \lambda \otimes P \text{ a.e.}$$

For this purpose we apply a variant of the Fubini theorem: g being nonnegative we may write

$$\int_{\mathbb{R}} g(x)\mathrm{d}x = \int_{\mathbb{R}} \mathrm{E}^{\mathcal{I}} g_0(x)\mathrm{d}x = \mathrm{E}^{\mathcal{I}} \int_{\mathbb{R}} g_0(x)\mathrm{d}x = 1 \text{ (a.s.)},$$

thus, for almost all ω, g is a density.

Applying the Scheffé lemma (Billingsley 1968) and (9.15) we obtain

$$\| f_{n,0} - g \|_{L^1(\lambda)} \to 0, \quad \text{a.s.},$$

and

$$\sup_{B \in \mathcal{B}_{\mathbb{R}}} |\mu_n(B) - \nu(B)| \to 0 \quad \text{a.s.},$$

where ν is the measure with density g.

But, the ergodic theorem entails

$$\mu_n(B) = \frac{1}{n} \int_0^n \mathbb{1}_B(X_t)\mathrm{d}t \to \mathrm{E}\mathbb{1}_B(X_0) = \mu(B), \quad \text{a.s.}, \tag{9.16}$$

$B \in \mathcal{B}_{\mathbb{R}}$.

From (9.16) we deduce that, if (B_j) is a sequence of Borel sets which generates $\mathcal{B}_{\mathbb{R}}$, there exists Ω_0, such that $P(\Omega_0) = 1$, and

$$\mu_n^{\omega}(B_j) \to \mu(B_j), \quad j \geq 0, \quad \omega \in \Omega_0,$$

hence, with probability 1, $\nu = \mu$, that is $f = g$, $\lambda \otimes P$ a.e., and (9.15) gives (9.10).

(9.11) and (9.13) are consequences of (9.10) and the Scheffé lemma. (9.12) follows from the dominated convergence theorem since $\| f_{n,0} - f \|_{L^1(\lambda)} \leq 2$.

Finally, if ℓ_T is square integrable, $\mathrm{E}|g_0(x)|^2 < \infty$ a.e., thus

$$\mathrm{E}|f_{n,0}(x) - g(x)|^2 \to 0,$$

hence (9.14) since $g = f$, $\lambda \otimes P$ a.e. ∎

Note that, in discrete time, the analogue of (9.13) does not hold. Davydov (2001) has shown that (9.13) is satisfied if and only if X has an occupation density.

With some additional assumptions it is possible to establish that the a.s. convergence is uniform over compact sets (see Bosq and Davydov 1999).

9.5 Rates of convergence

The main purpose of this section is to derive the parametric rate of the EDE.

Lemma 9.2
If Assumptions A9.1 hold, and $P_{X_t} = \mu$, $t \in \mathbb{R}$, *then*

$$\operatorname{Var} f_{T,0}(x) = \frac{1}{T^2} \int_{[0,T]^2} g_{s,t}(x,x) \mathrm{d}s \, \mathrm{d}t, \quad x \in \mathbb{R}, \tag{9.17}$$

where $g_{s,t} = f_{s,t} - f \otimes f$.

If, in addition, $g_{s,t} = g_{|t-s|,0} := g_{|t-s|}$, *we have*

$$\operatorname{Var} f_{T,0}(x) = \frac{2}{T} \int_0^T \left(1 - \frac{u}{T}\right) g_u(x,x) \mathrm{d}u, \quad x \in \mathbb{R}. \tag{9.18}$$

PROOF: *(Sketch)*
Let $K \in \mathcal{K}_1$, Theorem 9.2 yields $\mathrm{E}(Z_h^{(K)}(x))^2 \underset{h \to 0}{\longrightarrow} \mathrm{E}\ell_T^2(x)$; it follows that

$$\mathrm{E}\ell_T^2(x) = F_T(x,x) = \int_{[0,T]^2} f_{s,t}(x,x) \mathrm{d}s \, \mathrm{d}t.$$

On the other hand,

$$(\mathrm{E}Z_h^{(K)}(x))^2 \to (\mathrm{E}\ell_T(x))^2 = T^2 f(x) = \int_{[0,T]^2} f^2(x) \mathrm{d}s \, \mathrm{d}t,$$

hence (9.17).

(9.18) is an easy consequence of $g_{s,t} = g_{|t-s|}$ and (9.17). ■

Theorem 9.5
Suppose that Assumptions A9.1 hold, and P_{X_t} *does not depend on* t.

(1) If

$$G(x) := \overline{\lim_{T \to \infty}} \frac{1}{T} \int_{[0,T]^2} g_{s,t}(x,x) \mathrm{d}s \, \mathrm{d}t < \infty,$$

then

$$\overline{\lim_{T \to \infty}} \, T \cdot \operatorname{Var} f_{T,0}(x) \leq G(x).$$

(2) If $g_{s,t} = g_{|t-s|}$ *and*

$$\int_0^\infty |g_u(x,x)| \mathrm{d}u < \infty \tag{9.19}$$

then

$$T \cdot \operatorname{Var} f_{T,0}(x) \to \int_{-\infty}^{\infty} g_u(x,x)\mathrm{d}u. \tag{9.20}$$

PROOF:
It suffices to take $\overline{\lim}$ in (9.17) and lim in (9.18). ■

(9.20) shows that $f_{T,0}$ has the same asymptotic behaviour as the kernel estimator $f_T^{(K)}$ (see Theorem 7.2, Chapter 7). Note however that the parametric rate is obtained for a larger class of densities. In fact, whatever the choice of $h_T \to 0$, it is possible to construct a continuous density f_0 associated with a suitable process X for which

$$T \cdot \mathrm{E}(f_T^{(K)}(x) - f_0(x))^2 \to \infty$$

while

$$T \cdot \mathrm{E}(f_{T,0}(x) - f_0(x))^2 \to \int_{-\infty}^{+\infty} g_u(x,x)\mathrm{d}u.$$

Finally note that (9.19) is close to Castellana and Leadbetter's condition (see Theorem 7.2). It also implies condition A9.1(ii). Actually, under some mild additional conditions, it may be proved that the kernel estimator and the EDE converge at rate $1/T$ if and only if (9.19) holds. For details we refer to Bosq (1998), Chapter 6.

We now turn to uniform convergence. We need the following family of assumptions:

Assumptions 9.2 (A9.2)

(i) X is a strictly stationary GSM process,

(ii) X admits an occupation density ℓ_1 such that $\mathrm{E}[\exp c\ell_1(x)] < \infty$, $x \in \mathbb{R}$ $(c > 0)$,

(iii) $\inf_{a \le x \le b} \mathrm{E}|\ell_1(x) - f(x)|^2 > 0$, $a < b$,

(iv) $\sup_{a \le x,y \le b, |x-y| < \delta} |\ell_1(y) - \ell_1(x)| \le V_1 \delta^d$, $\delta > 0$, $(d > 0)$, with $V_1 \in L^1(p)$

(v) $\mathrm{E}\ell_T(\cdot)$ is continuous.

The Ornstein–Uhlenbeck process satisfies these conditions. The next statement furnishes a uniform rate.

Theorem 9.6
Under Assumptions A9.2, we have

$$\sup_{a \le x \le b} |f_{T,0}(x) - f(x)| = \mathcal{O}\left(\frac{\ln T}{\sqrt{T}}\right) a.s.$$

PROOF: *(Sketch)*
Again we may assume that T is an integer. Then, using (9.9) and an exponential type inequality under Cramér's conditions (see e.g. Bosq 1998, Theorem 1.4), one easily obtains

$$|f_{n,0}(x) - f(x)| = \mathcal{O}\left(\frac{\ln n}{\sqrt{n}}\right) \text{ a.s.},$$

$a \leq x \leq b$.

Similarly to Theorem 6.6, one deduces uniformity from A9.2(iv).
Note that regularity of f is hidden in A9.2(iv) since

$$|f(y) - f(x)| = |\mathrm{E}\ell_1(y) - \mathrm{E}\ell_1(x)| \leq \mathrm{E}|\ell_1(y) - \ell_1(x)|. \qquad \blacksquare$$

We finally state a result concerning limit in distribution.

Theorem 9.7
Suppose that

(*a*) *X is strictly stationary and α-mixing with $\alpha(u) = \mathcal{O}(u^{-\beta})$, $\beta > 1$,*

(*b*) *g_u exists for $u \neq 0$, is continuous over D, and $\int_{-\infty}^{+\infty} \| g_u \|_\infty \, \mathrm{d}u < \infty$,*

(*c*) *there exists $\delta > 2\beta/(2\beta - 1)$ such that $\mathrm{E}\ell_1^\delta(x) < \infty$, $x \in \mathbb{R}$;*

then

$$\sqrt{T}(f_{T,0}(x_1) - f(x_1), \ldots, f_{T,0}(x_k) - f(x_k))' \xrightarrow{\mathcal{D}} N_k \approx \mathcal{N}(0, \Gamma), \tag{9.21}$$
$$x_1, \ldots, x_k \in \mathbb{R}, k \geq 1, \textit{where } \Gamma = \left(\int_{-\infty}^{+\infty} g_u(x_i, x_j)\mathrm{d}u\right)_{1 \leq i,j \leq k}.$$

PROOF: *(Sketch)*
It suffices to prove (9.21) for $T = n$ and $k = 1$, since for $k > 1$ we can use the Cramér–Wold device (Billingsley 1968).
Then we have

$$\sqrt{n}(f_{n,0}(x) - f(x)) = \frac{1}{\sqrt{n}} \sum_{i=1}^{n} (\ell_{[i-1,i]}(x) - f(x)),$$

$x \in \mathbb{R}$, and the central limit theorem for α-mixing processes (Rio 2000) gives the desired result. $\blacksquare$

Under the same conditions one may derive a functional law of the iterated logarithm (cf. Bosq 1998).

9.6 Approximation of empirical density by common density estimators

The EDE has many interesting properties: it is unbiased, recursive, invariant and asymptotically efficient. However, due to its implicit form, it is difficult to compute, thus approximation is needed.

It is noteworthy that the usual nonparametric density estimators are good approximations of the EDE.

Concerning the kernel estimator (see Chapter 7), (9.2) yields

$$f_{T,h_T}^{(K)}(x) = \frac{1}{Th_T}\int_0^T K\left(\frac{x - X_t}{h_T}\right)\mathrm{d}t = \int_{-\infty}^{+\infty} \frac{1}{h_T} K\left(\frac{x-y}{h_T}\right)\frac{\ell_T(y)}{T}\mathrm{d}y, \quad x \in \mathbb{R},$$

that is $f_{T,h_T}^{(K)} = K_{h_T} * f_{T,0}$, where $K_{h_T} = h_T^{-1}K(\cdot/h_T)$, and K is a bounded density such that $\lim_{|x|\to\infty} xK(x) = 0$.

Suppose that T is fixed and write h instead of h_T. Then, if ℓ_T is continuous at x, the Bochner lemma 6.1, gives $\lim_{h\to 0} f_{T,h}^{(K)}(x) = f_{T,0}(x)$. If Assumptions A9.1 hold, (9.5) implies

$$\lim_{h\to 0} \sup_{a\le x\le b} \mathrm{E}\left| f_{T,h}^{(K)}(x) - f_{T,0}(x)\right|^2 = 0.$$

Now, if $\ell_T \in L^2(\lambda \otimes P)$, let $(e_j, j \ge 0)$ be an orthonormal basis of $L^2(\lambda)$. For almost all $\omega \in \Omega$ we have $\ell_T(\cdot, \omega) \in L^2(\lambda)$, hence

$$f_{T,0} = \sum_{j=0}^{\infty} \langle f_{T,0}, e_j\rangle_{L^2(\lambda)} e_j \text{ (a.s.)}$$

where convergence takes place in $L^2(\lambda)$. Then (9.2) entails

$$\langle f_{T,0}, e_j\rangle_{L^2(\lambda)} = \frac{1}{T}\int_{-\infty}^{+\infty} \ell_T(x)e_j(x)\mathrm{d}x = \frac{1}{T}\int_0^T e_j(X_t)\mathrm{d}t = \hat{a}_{jT},$$

$j \ge 0$, thus $f_{T,0} = \sum_{j=0}^{\infty} \hat{a}_{jT} e_j$ (a.s.). Note that this implies the nontrivial property

$$\sum_{j=0}^{\infty} \hat{a}_{jT}^2 < \infty \quad \text{(a.s.)}. \tag{9.22}$$

Note also that (9.22) is, in general, not satisfied in the discrete case.

Now, the projection estimator $f_{T,k_T}^{(e)}$ associated with (e_j) is defined by

$$f_{T,k_T}^{(e)} = \sum_{j=0}^{k_T} \hat{a}_{jT} e_j$$

(cf. Chapter 3) thus

$$f_{T,k_T}^{(e)} = \Pi^{k_T}(f_{T,0}) \tag{9.23}$$

where Π^{k_T} is the *orthogonal projector* of $\mathrm{sp}(e_j, 0 \le j \le k_T)$.

If T is fixed, and writing k for k_T, it follows that

$$\lim_{k\to\infty} \| f_{T,k}^{(e)} - f_{T,0} \| = 0 \text{ (a.s.)}$$

where $\| \cdot \|$ is the $L^2(\lambda)$-norm.

As stated in Chapter 3, approximation of EDE by the projection estimator allows us to obtain a parametric rate for the latter. The starting point is asymptotic behaviour of the mean integrated square error (MISE) of $f_{T,0}$:

Theorem 9.8

If X is strictly stationary, with an occupation density $\ell_T \in L^2(\lambda \otimes P)$ and such that the series

$$L = \sum_{k\in\mathbb{Z}} \int_{-\infty}^{+\infty} \mathrm{Cov}(\ell_1(x), \ell_{[k-1,k]}(x))\mathrm{d}x \tag{9.24}$$

is absolutely convergent, then

$$\lim_{T\to\infty} T \cdot \mathrm{E} \| f_{T,0} - f \|^2 = L. \tag{9.25}$$

PROOF:

Note first that f exists (Lemma 9.1) and is in $L^2(\lambda)$ since

$$\int_{-\infty}^{+\infty} f^2(x)\mathrm{d}x = \int_{-\infty}^{+\infty} (\mathrm{E}\ell_1(x))^2\mathrm{d}x \le \int_{-\infty}^{+\infty} \mathrm{E}\ell_1^2(x)\mathrm{d}x < \infty.$$

Now take $T = n$ and write

$$n\mathrm{E} \| f_{n,0} - f \|^2 = \sum_{k=-(n-1)}^{n-1} \left(1 - \frac{|k|}{n}\right) \int_{-\infty}^{+\infty} \mathrm{Cov}(\ell_1(x), \ell_{[k-1,k]}(x))\mathrm{d}x;$$

then (9.25) follows by dominated convergence with respect to the counting measure on $\mathbb{Z}$. ■

Applying the Davydov inequality, one sees that (9.24) holds provided $\int_{\mathbb{R}}[\mathrm{E}\ell^r(x)]^{2/r}\mathrm{d}x < \infty$, $(r > 2)$, and X is α-mixing with $\sum_k [\alpha(k)]^{(r-2)/r} < \infty$. Also note that, if in addition, $\int_{\mathbb{R}^2} |g_u(x,x)|\mathrm{d}u\,\mathrm{d}x < \infty$ then

$$T \cdot \mathrm{E} \parallel f_{T,0} - f \parallel^2 \to \int_{\mathbb{R}^2} g_u(x,x)\mathrm{d}u\,\mathrm{d}x,$$

thus $L = \int_{\mathbb{R}^2} g_u(x,x)\mathrm{d}u\,\mathrm{d}x$.

The parametric rate for $f^{(e)}_{T,k_T}$ follows from Theorem 9.8.

Corollary 9.1

If conditions in Theorem 9.8 hold and (k_T) is such that

$$\sum_{j>k_T} \left(\int_{-\infty}^{+\infty} f(x)e_j(x)\mathrm{d}x \right)^2 = o\left(\frac{1}{T}\right) \tag{9.26}$$

then

$$\overline{\lim}_{T\to\infty} T \cdot \mathrm{E} \parallel f^{(e)}_{T,k_T} - f \parallel^2 \leq L.$$

Proof:

Write

$$\mathrm{E} \parallel f^{(e)}_{T,k_T} - f \parallel^2 = \mathrm{E} \parallel f^{(e)}_{T,k_T} - \Pi^{k_T} f \parallel^2 + \parallel \Pi^{k_T} f - f \parallel^2,$$

from (9.23) it follows that

$$\begin{aligned}\mathrm{E} \parallel f^{(e)}_{T,k_T} - f \parallel^2 &= \mathrm{E} \parallel \Pi^{k_T}(f_{T,0} - f) \parallel^2 + o\left(\frac{1}{T}\right)\\ &\leq \mathrm{E} \parallel f_{T,0} - f \parallel^2 + o\left(\frac{1}{T}\right)\end{aligned}$$

and (9.25) gives the conclusion. ■

A practical choice of k_T is possible if one has some knowledge concerning the Fourier coefficients of f. For example, if $|\int fe_j| = \mathcal{O}(j^{-\beta})$ $(\beta > 1/2)$ the choice $k_T \simeq T^\delta$, where $\delta > 1/(2\beta - 1)$, entails (9.26). Note that, in such a situation, construction of an adaptive estimator $f^{(e)}_{T,\hat{k}_T}$ (cf. Chapter 3) is not necessary since the nonadaptive estimator reaches the superoptimal rate even on $\mathcal{F}_1$.

However $\hat{f}_T = f^{(e)}_{T,\hat{k}_T}$ (see Section 3.7) has an interesting property if $f \in \mathcal{F}_0$: under the assumptions in Theorem 3.12 we have

$$\lim_{T\to\infty} T \cdot \mathrm{E} \parallel \hat{f}_T - f \parallel^2 = 2\sum_{j=0}^{K(f)} \int_0^\infty \mathrm{Cov}(e_j(X_0), e_j(X_u))\mathrm{d}u. \tag{9.27}$$

Now

$$T \cdot \mathrm{E} \parallel f_{T,0} - f \parallel^2 = T \sum_{j=0}^{\infty} \mathrm{Var}\, \hat{a}_{jT}$$
$$= \sum_{j=0}^{\infty} 2 \int_0^T \left(1 - \frac{u}{T}\right) \mathrm{Cov}(e_j(X_0), e_j(X_u)) \mathrm{d}u \qquad (9.28)$$

and, since $\int_0^\infty |\mathrm{Cov}(e_j(X_0), e_j(X_u))| \mathrm{d}u < \infty$, the dominated convergence theorem yields

$$T \cdot \mathrm{Var}\, \hat{a}_{jT} \to 2 \int_0^\infty \mathrm{Cov}(e_j(X_0), e_j(X_u)) \mathrm{d}u, \, j \geq 0.$$

Applying the Fatou lemma for the counting measure, one finds

$$\liminf_{T \to \infty} T \cdot \mathrm{E} \parallel f_{T,0} - f \parallel^2 \geq 2 \sum_{j=0}^{\infty} \mathrm{Cov}(e_j(X_0), e_j(X_u)) \mathrm{d}u.$$

Since it is easy to construct examples where $\int_0^\infty \mathrm{Cov}(e_j(X_0), e_j(X_u)) \mathrm{d}u > 0$ for some $j > K(f)$, (9.27) and (9.28) show that $\hat{f}_T$ is asymptotically strictly better than $f_{T,0}$ for $f \in \mathcal{F}_0$, with respect to the MISE.

Discrete time data

Another type of approximation appears if only discrete time data are available. Suppose that the observations are $(X_{i\tau/n}, 1 \leq i \leq n)$ and consider the kernel estimator

$$\bar{f}_{n,h_n}^{(K)}(x) = \frac{1}{n} \sum_{i=1}^{n} K_{h_n}(x - X_{i\tau/n}), \quad x \in \mathbb{R}.$$

We have the following approximation result.

Theorem 9.9

Let $X = (X_t, t \in \mathbb{R})$ be a real measurable strictly stationary process with a local time ℓ_T and such that

(*a*) $\mathrm{E}(\sup_{|x-y|<\delta} |\ell_T(y) - \ell_T(x)|) \leq c_T \delta^\lambda$, $\delta > 0$, $(\lambda > 0)$,

and

(*b*) $\mathrm{E}|X_t - X_s| \leq d_T |t - s|^\gamma$; $(s,t) \in [0,T]^2 (\gamma > 0)$, *then, if K is a density with bounded derivative and such that* $\int_{\mathbb{R}} |u|^{\lambda} K(u)\mathrm{d}u < \infty$, *we have*

$$\mathrm{E} \parallel \bar{f}_{n,h_n}^{(K)} - f_{T,0} \parallel_\infty \leq \frac{d_T \parallel K' \parallel_\infty}{\gamma+1} \frac{T^\gamma}{n^\gamma h_n^2} + c_T \int_{\mathbb{R}} |u|^{\lambda} K(u)\mathrm{d}u \cdot \frac{h_n^{\lambda}}{T};$$

consequently, for T fixed, if $n \to \infty$ *and* $h_n \simeq n^{-\gamma/(2+\lambda)}$ *it follows that*

$$\mathrm{E} \parallel \bar{f}_{n,h_n}^{(K)} - f_{T,0} \parallel_\infty = \mathcal{O}(n^{-\gamma\lambda/(2+\lambda)}).$$

PROOF:
See Bosq (1998). ■

Notes

Nguyen and Pham (1980) and Doukhan and León (1994) have noticed the role played by the occupation density in density estimation. The fact that the projection estimator is the projection of EDE has been noticed by Frenay (2001).

Most of the results in this chapter come from Bosq and Davydov (1999) and Bosq and Blanke (2004).

In the context of diffusion processes, Kutoyants (1997a) has defined and studied an unbiased density estimator f_T^* based on *local time*.

The local time L_T is the occupation density associated with the quadratic variation of the process (see Chung and Williams 1990). In particular, if X is a stationary diffusion satisfying the stochastic differential equation

$$\mathrm{d}W_t = S(X_t)\mathrm{d}t + \sigma(X_t)\mathrm{d}W_t \quad (\sigma(\cdot) > 0),$$

we have

$$L_T(x) = \ell_T(x)\sigma^2(x), \quad x \in \mathbb{R}$$

and it can be shown that

$$\mathrm{E}(f_T^*(x) - f_{T,0}(x))^2 = \mathcal{O}\left(\frac{1}{T^2}\right)$$

thus, $f_{T,0}$ and f_T^* have the same asymptotic behaviour in mean-square.

Kutoyants (1998) has shown that the rate $1/T$ is minimax and that the constant $\int_{-\infty}^{+\infty} g_u(x,x)\mathrm{d}u$ is minimax. An estimation of this quantity appears in Blanke and Merlevède (2000).

Extensions

Extension to multidimensional processes is not possible in general since existence of occupation density in $\mathbb{R}^d (d > 1)$ means that X has very exotic sample paths.

Regression estimation

Serot (2002) has defined a *conditional occupation density* and used it for the construction of a regression estimator.

Part V

Linear Processes in High Dimensions

10

Functional linear processes

10.1 Modelling in large dimensions

Let $\xi = (\xi_t, t \in \mathbb{R})$ be a real continuous time process. In order to study the behaviour of ξ over time intervals of length δ, one may set

$$X_n(s) = \xi_{n\delta+s}, \quad 0 \le s \le \delta; \; n \in \mathbb{Z}. \tag{10.1}$$

This construction generates a sequence $(X_n, n \in \mathbb{Z})$ of random variables with values in a suitable function space, say F.

For example, if one observes temperature in continuous time during N days, and wants to predict evolution of temperature during the $(N+1)$th day, the problem becomes: predict X_{N+1} from the data $X_1, \ldots, X_N$. Here one may choose $F = C[0, \delta]$, where δ is length of one day.

Another example of modelling in large dimensions is the following: consider an economic variable associated with individuals. At instant n, the variable associated with the individual i is $X_{n,i}$. In order to study global evolution of that variable for a large number of individuals, and during a long time, it is convenient to set

$$X_n = (X_{n,i}, \; i \ge 1), \; n \in \mathbb{Z},$$

which defines a process $X = (X_n, n \in \mathbb{Z})$ with values in some sequence space F.

These examples show that infinite-dimensional modelling is interesting in various situations. Clearly the model is useful for applications, especially if the X_n's are correlated.

In this chapter we study functional linear processes (FLP), a model which is rather simple and general enough for various applications.

Inference and Prediction in Large Dimensions D. Bosq and D. Blanke

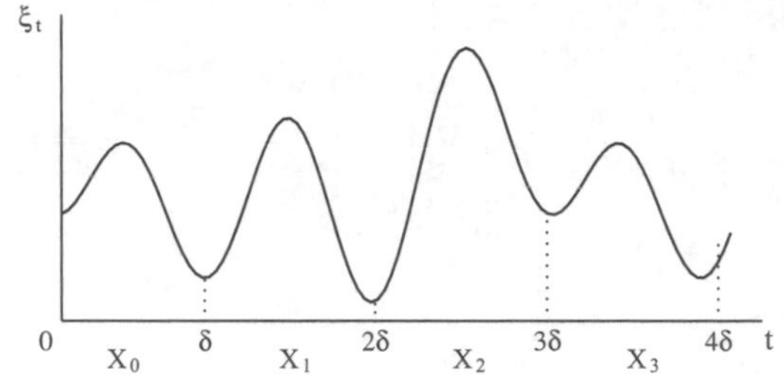

Figure 10.1 Representation of a process as a sequence of F-random variables.

We begin with a presentation of linear prediction and Wold decomposition in Hilbert spaces. This allows us to give a general definition of FLP in such spaces. Concerning properties of FLP we focus on autoregressive and moving average processes in Hilbert spaces, two simple models that have been used in practice. Some extensions to Banach spaces are indicated.

10.2 Projection over linearly closed spaces

Let H be a real separable Hilbert space with its norm $\|\cdot\|$ and its scalar product $\langle\cdot,\cdot\rangle$. $\mathcal{L}$ denotes the space of continuous linear operators from H to H; it is equipped with its usual norm

$$\| \ell \|_{\mathcal{L}} = \sup_{\|x\|\leq 1} \| \ell(x) \|, \ \ell \in \mathcal{L}.$$

Consider the space $L_H^2 := L_H^2(\Omega, \mathcal{A}, P)$ of (classes of) random variables X, defined on the probability space $(\Omega, \mathcal{A}, P)$, with values in H, and such that $\mathrm{E} \| X \|^2 < \infty$. The scalar product

$$\langle X, Y \rangle_{L_H^2} = \mathrm{E}\langle X, Y \rangle; \ X, Y \in L_H^2$$

provides a structure of Hilbert space to L_H^2.

If X and Y are in L_H^2, the *cross-covariance operator* of X and Y is defined as

$$C_{X,Y}(x) = \mathrm{E}\,[\langle X - \mathrm{E}\,X, x\rangle (Y - \mathrm{E}\,Y)], \ x \in H,$$

where $\mathrm{E}\,X$ and $\mathrm{E}\,Y$ are expectations of X and Y (i.e. $\mathrm{E}\,X$ is the unique element of H such that $\langle \mathrm{E}\,X, a\rangle = \mathrm{E}\langle X, a\rangle$ for all a in H).

The *covariance operator* $C_{X,X}$ of X is denoted by C_X. In the following we consider *zero-mean* H-valued random variables, except where otherwise stated.

Now, a linear subspace $\mathcal{G}$ of L_H^2 is said to be *linearly closed* (LCS) if:

(a) $\mathcal{G}$ is closed in L_H^2,

and

(b) $X \in \mathcal{G}$, $\ell \in \mathcal{L}$ implies $\ell(X) \in \mathcal{G}$.

Finally, X and Y in L_H^2 are said to be

- *weakly orthogonal* $(X \perp Y)$ if $\mathrm{E}\langle X, Y\rangle = 0$,
- *strongly orthogonal* $(X \perp\!\!\!\!\perp Y)$ if $C_{X,Y} = 0$.

Stochastic independence implies strong orthogonality, which in turn implies weak orthogonality. Weak orthogonality does not imply strong orthogonality, except if H is one-dimensional. However, if $\mathcal{G}$ is an LCS, $Y \perp \mathcal{G}$ yields $Y \perp\!\!\!\!\perp \mathcal{G}$.

We now point out some properties of the orthogonal projector Π^X of the LCS $\mathcal{G}_X$ generated by $X \in L_H^2$:

$$\mathcal{G}_X = \overline{\mathrm{sp}}\{\ell(X),\ \ell \in \mathcal{L}\}.$$

Note that $\mathcal{G}_X$ is, in general, infinite-dimensional and that elements of $\mathcal{G}_X$ are not necessarily of the form $\ell(X)$, $\ell \in \mathcal{L}$. The following statement summarizes some useful properties of Π^X.

Proposition 10.1

(a) *For each Y in L_H^2, there exists a measurable mapping λ from H to H such that*

$$\Pi^X(Y) = \lambda(X).$$

(b) *There exists a sequence (ℓ_k) in $\mathcal{L}$ and a linear subspace $\mathcal{V}$ of H such that $P_x(\mathcal{V}) = 1$ and*

$$\ell_k(x) \to \lambda(x),\ x \in \mathcal{V}. \tag{10.2}$$

Moreover one may set

$$\lambda(x) = 0,\ x \notin \mathcal{V}, \tag{10.3}$$

then (10.2) *and* (10.3) *guarantee uniqueness of λ (P_X a.s.).*

(c) *λ is linear over $\mathcal{V}$, and such that*

$$\begin{aligned} \lambda(\mathrm{E}X) &= \mathrm{E}(\lambda(X)),\ \textit{(even if } \mathrm{E}X \neq 0),\\ C_{Z,\lambda(X)} &= \lambda C_{Z,X},\ Z \in L_H^2,\\ C_{X,Y} &= C_{X,\lambda(X)} = \lambda C_X,\\ C_{\lambda(X)}(y) &= (\lambda C_X \lambda^*)(y),\ \textit{provided } \lambda^*(y) \in H, \end{aligned} \tag{10.4}$$

where λ^* is the adjoint of λ.

PROOF:
See Bosq (2007). ■

Notice that (10.4) entails uniqueness of λ on $C_X(H)$. However, this property turns out to be trivial if $P_x(C_X(H)) = 0$ which may be the case if X is Gaussian (see Fortet 1995). Proposition 10.1 shows that, despite $\lambda \notin \mathcal{L}$ (in general), it possesses some properties of elements in $\mathcal{L}$. For explicit $\lambda \notin \mathcal{L}$ we refer to examples 10.2, 10.3 and 10.4 below.

Now it is of interest to give conditions that yield existence of $\ell \in \mathcal{L}$ such that

$$\ell(X) = \lambda(X) \text{ a.s.}$$

For this purpose we need a lemma concerning continuous linear operators. First we give a definition of dominance: let A and B be in $\mathcal{L}$; A is said to be dominated by B ($A \prec B$) if and only if

$$(\exists \alpha \geq 0) : \| A(x) \| \leq \alpha \| B(x) \|, \; x \in H. \tag{10.5}$$

Lemma 10.1
$A \prec B$ if and only if there exists $R \in \mathcal{L}$ such that

$$A = RB. \tag{10.6}$$

Moreover, by putting $R = 0$ on $\overline{B(H)}^{\perp}$ one ensures uniqueness of R.

PROOF:
Suppose that $A \prec B$ and put

$$f_s(B(x)) = \langle s, A(x) \rangle; \; s, \, x \in H, \tag{10.7}$$

then, from (10.5), it follows that

$$| f_s(B(x))| \leq \| s \| \| A(x) \| \leq \alpha \| s \| \| B(x) \|. \tag{10.8}$$

Thus, $y \mapsto f_s(y)$ is a continuous linear functional on $B(H)$ that can be extended by continuity on $\overline{B(H)}$.

Now, by using the Riesz theorem on $\overline{B(H)}$ (see Vakhania *et al.* 1987) one may claim existence (and uniqueness) of $z_s \in \overline{B(H)}$ such that

$$f_s(B(x)) = \langle z_s, B(x) \rangle$$

and (10.8) yields

$$\| z_s \| \leq \alpha \| s \|. \tag{10.9}$$

This allows us to define a continuous linear mapping from H to $\overline{B(H)}$ by setting $R^*(s) = z_s$, and (10.9) implies $\| R^* \|_{\mathcal{L}} \leq \alpha$.

Now, for every $(s, x) \in H^2$, we have

$$f_s(B(x)) = \langle z_s, B(x) \rangle = \langle R^*(s), B(x) \rangle = \langle (B^* R^*)(s), x \rangle,$$

and, from (10.7), it follows that $\langle A^*(s), x \rangle = \langle (B^* R^*)(s), x \rangle$, which implies $A = RB$ with $R : \overline{B(H)} \to H$ defined by

$$\langle R(x), y \rangle = \langle x, R^*(y) \rangle, \ x \in \overline{B(H)}, \ y \in H.$$

R can be extended to H by putting

$$R(x) = 0, \ x \in \overline{B(H)}^{\perp}.$$

Since $A(x) = R(B(x))$ with $B(x) \in B(H)$, we obtain

$$A(x) = (RB)(x), \ x \in H.$$

Concerning uniqueness, if $A = R'B$ where $R' \in \ell$ and $R' = 0$ over $\overline{B(H)}^{\perp}$, then $R'B = RB$, that is $R' = R$ over $B(H)$. By continuity it follows that $R' = R$ over $\overline{B(H)}$ and finally $R' = R$ over H.

Conversely, if (10.6) holds, we have

$$\| A(x) \| = \| (RB)(x) \| \leq \| R \|_{\mathcal{L}} \| B(x) \|, \ x \in H,$$

which gives (10.5) with $\alpha = \| R \|_{\mathcal{L}}$. ■

As an application we have the desired property:

Theorem 10.1
The following conditions are equivalent

$$C_{X,Y} \prec C_X \tag{10.10}$$

and

$$(\exists \ell \in \mathcal{L}) \ : \ \Pi^X(Y) = \ell(X) \ (a.s.). \tag{10.11}$$

PROOF:
If (10.10) is valid, Lemma 10.1 implies existence of $\ell \in \mathcal{L}$ such that $C_{X,Y} = \ell C_X$, and since $C_{X,\ell(X)} = \ell C_X$, one obtains $C_{X,Y-\ell(X)} = 0$ which means that $Y - \ell(X) \perp \mathcal{G}_X$, hence (10.11).

Conversely (10.11) entails $C_{X,Y-\ell(X)} = C_{X,Y-\Pi^X(Y)} = 0$, therefore $C_{X,Y} = C_{X,\ell(X)} = \ell C_X$ and (10.10) follows from Lemma 10.1. ■

Note that (10.10) may be written under the form $C_{X,\Pi^X(Y)} \prec C_X$.

If H is infinite-dimensional, (10.10) is, in general, not satisfied, as shown by the following example.

Example 10.1
Let $(X_j,\ j \geq 1)$ be a sequence of real independent, not degenerate, random variables such that

$$\sum_{j=1}^{\infty} \mathrm{E}\, X_j^2 \cdot \max(1, \alpha_j^2) < \infty$$

where $(\alpha_j,\ j \geq 1)$ is a sequence of real coefficients.

Set $X = \sum_{j=1}^{\infty} X_j v_j$ and $Y = \sum_{j=1}^{\infty} \alpha_j X_j v_j$ where $(v_j,\ j \geq 1)$ is an orthonormal system in H. Then

$$C_X(v_j) = (\mathrm{E}\, X_j^2) v_j \quad \text{and} \quad C_{X,Y}(v_j) = C_{Y,X}(v_j) = \alpha_j (\mathrm{E}\, X_j^2) v_j.$$

It is easy to see that (10.10) holds if and only if $\sup_{j \geq 1} |\alpha_j| < \infty$. In that case $\alpha = \sup_{j \geq 1} |\alpha_j|$. ◇

Note that $C_{Y,X} \prec C_X^{1/2}$ is satisfied (cf. Baker 1973). We now extend Proposition 10.1 and Theorem 10.1 to more general LCS.

Corollary 10.1
Let (Z_n) be a sequence of strongly orthogonal H-random variables and $Y \in L_H^2$, then

(a) *there exists a sequence (λ_n) of measurable mappings from H to H such that*

$$\Pi^{(Z_n)}(Y) = \sum_{n \geq 1} \lambda_n(Z_n) \tag{10.12}$$

where $\Pi^{(Z_n)}$ denotes the orthogonal projector of the LCS $\mathcal{G}_{(Z_n)}$ generated by (Z_n). The series converges in L_H^2.

(b) *Each λ_n satisfies*

$$\lambda_n(\mathrm{E} Z_n) = \mathrm{E}\left[\lambda_n(Z_n)\right] \ (\textit{even if } \mathrm{E}\, Z_n \neq 0),$$

$$C_{Z_n,Y} = C_{Z_n,\lambda_n(Z_n)} = \lambda C_{Z_n},$$

$$C_{\lambda_n(Z_n)}(y) = (\lambda_n C_{Z_n} \lambda_n^*)(y),$$

provided $\lambda_n^*(y) \in H$.

(c) The following conditions are equivalent:

$$C_{Z_n,Y} \prec C_{Z_n},\ n \geq 1$$
$$\Pi^{(Z_n)}(Y) = \sum_{n\geq 1} \ell_n(Z_n)$$

where $\ell_n \in \mathcal{L}$, $n \geq 1$.

PROOF:
It is easy to verify that

$$\mathcal{G}_{(Z_n,\ n\geq 1)} = \mathcal{G}_{Z_1} \oplus \mathcal{G}_{Z_2} \cdots \oplus \mathcal{G}_{Z_n} \oplus \cdots,$$

consequently $\Pi^{(Z_n,n\geq 1)} = \sum_{n\geq 1} \Pi^{Z_n}$. Then it suffices to apply Proposition 10.1 and Theorem 10.1 to obtain the claimed results. ■

If (Z_n) is not strongly orthogonal, more intricate results may be obtained but we do not need them in the following.

10.3 Wold decomposition and linear processes in Hilbert spaces

An H-valued (zero-mean) process $X = (X_n,\ n \in \mathbb{Z})$ is said to be *weakly stationary* (WSH) if $\mathrm{E} \parallel X_n \parallel^2 < \infty$, $n \in \mathbb{Z}$ and

$$C_{X_{n+h},X_{m+h}} = C_{X_n,X_m},\ h,\ n,\ m \in \mathbb{Z}.$$

Let X be weakly stationary and let $\mathcal{M}_n$ be the LCS generated by $(X_s,\ s \leq n)$. Set

$$\varepsilon_n = X_n - \Pi^{\mathcal{M}_{n-1}}(X_n),\ n \in \mathbb{Z} \tag{10.13}$$

where $\Pi^{\mathcal{M}_{n-1}}$ is the orthogonal projector of $\mathcal{M}_{n-1}$. Then X is said to be *regular* if

$$\sigma^2 := \mathrm{E} \parallel \varepsilon_n \parallel^2 > 0.$$

In that case (ε_n) is an *H-white noise* (i.e. $C_{\varepsilon_n} = C_{\varepsilon_0} \neq 0$, $\mathrm{E}\varepsilon_n = 0$, $\varepsilon_n \perp \varepsilon_m$, $n \neq m$; $n, m \in \mathbb{Z}$). Moreover $\varepsilon_n \in \mathcal{M}_n$ and $\varepsilon_n \perp \mathcal{M}_{n-1}$: (ε_n) is the *innovation* process of X.

Now, if $\mathcal{J}_n$ is the LCS generated by $(\varepsilon_s,\ s \leq n)$, the *Wold decomposition* of X is defined by

$$\begin{aligned} X_n &= \Pi^{\mathcal{J}_n}(X_n) + \Pi^{\mathcal{J}_n^\perp}(X_n) \\ &=: Y_n + Z_n,\ n \in \mathbb{Z}. \end{aligned}$$

Properties of this decomposition are similar to those in the real case (see, for example Brockwell and Davis 1991). In particular one has

$$Z_n \perp \varepsilon_s;\ n, s \in \mathbb{Z} \quad \text{and} \quad Z_n \in \bigcap_{j=0}^{\infty} \mathcal{M}_{n-j},\ n \in \mathbb{Z}.$$

We are know in a position to define *linear process in H* (LPH).

Definition 10.1
A WSH regular process X is an LPH if one of the three following equivalent conditions is satisfied:

$$X_n = \Pi^{\mathcal{J}_n}(X_n),\ n \in \mathbb{Z} \tag{10.14}$$

$$\mathcal{M}_n = \mathcal{J}_n,\ n \in \mathbb{Z} \tag{10.15}$$

$$\Pi^{\mathcal{M}_{n-1}}(X_n) = \Pi^{\mathcal{J}_{n-1}}(X_n),\ n \in \mathbb{Z}. \tag{10.16}$$

Let us show that these conditions are equivalent. First, since $\varepsilon_n \perp \mathcal{M}_{n-1}$, we have

$$\Pi^{\varepsilon_n}(X_n) = \varepsilon_n + \Pi^{\varepsilon_n}\Pi^{\mathcal{M}_{n-1}}(X_n) = \varepsilon_n,$$

thus Corollary 10.1 gives

$$\begin{aligned} \Pi^{\mathcal{J}_n}(X_n) &= \varepsilon_n + \sum_{j=1}^{\infty} \Pi^{\varepsilon_{n-j}}(X_n) \\ &= \varepsilon_n + \Pi^{\mathcal{J}_{n-1}}(X_n). \end{aligned}$$

Now, if (10.14) holds, it follows that

$$X_n = \varepsilon_n + \Pi^{\mathcal{J}_{n-1}}(X_n).$$

Comparing with (10.13) one obtains (10.16).

If (10.16) takes place we may write

$$\Pi^{\mathcal{J}_{n-1}}(X_n) = \Pi^{\mathcal{M}_{n-1}}(X_n) = X_n - \varepsilon_n,$$

thus

$$X_n = \varepsilon_n + \Pi^{\mathcal{J}_{n-1}}(X_n) = \Pi^{\mathcal{J}_n}(X_n)$$

which is (10.14).

Obviously (10.15) yields (10.16). Now, assume (10.16); since (10.13) entails

$$\varepsilon_s = X_s - \Pi^{\mathcal{M}_{n-1}}(X_s), \ s \leq n-1$$

we have $\mathcal{J}_{n-1} \subset \mathcal{M}_{n-1}$.

On the other hand, since (10.16) implies (10.14) one has

$$X_s = \Pi^{\mathcal{J}_s}(X_s) \in \mathcal{J}_{n-1}, \ s \leq n-1,$$

then $\mathcal{M}_{n-1} \subset \mathcal{J}_{n-1}$ and, finally, (10.15) holds. ■

By using (10.12) one may write (10.14) under the form

$$X_n = \varepsilon_n + \sum_{j=1}^{\infty} \lambda_{j,n}(\varepsilon_{n-j}), \ n \in \mathbb{Z}. \tag{10.17}$$

Now, if (ε_n) is a *strong H-white noise* (i.e. the ε_n's are i.i.d.) it is clearly possible to choose versions of the $\lambda_{j,n}$'s that do not depend on n; then

$$X_n = \varepsilon_n + \sum_{j=1}^{\infty} \lambda_j(\varepsilon_{n-j}), \ n \in \mathbb{Z}.$$

In the general case, one has

$$X_n = \varepsilon_n + \sum_{j=1}^{\infty} \Pi^{\varepsilon_{n-j}}(X_n), \ n \in \mathbb{Z}$$

and $\Pi^{\varepsilon_{n-j}}(X_n)$ only depends on C_{ε_0}, C_{ε_0, X_j} and X_n.

On the other hand, if $C_{\varepsilon_{n-j}, X_n} \prec C_{\varepsilon_{n-j}}$, $j \geq 1$, then Corollary 10.1(c) and stationarity of X yield

$$X_n = \varepsilon_n + \sum_{j=1}^{\infty} \ell_j(\varepsilon_{n-j}), \ n \in \mathbb{Z}$$

where $\ell_j \in \mathcal{L}_j$, $j \geq 1$, which means that X is an *H-linear process in the 'classical sense'* (see Bosq 2000, and references therein).

Finally, if no prediction property is involved, we will say that X is a linear process in the *wide sense* (LPHWS) if there exists an H-white noise (ε_n) such that (10.17) holds.

We now give a simple example of an H-linear process.

Example 10.2
Consider a sequence of real linear processes defined as

$$X_{n,k} = \varepsilon_{n,k} + \sum_{j=1}^{\infty} a_{j,k}\varepsilon_{n-j,k}, \; n \in \mathbb{Z}, \; k \geq 1$$

where $(\varepsilon_{n,k}, \; n \in \mathbb{Z})$ is a white noise, $\mathrm{E}\,\varepsilon_{n,k}^2 = \sigma_k^2$ and $A_k^2 = 1 + \sum_{j=1}^{\infty} a_{j,k}^2 < \infty$, $k \geq 1$. Suppose that $(\varepsilon_{n,k}, \; n \in \mathbb{Z}, \; k \geq 1)$ is a family of independent random variables and

$$\sum_{k=1}^{\infty} \sigma_k^2 A_k^2 < \infty. \tag{10.18}$$

Now set

$$X_n = (X_{n,k}, \; k \geq 1), \; \varepsilon_n = (\varepsilon_{n,k}, \; k \geq 1), \; n \in \mathbb{Z}.$$

From (10.18) it follows that

$$\sum_{k=1}^{\infty} X_{n,k}^2 < \infty \quad \text{and} \quad \sum_{k=1}^{\infty} \varepsilon_{n,k}^2 < \infty \text{ (a.s.)}.$$

So, (X_n) and (ε_n) are sequences of ℓ^2-valued random variables where, as usual, $\ell^2 = \{(x_k), \; \sum_{k=1}^{\infty} x_k^2 < \infty\}$.

Now we have

$$X_n = \varepsilon_n + \sum_{j=1}^{\infty} \lambda_j(\varepsilon_{n-j}), \; n \in \mathbb{Z}$$

where λ_j is defined by

$$\lambda_j((x_k)) = (a_{j,k}x_k)$$

as soon as

$$\sum_{k=1}^{\infty} a_{j,k}^2 x_k^2 < \infty,$$

thus, from (10.18), $\lambda_j(\varepsilon_{n-j})$ is well defined.

Note that λ_j is bounded if and only if $\sup_{k\geq 1} |a_{j,k}| < \infty$. Thus if, for example, $\sigma_k^2 = k^{-4}$ and $a_{j,k} = k/j$, $k \geq 1$, $j \geq 1$, (10.18) is satisfied but λ_j is unbounded for every j.

Finally, note that, if $(\varepsilon_{n,k})$ is the innovation of $(X_{n,k})$ for all k, one may verify that (ε_n) is the innovation of (X_n). ◇

10.4 Moving average processes in Hilbert spaces

We now particularize our model by considering moving averages. From a statistical point of view these processes are more tractable than general linear processes since they only depend on a finite number of parameters.

A *moving average of order q in H* (MAH(q)) is a linear process such that

$$\Pi^{\mathcal{M}_{n-1}}(X_n) = \Pi^{\{\varepsilon_{n-j},\, 1\leq j\leq q\}}(X_n), \; n \in \mathbb{Z} \tag{10.19}$$

and

$$\mathrm{E} \parallel \Pi^{\varepsilon_{n-q}}(X_n) \parallel > 0, \; n \in \mathbb{Z}. \tag{10.20}$$

More generally (X_n) is a moving average in the *wide sense* if there exists an H-white noise (ε_n) such that (10.19) holds for some q (thus (10.20) may not hold).

Clearly results and discussions in Section 10.3 remain valid here. We now give a special criterion for a process to be MAH.

Proposition 10.2
Let X be a regular WSH, then it is an MAH(q) if and only if

$$C_q \neq 0; \; C_h = 0, \; h > q. \tag{10.21}$$

PROOF:
If (X_n) is an MAH(q), (10.13) and (10.19) give

$$X_n = \varepsilon_n + \Pi^{\{\varepsilon_{n-j},\, 1\leq j\leq q\}}(X_n), \; n \in \mathbb{Z},$$

from (10.15) it follows that

$$X_n \perp \mathcal{J}_{n-h} = \mathcal{M}_{n-h}, \; h > q,$$

which implies

$$C_h = C_{X_n, X_{n-h}} = 0, \; h > q.$$

Moreover $C_q \neq 0$; if not, one has $X_n \perp X_{n-q}$, but

$$X_{n-q} = \varepsilon_{n-q} + \Pi^{\{\varepsilon_{n-q-j},\, 1 \leq j \leq q\}}(X_{n-q}),$$

and, since $X_n \perp \varepsilon_{n-q-j}$, $1 \leq j \leq q$, this implies $X_n \perp \varepsilon_{n-q}$, thus $\Pi^{\varepsilon_{n-q}}(X_n) = 0$ which contradicts (10.20).

Conversely, suppose that (10.21) holds. Since

$$\mathcal{J}_{n-1} = \mathcal{G}_{(\varepsilon_{n-1},\ldots,\varepsilon_{n-q})} \oplus \mathcal{J}_{n-q-1},$$

(10.15) yields

$$\mathcal{M}_{n-1} = \mathcal{G}_{(\varepsilon_{n-1},\ldots,\varepsilon_{n-q})} \oplus \mathcal{M}_{n-q-1},$$

therefore

$$\Pi^{\mathcal{M}_{n-1}}(X_n) = \Pi^{\{\varepsilon_{n-j},\, 1 \leq j \leq q\}}(X_n) + \Pi^{\mathcal{M}_{n-q-1}}(X_n),$$

and since $C_h = 0$, $h > q$ entails $X_n \perp X_{n-h}$, $h > q$, we have $\Pi^{\mathcal{M}_{n-q-1}}(X_n) = 0$, hence (10.19).

On the other hand

$$\begin{aligned} C_{\varepsilon_{n-q},X_n} &= C_{X_{n-q},X_n} + C_{\Pi^{\mathcal{M}_{n-q-1}}(X_{n-q}),X_n} \\ &= C_q \neq 0 \end{aligned}$$

and $C_{\varepsilon_{n-q},X_n} = C_{\varepsilon_{n-q},\Pi^{\varepsilon_{n-q}}(X_n)}$, thus (10.20) holds and the proof is now complete. ■

The next statement connects MAH with real moving averages.

Proposition 10.3
Let (X_n) be an MAH(q) defined by

$$X_n = \varepsilon_n + \sum_{j=1}^{q} \lambda_j(\varepsilon_{n-j}), \ n \in \mathbb{Z}.$$

Suppose that there exists $v \in H$ such that

$$\lambda_j^*(v) = \alpha_j v; \ (\alpha_q \neq 0), \ 1 \leq j \leq q$$

then $(\langle X_n, v\rangle, \ n \in \mathbb{Z})$ is an MA(q) in the wide sense provided $\mathrm{E}(\langle \varepsilon_n, v\rangle^2) \neq 0$.

PROOF:
We have the relation

$$\langle X_n, v\rangle = \langle \varepsilon_n, v\rangle + \sum_{j=1}^{q} \alpha_j \langle \varepsilon_{n-j}, v\rangle, \ n \in \mathbb{Z},$$

and $(\langle \varepsilon_n, v\rangle,\ n \in \mathbb{Z})$ is clearly a real white noise. ■

Note that, if $1 + \sum_{j=1}^{q} \alpha_j z^j \neq 0$ for $|z| \leq 1$, $(\langle X_n, v\rangle,\ n \in \mathbb{Z})$ becomes an MA(q) in the strict sense (see Brockwell and Davis 1991).

We now point out two examples of MAH.

Example 10.3
Consider the Hilbert space $H = L^2(\mathbb{R}_+, \mathcal{B}_{\mathbb{R}_+}, \mu)$ where $\mathcal{B}_{\mathbb{R}_+}$ is the Borel σ-field of $\mathbb{R}_+$ and μ is the standard exponential probability measure with density e^{-t}, $t \geq 0$.

Let $\mathcal{C}$ be the subspace of elements of H that admit a version φ with a uniformly continuous derivative over $\mathbb{R}_+$ (here $\varphi'(0) = \lim_{h\to 0(+)}(\varphi(h) - \varphi(0))/h$. Then, the mapping $T\ :\ \varphi \mapsto \varphi'$ is not continuous on $\mathcal{C}$. However the family of mappings defined as

$$T_h(\varphi)(\cdot) = \frac{\varphi(\cdot + h) - \varphi(\cdot)}{h}, \ \varphi \in H, \ h > 0$$

is a well defined subfamily of $\mathcal{L}$ such that

$$\mathrm{E} \parallel T_h(\varepsilon_n) - T(\varepsilon_n) \parallel^2 \underset{h\to 0}{\longrightarrow} 0, \ n \in \mathbb{Z} \tag{10.22}$$

where (ε_n) is any H-white noise with sample paths in $\mathcal{C}$. Now set

$$X_n(t) = \varepsilon_n(t) + c\,\varepsilon'_{n-1}(t), \ t \geq 0, \ n \in \mathbb{Z}, \ (c \in \mathbb{R}),$$

then (X_n) is an MAH(1) in the wide sense since

$$X_n = \varepsilon_n + \lambda(\varepsilon_{n-1}), \ n \in \mathbb{Z}$$

where $\lambda\ :\ \varphi \mapsto c\,\varphi'$ is unbounded, but (10.22) implies $\lambda(\varepsilon_{n-1}) \in \mathcal{G}_{\varepsilon_{n-1}}$, thus $\lambda(\varepsilon_{n-1}) = \Pi^{\varepsilon_{n-1}}(X_n)$. ◇

Example 10.4 *(Truncated Ornstein–Uhlenbeck process)*
Consider the real continuous time process

$$\xi_t = \int_{\{t\}-q}^{t} \mathrm{e}^{-\theta(t-s)} \mathrm{d}W(s), \ t \in \mathbb{R} \ (\theta > 0)$$

where W is a bilateral Wiener process, $q \geq 1$, an integer, and $\{t\}$ the smallest integer $\geq t-1$. $(\xi_t,\ t \in \mathbb{R})$ is a fixed continuous version of the stochastic integral.

Let us set

$$X_n(t) = \xi_{n+t},\ 0 \leq t \leq 1;\ n \in \mathbb{Z},$$

then $(X_n(t),\ 0 < t < 1)$ has continuous sample paths.

Now let $(e_j,\ j \geq 0)$ be an orthonormal basis of $H = L^2(\mu + \delta_{(1)})$ where μ is Lebesgue measure on $[0, 1]$. For convenience we take $e_0 = \mathbb{1}_{\{1\}}$ and $(e_j,\ j \geq 1)$ a version of an orthonormal basis of $L^2(\mu)$ satisfying $e_j(1) = 0,\ j \geq 1$.

Then one may identify X_n with an $L^2(\mu + \delta_{(1)})$-valued random variable by putting

$$X_n(\cdot) = X_n(1)\mathbb{1}_{\{1\}}(\cdot) + \sum_{j \geq 1} \left[\int_0^1 X_n(s) e_j(s)\,\mathrm{d}s \right] e_j(\cdot).$$

We now introduce the operator ℓ_θ defined by

$$\ell_\theta(x)(t) = \mathrm{e}^{-\theta t} x(1),\ x \in H.$$

We have

$$\| \ell_\theta^j \|_{\mathcal{L}} = \left(\frac{1 - \mathrm{e}^{-2\theta} + 2\theta \mathrm{e}^{-2\theta}}{2\theta} \right)^{1/2} \mathrm{e}^{-\theta(j-1)},\ j \geq 1,$$

and (X_n) has decomposition

$$X_n = \sum_{j=0}^{q} \ell_\theta^j(\varepsilon_{n-j}),\ n \in \mathbb{Z}, \tag{10.23}$$

where

$$\varepsilon_n(t) = \int_n^{n+t} \mathrm{e}^{-\theta(n+t-s)} \mathrm{d}W(s),\ 0 \leq t \leq 1.$$

Thus, (X_n) is MAH(q) and one can show that (ε_n) is the innovation of (X_n).

Note that, since

$$\ell_\theta^*(\mathbb{1}_{\{1\}}) = \mathrm{e}^{-\theta} \mathbb{1}_{\{1\}},$$

it follows that $X_n(1) = \langle X_n, \mathbb{1}_{\{1\}} \rangle$, $n \in \mathbb{Z}$ is a real MA(q) such that

$$X_n(1) = \sum_{j=0}^{q} \mathrm{e}^{-\theta j} \varepsilon_{n-j}(1),\ n \in \mathbb{Z},$$

(Proposition 10.3).

If $q = 1$, (10.23) takes the form

$$X_n = \varepsilon_n + \ell_\theta(\varepsilon_{n-1}), \ n \in \mathbb{Z},$$

and, despite the fact that we may have $\| \ell_\theta \| \geq 1$, it follows that

$$\varepsilon_n = \sum_{j=0}^{\infty} (-1)^j \ell_\theta^j (X_{n-j}), \ n \in \mathbb{Z}$$

where convergence takes place in L_H^2. ◇

10.5 Autoregressive processes in Hilbert spaces

Let $X = (X_n, \, n \in \mathbb{Z})$ be a regular WSH process, set

$$\mathcal{M}_n^k = \mathcal{G}_{(X_n,\ldots,X_{n-k})}, \ n \in \mathbb{Z}, \ k \geq 0$$

where, as above, $\mathcal{G}_{(X_n,\ldots,X_{n-k})}$ is the LCS generated by $X_n, \ldots, X_{n-k}$.

Then X is said to be an *autoregressive Hilbertian process of order* p (ARH(p)) if

$$\Pi^{\mathcal{M}_{n-1}}(X_n) = \Pi^{\mathcal{M}_{n-1}^p}(X_n)$$

and, if $p > 1$,

$$\mathrm{E}\left\| \Pi^{\mathcal{M}_{n-1}^p}(X_n) - \Pi^{\mathcal{M}_{n-1}^{p-1}}(X_n) \right\| > 0.$$

From now on we suppose that $p = 1$. Actually, it is easy to verify that, if X is ARH(p), then $((X_n, \ldots, X_{n-p+1}), \ n \in \mathbb{Z})$ is ARH^p(1), where $H^p = H \otimes \cdots \otimes H$.

Applying Proposition 10.1, one may characterize an ARH(1) by a relation of the form

$$X_n = \lambda_n(X_{n-1}) + \varepsilon_n, \ n \in \mathbb{Z}.$$

If X is strictly stationary, it is possible to choose $\lambda_n = \lambda$ not depending on n. Now, if

$$C_{X_{n-1},X_n} \prec C_{X_{n-1}},$$

Theorem 10.1 yields existence of an element in $\mathcal{L}$, say ρ, such that

$$X_n = \rho(X_{n-1}) + \varepsilon_n, \ n \in \mathbb{Z}. \tag{10.24}$$

We will say that ρ is the *autocorrelation operator* of X, even if different definitions may be considered (cf. Baker 1973). Note that ρ is unique over

$$S = \overline{\text{sp}} \bigcup_{n \in \mathbb{Z}} \{S_{X_n} \cup S_{\varepsilon_n}\}$$

where S_Z denotes support of P_Z.

Conversely, given a white noise (ε_n) and $\rho \in \mathcal{L}$, one wants to show existence of (X_n) satisfying (10.24). For this purpose we introduce the following conditions

$\boldsymbol{c}_0$: *There exists an integer* $j_0 \geq 1$ *such that* $\| \rho^{j_0} \|_{\mathcal{L}} < 1$,

and

$\boldsymbol{c}_1$: *There exist* $a > 0, 0 < b < 1$ *such that* $\| \rho^{j_0} \|_{\mathcal{L}} \leq ab^j, \quad j \geq 0$.

The next lemma is simple but somewhat surprising.

Lemma 10.2
(c_0) *and* (c_1) *are equivalent.*

PROOF:
Obviously (c_1) entails (c_0). If (c_0) holds, in order to establish (c_1) we may and do suppose that $j > j_0$ and $0 < \| \rho^{j_0} \|_{\mathcal{L}} < 1$. Dividing such a j by j_0, one obtains

$$j = j_0 q + r$$

where $q \geq 1$ and $r \in [0, j_0[$ are integers.

Now, since

$$\| \rho^j \|_{\mathcal{L}} \leq \| \rho^{j_0} \|_{\mathcal{L}}^q \|\rho^r \|_{\mathcal{L}},$$

$q > jj_0^{-1} - 1$ and $0 < \| \rho^{j_0} \|_{\mathcal{L}} < 1$, it follows that

$$\| \ell^j \|_{\mathcal{L}} \leq ab^j, \ j > j_0$$

where $a = \| \ell^{j_0} \|_{\mathcal{L}}^{-1} \max_{0 \leq r < j_0} \| \ell^r \|_{\mathcal{L}}$ and $b = \| \ell^{j_0} \|_{\mathcal{L}}^{1/j_0} < 1$. ■

This elementary lemma shows that the 'natural' condition $\sum_{j=0}^{\infty} \| \rho^j \|_{\mathcal{L}} < \infty$ for constructing X_n (see Proposition 10.4 below) is satisfied as soon as $\| \rho^{j_0} \|_{\mathcal{L}} < 1$ for some $j_0 \geq 1$. Finally, observe that (c_0) (or (c_1)) does not imply $\| \rho \|_{\mathcal{L}} < 1$, contrary to the one-dimensional case (see Example 10.5).

The next statement gives existence and uniqueness of (X_n).

Proposition 10.4
If (c_0) *holds,* (10.24) *has a unique stationary solution given by*

$$X_n = \sum_{j=0}^{\infty} \rho^j(\varepsilon_{n-j}), \; n \in \mathbb{Z} \tag{10.25}$$

where the series converges almost surely and in L^2_H*. Moreover* (ε_n) *is the innovation of* (X_n).

PROOF:
Put $\sigma^2 = \mathrm{E} \parallel \varepsilon_n \parallel^2 > 0$, $n \in \mathbb{Z}$; we have

$$\sum_{j=0}^{\infty} \parallel \rho^j(\varepsilon_{n-j}) \parallel_{L^2_H} \leq \sigma \sum_{j=0}^{\infty} \parallel \rho^j \parallel_{\mathcal{L}}$$

and Lemma 10.2 gives convergence in L^2_H.

Now, since

$$\mathrm{E}\left(\sum_{j=0}^{\infty} \parallel \rho^j \parallel_{\mathcal{L}} \parallel \varepsilon_{n-j} \parallel \right) < \infty$$

it follows that

$$\sum_{j=0}^{\infty} \parallel \rho^j \parallel_{\mathcal{L}} \parallel \varepsilon_{n-j} \parallel < \infty \text{ a.s.}$$

hence the series in (10.25) converges almost surely.

Finally (X_n) is clearly a stationary solution of (10.24).

Conversely, if (Y_n) denotes a stationary solution of (10.24) a straightforward induction gives

$$Y_n = \sum_{j=0}^{k} \rho^j(\varepsilon_{n-j}) + \rho^{k+1}(Y_{n-k-1}), \; k \geq 1,$$

therefore

$$\mathrm{E}\left\| Y_n - \sum_{j=0}^{k} \rho^j(\varepsilon_{n-j}) \right\|^2 \leq \parallel \rho^{k+1} \parallel_{\mathcal{L}}^2 \mathrm{E} \parallel Y_{n-k-1} \parallel^2 .$$

By stationarity $\mathrm{E}\| Y_{n-k-1} \|^2$ remains constant and Lemma 10.2 yields $\| \rho^{k+1} \|_{\mathcal{L}}^2 \to 0$ as $k \to \infty$. Consequently

$$Y_n = \sum_{j=0}^{\infty} \rho^j(\varepsilon_{n-j}), \; n \in \mathbb{Z} \; (a.s.)$$

which proves uniqueness (a.s.).

Now, concerning (ε_n), we have

$$\varepsilon_n = X_n - \rho(X_{n-1}) \in \mathcal{M}_n$$

and, since $\varepsilon_n \perp \rho^j(\varepsilon_{m-j})$, $j \geq 0$, $m < n$, then $\varepsilon_n \perp X_m = \sum_{j=0}^{\infty} \rho^j(\varepsilon_{m-j})$, $m < n$, which implies $\varepsilon_n \perp \mathcal{M}_m$, $m < n$, thus (ε_n) is the innovation of (X_n). ■

A simple example of ARH(1) process is the following.

Example 10.5 *(Ornstein–Uhlenbeck process)*
Let $\xi = (\xi_t, \; t \in \mathbb{R})$ be a measurable version of the Ornstein–Uhlenbeck process:

$$\xi_t = \int_{-\infty}^{t} \mathrm{e}^{-\theta(t-s)} \mathrm{d}W(s), \; t \in \mathbb{R}, \; (\theta > 0),$$

where W is a standard bilateral Wiener process. Setting

$$X_n(t) = \xi_{n+t}, \; 0 \leq t \leq 1, \; n \in \mathbb{Z}$$

one defines a sequence of random variables with values in the Hilbert space $H = L^2([0, 1], \mathcal{B}_{[0,1]}, \mu + \delta_{(1)})$ where μ is Lebesgue measure on [0,1] (Example 10.4).

Now define $\rho_\theta : H \to H$ by putting

$$\rho_\theta(x)(t) = \mathrm{e}^{-\theta t} x(1), \; t \in [0, 1], \; x \in H,$$

and define the H-white noise

$$\varepsilon_n(t) = \int_{n}^{n+t} \mathrm{e}^{-\theta(n+t-s)} \mathrm{d}W(s), \; t \in [0, 1[, \; n \in \mathbb{Z}$$

and $\varepsilon_n(1) = X_n(1) - \mathrm{e}^{-\theta} X_{n-1}(1)$, $n \in \mathbb{Z}$. Then we have

$$\begin{aligned} X_n(t) &= \int_{-\infty}^{n+t} \mathrm{e}^{-\theta(n+t-s)} \mathrm{d}W(s) \\ &= \mathrm{e}^{-\theta t} \int_{-\infty}^{(n-1)+1} \mathrm{e}^{-\theta((n-1)+1-s)} \mathrm{d}W(s) + \int_{n}^{n+t} \mathrm{e}^{-\theta(n+t-s)} \mathrm{d}W(s), \end{aligned}$$

$0 \leq t \leq 1$, consequently $X_n = \rho_\theta(X_{n-1}) + \varepsilon_n$, $n \in \mathbb{Z}$.

Now

$$\| \rho_\theta \|_{\mathcal{L}}^2 = \int_0^1 \mathrm{e}^{-2\theta t} \mathrm{d}(\mu + \delta_{(1)})(t) = \frac{1 - \mathrm{e}^{-2\theta}}{2\theta} + \mathrm{e}^{-2\theta} =: \nu(\theta),$$

and, more generally,

$$\| \rho_\theta^j \|_{\mathcal{L}}^2 = \mathrm{e}^{-2\theta(j-1)} \nu(\theta), \ j \geq 1;$$

then condition (c_0) holds and (X_n) is ARH(1) with innovation (ε_n). Note that, if $0 < \theta \leq 1/2$, one has $\| \rho_\theta \|_{\mathcal{L}} \geq 1$, hence $j_0 > 1$.

Also note that this autoregressive process is somewhat degenerate since the range of ρ_θ is one-dimensional. This suggests that the ARH(1) model has a good amount of generality.

We will see later that ξ can also be associated with an AR process in some suitable Banach space. ◇

In some special cases, (c_0) is not necessary for obtaining (10.25). We give an example.

Example 10.6

Let ρ be defined as

$$\rho(x) = \alpha \langle x, e_1 \rangle e_1 + \beta \langle x, e_3 \rangle e_2, \ x \in H$$

where $\{e_1, e_2, e_3\}$ is an orthonormal system in H and $0 < \beta < 1 \leq \alpha$.

Here we have

$$\sum_{j=0}^{\infty} \| \rho^j \|_{\mathcal{L}}^2 = \sum_{j=0}^{\infty} \alpha^{2j} = \infty.$$

However, if (ε_n) is a strong white noise such that

$$P(\langle \varepsilon_n, e_1 \rangle = 0) = 1, \ n \in \mathbb{Z},$$

one obtains $\rho(\varepsilon_n) = \beta \langle \varepsilon_n, e_3 \rangle e_2$ (a.s.) and $\rho^j(\varepsilon_n) = 0$, $j \geq 2$; thus (10.24) has a stationary solution given by $X_n = \varepsilon_n + \rho(\varepsilon_{n-1})$, $n \in \mathbb{Z}$ and (ε_n) is the innovation of (X_n).

Here (X_n) is at the same time ARH(1) and MAH(1). This kind of phenomenon cannot occur in a univariate context.

Clearly the above example can be extended by taking ρ *nilpotent* (i.e. there exists $j \geq 2$ such that $\rho^j = 0$). ◇

An ARH(1) that satisfies (c_0) will be said to be *standard*. The next result is a characterization of standard ARH(1).

Proposition 10.5
Let X be a zero-mean, not degenerate WSH such that $C_h = \rho^h C_0$, $h \geq 0$ where $\rho \in \mathcal{L}$ and $\| \rho^{j_0} \|_{\mathcal{L}} < 1$ for some $j_0 \geq 1$.

Then X is a standard ARH(1) with autocorrelation operator ρ.

PROOF:
Set $\varepsilon_n = X_n - \rho(X_{n-1})$, $n \in \mathbb{Z}$, then, if $h \geq 1$ and $x \in H$, we have

$$\begin{aligned} \mathrm{E}\left(\langle \varepsilon_n, x\rangle \varepsilon_{n+h}\right) &= \mathrm{E}\left(\langle X_n - \rho(X_{n-1}), x\rangle (X_{n+h} - \rho(X_{n+h-1}))\right) \\ &= (\rho^h C_0 - \rho\, \rho^{h-1} C_0 - \rho^{h+1} C_0 \rho^* + \rho\, \rho^h C_0 \rho^*)(x) = 0 \end{aligned}$$

and

$$\mathrm{E}\left(\langle \varepsilon_n, x\rangle \varepsilon_n\right) = (C_0 - \rho C_0 \rho^*)(x).$$

So, (ε_n) is a white noise, provided $C_0 \neq \rho C_0 \rho^*$. But, $C_0 = \rho C_0 \rho^*$ yields $\varepsilon_n = 0$ a.s., hence $X_n = \rho(X_{n-1})$, $n \in \mathbb{Z}$; consequently $X_n = \rho^n(X_0)$, $n \geq 0$ and $\| X_n \| \leq \| \rho^n \|_{\mathcal{L}} \| X_0 \|$ which implies $\mathrm{E} \| X_n \|^2 \to 0$ as $n \to \infty$ and, since $\mathrm{E} \| X_n \|^2$ is fixed, one obtains $X_n = 0$ a.s., a contradiction.
Finally (ε_n) is a white noise and (X_n) is an ARH(1) with innovation (ε_n) and autocorrelation operator ρ. ■

The next statement is similar to Proposition 10.3.

Proposition 10.6
Let (X_n) be a standard ARH(1). Then, if there exists $v \in H$ such that

$$\rho^*(v) = \alpha v \ (0 < |\alpha| < 1),$$

$(\langle X_n, v\rangle,\ n \in \mathbb{Z})$ is degenerate or is AR(1).

PROOF:
From (10.24) it follows that

$$\langle X_n, v\rangle = \alpha \langle X_{n-1}, v\rangle + \langle \varepsilon_n, v\rangle,\ n \in \mathbb{Z}. \tag{10.26}$$

If $\langle \varepsilon_n, v\rangle = 0$ (a.s.) we have

$$\mathrm{E}\langle X_n, v\rangle^2 = \alpha^{2n} \mathrm{E}\left(\langle X_0, v\rangle^2\right)$$

and by stationarity, $\langle X_n, v\rangle = 0$, $n \in \mathbb{Z}$ (a.s.).

If not, $(\langle \varepsilon_n, v \rangle)$ is a real white noise and (10.26) shows that $(\langle X_n, v \rangle)$ is AR(1). ∎

Applying Proposition 10.6 to Example 10.5 one obtains that $(X_n(1))$ is an AR(1) since

$$X_n(1) = \langle X_n, \mathbb{1}_{\{1\}} \rangle \text{ and } \rho_\theta^*(\mathbb{1}_{\{1\}}) = \mathrm{e}^{-\theta}\mathbb{1}_{\{1\}}.$$

Thus

$$X_n(1) = \mathrm{e}^{-\theta} X_{n-1}(1) + \varepsilon_n(1), \; n \in \mathbb{Z}.$$

We now specify the Markovian character of ARH(1). We first introduce Markov processes by using LCS.

Definition 10.2
Let (X_n) be a second order H-process. It is said to be Markovian in the wide sense *(MWS) if*

$$\Pi^{\{X_{r_1},\ldots,X_{r_k},X_s\}}(X_t) = \Pi^{X_s}(X_t), \tag{10.27}$$

$k \geq 1, \; r_1 < \cdots < r_k < s < t.$
In order to extend a well known characterization theorem (see Rao 1984), we set

$$\lambda_{s,t} = \lambda_{X_t}^{X_s}, \; s < t$$

where $\lambda_{X_t}^{X_s}$ is a measurable mapping λ such that

$$\lambda(X_s) = \Pi^{X_s}(X_t).$$

Proposition 10.7

(1) (X_n) is MWS if and only if

$$\lambda_{r,t} = \lambda_{s,t}\lambda_{r,s} \quad \textit{over } C_{X_r}(H), \; r < s < t.$$

(2) Let (X_n) be a strictly stationary second order H-process. Then, it is MWS if and only if

$$\lambda_k = \lambda_1^k \textit{ over } C_{X_0}(H), \; k \geq 1$$

where $\lambda_k = \lambda_{s,s+k}, \; s \in \mathbb{Z}, \; k \geq 1.$

For a proof we refer to Bosq (2007).

The following statement connects MWS and ARH.

Proposition 10.8

(a) A regular strictly stationary MWS is ARH(1).

(b) A strictly stationary ARH(1) is MWS.

(c) A standard ARH(1) is MWS.

PROOF:

(a) Let (X_n) be a regular strictly stationary MWS. Using Fortet (1995, p. 67) and (10.27) it is easy to verify that

$$\Pi^{\mathcal{M}_{n-1}}(X_n) = \Pi^{X_{n-1}}(X_n)$$

which means that (X_n) is ARH(1).

(b) If (X_n) is a strictly stationary ARH(1), and $r_1 < \cdots < r_k < t-1$, we have

$$\mathcal{G}_{X_{t-1}} \subset \mathcal{G}_{(X_{t-1}, X_{r_k}, \ldots, X_{r_1})} \subset \mathcal{M}_{t-1}$$

and $\Pi^{X_{t-1}}(X_t) = \Pi^{\mathcal{M}_{t-1}}(X_t)$; it follows that $\Pi^{X_{t-1}}(X_t) = \Pi^{(X_{t-1}, X_{r_k}, \ldots, X_{r_1})}(X_t)$ which is (10.27) for $s = t-1$.

In order to establish a similar result for $s < t-1$ we first note that, if $\mathcal{G}$ is an LCS and $\ell \in \mathcal{L}$ then

$$\Pi^{\mathcal{G}} \ell(Y) = \ell \Pi^{\mathcal{G}}(Y), \ \ Y \in L^2_H$$

since $Y - \Pi^{\mathcal{G}} Y \perp \mathcal{G}$ implies $\ell(Y) - \ell \Pi^{\mathcal{G}}(Y) \perp \mathcal{G}$, thus $\ell \Pi^{\mathcal{G}}(Y)$ is the orthogonal projection of $\ell(Y)$ on $\mathcal{G}$. Now we take $s = t-2$. By strict stationarity one may choose λ and $(\ell_k) \subset \mathcal{G}$ such that

$$\Pi^{X_{n-1}}(X_n) = \lambda(X_{n-1}), \ \ n \in \mathbb{Z},$$

and

$$\ell_k(X_{n-1}) \underset{L^2_H}{\overset{\text{a.s.}}{\rightarrow}} \lambda(X_{n-1}), \ \ n \in \mathbb{Z}. \tag{10.28}$$

From (10.28) we infer that

$$\Pi^{\mathcal{M}_{t-2}} \ell_k(X_{t-1}) \underset{L^2_H}{\overset{\text{a.s.}}{\rightarrow}} \Pi^{\mathcal{M}_{t-2}} \lambda(X_{t-1}), \tag{10.29}$$

which means that $\Pi^{\mathcal{M}_{t-2}}(X_{t-1}) \in \mathcal{V}$ (a.s.) (see Proposition 10.1(b)), therefore

$$\ell_k[\Pi^{\mathcal{M}_{t-2}}] \underset{\text{a.s.}}{\rightarrow} \lambda[\Pi^{\mathcal{M}_{t-2}}(X_{t-1})]. \tag{10.30}$$

Comparing (10.29) and (10.30) we get

$$\Pi^{\mathcal{M}_{t-2}}[\lambda(X_{t-1})] = \lambda[\Pi^{\mathcal{M}_{t-2}}(X_{t-1})].$$

Now, since (X_n) is ARH(1),

$$\Pi^{\mathcal{M}_{t-2}}(X_t) = \Pi^{\mathcal{M}_{t-2}}(\lambda(X_{t-1}) + \varepsilon_t) = \Pi^{\mathcal{M}_{t-2}}[\lambda(X_{t-1})];$$

on the other hand

$$\Pi^{\mathcal{M}_{t-2}}(X_{t-1}) = \Pi^{X_{t-2}}(X_{t-1}) = \lambda(X_{t-2}),$$

then

$$\Pi^{\mathcal{M}_{t-2}}(X_t) = \lambda^2(X_{t-2}),$$

and the same reasoning as above shows that

$$\Pi^{\{X_{t-2}, X_{r_k}, \ldots, X_{r_1}\}}(X_t) = \Pi^{X_{t-2}}(X_t),$$

$k \geq 1$, $r_1 < \cdots < r_k < t-2$.

Finally an easy induction gives (10.27) for every $s < t$.

(c) If (X_n) is standard, one has

$$X_n = \varepsilon_n + \rho(\varepsilon_{n-1}) + \cdots + \rho^{m-1}(\varepsilon_{n-m+1}) + \rho^m(X_{n-m}),\ m \geq 1.$$

It follows that

$$\Pi^{\mathcal{M}_{n-m}}(X_n) = \rho^m(X_{n-m}) = \Pi^{X_{n-m}}(X_n),$$

and

$$\Pi^{\{X_{r_1}, \ldots, X_{r_k}, X_s\}} = \rho^{t-s}(X_s),\ r_1 < \cdots < r_k < s < t. \qquad \blacksquare$$

The next proposition is useful for estimation and prediction.

Proposition 10.9
If (X_n) is a standard ARH(1), we have

$$C = \rho C \rho^* + C_{\varepsilon_0} = \sum_{j=0}^{\infty} \rho^j C_{\varepsilon_0} \rho *^j$$

*where the series converges in the space $\mathcal{N}$ of *nuclear operators* on H, and*

$$C_{X_n, X_m} = \rho^{(n-m)} C, \; m > n,$$

in particular $D = \rho C$ where $D = C_{X_0, X_1}$.

PROOF:
Straightforward and therefore omitted. ■

We now consider the special case where ρ is *symmetric* and *compact*. This means that

$$\langle \rho(x), y \rangle = \langle x, \rho(y) \rangle, \; x, y \in H$$

and that there exist (e_j), an orthonormal basis of H and $|\alpha_j| \searrow 0$ such that

$$\rho(x) = \sum_{j=1}^{\infty} \alpha_j \langle x, e_j \rangle e_j, \; x \in H,$$

or equivalently $\rho = \sum_{j=1}^{\infty} \alpha_j e_j \otimes e_j$.

For convenience we will suppose that $\alpha_j \mathrm{E}\left(\langle \varepsilon_0, e_j \rangle^2\right) \neq 0, j \geq 1$. Then we have:

Proposition 10.10
Let (ε_n) be an H-white noise and ρ be a symmetric compact operator. Then the equation

$$X_n = \rho(X_{n-1}) + \varepsilon_n, \; n \in \mathbb{Z} \tag{10.31}$$

has a stationary solution with innovation (ε_n) if and only if

$$\| \rho \|_{\mathcal{L}} < 1. \tag{10.32}$$

PROOF:
We have $\| \rho^j \|_{\mathcal{L}} = \| \rho \|_{\mathcal{L}}^j = |\alpha_1|^j, j \geq 1$ thus, (c_0) is satisfied iff (10.32) holds.

Now, if $\| \rho \|_{\mathcal{L}} < 1$, Proposition 10.4 entails existence of a unique stationary solution of (10.31) with innovation (ε_n).

Conversely, if (X_n) is a stationary solution of (10.31) with innovation (ε_n), it follows that

$$\mathrm{E}\left(\langle X_n, e_1\rangle^2\right) = \alpha_1^2 \mathrm{E}\left(\langle X_{n-1}, e_1\rangle^2\right) + \mathrm{E}(\langle \varepsilon_n, e_1\rangle^2),$$

thus

$$(1-\alpha_1^2)\mathrm{E}(\langle X_n, e_1\rangle^2) = \mathrm{E}(\langle \varepsilon_n, e_1\rangle^2) > 0,$$

therefore

$$\| \rho \|_{\mathcal{L}} = |\alpha_1| < 1.$$

■

The next statement shows that, if (X_n) is associated with a symmetric compact operator, it may be interpreted as a sequence of real autoregressive processes.

Proposition 10.11
If $\rho = \sum_1^\infty \alpha_j e_j \otimes e_j$ is symmetric compact, then (X_n) is an ARH(1) associated with $(\rho, (\varepsilon_n))$, iff $(\langle X_n, e_j\rangle,\ n \in \mathbb{Z})$ is an AR(1) associated with $(\alpha_j,\ (\langle \varepsilon_n, e_j\rangle))$ for each $j \geq 1$.

PROOF:
If (X_n) is ARH(1) the result follows from Proposition 10.6 applied to $v = e_j, j \geq 1$.

Conversely, if $\langle X_n, e_j\rangle = \alpha_j \langle X_{n-1}, e_j\rangle + \langle \varepsilon_n, e_j\rangle$, $n \in \mathbb{Z}$ where $|\alpha_j| < 1$ and $\mathrm{E}\left(\langle \varepsilon_n, e_j\rangle^2\right) > 0$, we may write

$$\begin{aligned}
\langle X_n, x\rangle &= \sum_{j=1}^{\infty} \langle X_n, e_j\rangle \langle x, e_j\rangle \\
&= \sum_{j=1}^{\infty} \alpha_j \langle X_{n-1}, e_j\rangle \langle x, e_j\rangle + \sum_{j=1}^{\infty} \langle \varepsilon_n, e_j\rangle \langle x, e_j\rangle \\
&= \langle \rho(X_{n-1}), x\rangle + \langle \varepsilon_n, x\rangle,\ x \in H,\ n \in \mathbb{Z}
\end{aligned}$$

thus $X_n = \rho(X_{n-1}) + \varepsilon_n$, $n \in \mathbb{Z}$ with $\| \rho \|_{\mathcal{L}} = |\alpha_1| < 1$, hence

$$X_n = \sum_{j=0}^{\infty} \rho^j(\varepsilon_{n-j}),$$

which shows that (ε_n) is the innovation of (X_n). ■

Example 10.7
Let $(X_{n,j},\ n \in \mathbb{Z})$ be a countable family of independent AR(1) processes defined by

$$X_{n,j} = \alpha_j X_{n-1,j} + \varepsilon_{n,j},\ n \in \mathbb{Z},\ j \geq 1,$$

where $|\alpha_j| < 1, j \geq 1, |\alpha_j| \downarrow 0$ as $j \uparrow \infty$ and

$$\mathrm{E}\,(\varepsilon_{n,j}^2) = \sigma_j^2 > 0 \quad \text{with} \quad \sum_{j=1}^{\infty} \sigma_j^2 < \infty.$$

Owing to the relations

$$\begin{aligned}\mathrm{E}\left(\sum_{j=1}^{\infty} X_{n,j}^2\right) &= \sum_{j=1}^{\infty} \mathrm{E}\left(\sum_{k=0}^{\infty} \alpha_j^k \varepsilon_{n-k,j}\right)^2 \\ &= \sum_{j=1}^{\infty} \frac{\sigma_j^2}{1-\alpha_j^2} < \frac{1}{1-|\alpha_1|^2} \sum_{j=1}^{\infty} \sigma_j^2 < \infty\end{aligned}$$

we may claim that $X_n = (X_{n,j},\ j \geq 1),\ n \in \mathbb{Z}$ defines a sequence of ℓ^2-valued random variables. Moreover, Proposition 10.11 yields

$$X_n = \rho(X_{n-1}) + \varepsilon_n,\ n \in \mathbb{Z}$$

where (ε_n) is the ℓ^2-white noise defined as $\varepsilon_n = (\varepsilon_{n,j},\ j \geq 1),\ n \in \mathbb{Z}$ and ρ the symmetric compact operator on ℓ^2 given by

$$\rho = \sum_{j=1}^{\infty} \alpha_j e_j \otimes e_j$$

where (e_j) is the standard orthonormal basis of ℓ^2:

$$e_1 = (1, 0, 0, \ldots),\ e_2 = (0, 1, 0, \ldots), \ldots$$

◇

10.6 Autoregressive processes in Banach spaces

Let B be a separable Banach space, with its norm $\|\cdot\|$ and its topological dual space B^*. The natural uniform norm on B^* is defined as

$$\| x^* \|_{B^*} = \sup_{\|x\| \leq 1} |x^*(x)|,\ x^* \in B^*.$$

We again denote by $\mathcal{L}$ the space of bounded linear operators from B to B, equipped with its classical norm $\| \cdot \|_{\mathcal{L}}$.

A *strong white noise on B* (SBWN) is a sequence $\varepsilon = (\varepsilon_n, n \in \mathbb{Z})$ of i.i.d. B-valued random variables such that $0 < \mathrm{E} \| \varepsilon_n \|^2 < \infty$ and $\mathrm{E}\, \varepsilon_n = 0$.

Example 10.8
Take $B = C[0, 1]$, consider a bilateral Wiener process W and set

$$\varepsilon_n^{(\varphi)}(t) = \int_n^{n+t} \varphi(n + t - s)\, \mathrm{d}W(s),\ \ 0 \le t \le 1,\ n \in \mathbb{Z}$$

where φ is a square integrable real function on [0,1] with $\int_0^1 \varphi^2(u)\mathrm{d}u > 0$.

Then a continuous version of $(\varepsilon_n^{(\varphi)})$ defines an SBWN. This is an easy consequence of the fact that W has stationary independent increments. ◇

We now define ARB processes.

Definition 10.3
Let $X = (X_n, n \in \mathbb{Z})$ be a strictly stationary sequence of B-random variables such that

$$X_n = \rho(X_{n-1}) + \varepsilon_n,\ \ n \in \mathbb{Z} \tag{10.33}$$

where $\rho \in \mathcal{L}$ and $\varepsilon = (\varepsilon_n)$ is an SBWN.

Then (X_n) is said to be a (strong) autoregressive process of order 1 in B (ARB(1)) associated with (ρ, ε).

The next statement deals with existence and uniqueness of X.

Proposition 10.12
If ρ satisfies (c_0), (10.33) has a unique strictly stationary solution given by

$$X_n = \sum_{j=0}^{\infty} \rho^j(\varepsilon_{n-j}),\ \ n \in \mathbb{Z}$$

where the series converges in $L_B^2(\Omega, \mathcal{A}, P)$ and with probability one.

PROOF:
Similar to the proof of Proposition 10.4 and therefore omitted. ■

The following properties of a standard ARB(1) (i.e. an ARB(1) with ρ satisfying (c_0)) are easy to derive:

(a) Define the covariance operator of X_0 by putting

$$C(x^*) = \mathrm{E}(x^*(X_0)X_0), \; x^* \in B^*,$$

and the cross-covariance operator of (X_0, X_1) by

$$D(x^*) = \mathrm{E}\,(x^*(X_0)X_1), \; x^* \in B^*,$$

then $D = \rho C$.

(b) Let $\mathcal{B}_n = \sigma(X_i, \; i \leq n) = \sigma(\varepsilon_i, \; i \leq n)$ then the conditional expectation given $\mathcal{B}_{n-1}$ is $\mathrm{E}^{\mathcal{B}_{n-1}}(X_n) = \rho(X_{n-1})$, so $\varepsilon_n = X_n - \mathrm{E}^{\mathcal{B}_{n-1}}(X_n)$, $n \in \mathbb{Z}$ and (ε_n) may be called the *innovation process* of (X_n).

(c) If there exists $v^* \in B^*$ and $\alpha \in]-1, +1[$ such that

$$\rho^*(v^*) = \alpha v^* \tag{10.34}$$

then $(v^*(X_n), \; n \in \mathbb{Z})$ is an AR(1), possibly degenerate.

Example 10.9 *(Ornstein–Uhlenbeck process)*
We consider again Example 10.5, but in a Banach space context. Taking $B = C([0, h])$ $(h > 0)$, one may set

$$X_n(t) = \xi_{nh+t}, \; 0 \leq t \leq h; \; n \in \mathbb{Z},$$

and

$$\varepsilon_n(t) = \int_{nh}^{nh+t} \mathrm{e}^{-\theta(nh+t-s)} \mathrm{d}W(s), \; t \in [0, h], \; n \in \mathbb{Z},$$

where $t \mapsto \varepsilon_n(t)$ is a continuous version of the stochastic integral.

Now, consider the operator $\rho_\theta \, : \, B \to B$ defined by

$$\rho_\theta(x)(t) = \mathrm{e}^{-\theta t} x(h), \; 0 \leq t \leq h; \; x \in C([0, h]),$$

then $\| \rho_\theta \|_{\mathcal{L}} = 1$ and, more generally

$$\| \rho_\theta^j \|_{\mathcal{L}} = \mathrm{e}^{-\theta(j-1)h}, \; j \geq 1.$$

Then, (X_n) is an ARB(1) process such that

$$X_n = \rho_\theta(X_{n-1}) + \varepsilon_n, \ n \in \mathbb{Z},$$

therefore

$$\mathrm{E}^{\mathcal{B}_{n-1}}(X_n)(t) = \mathrm{e}^{-\theta t} X_{n-1}(h) = \mathrm{e}^{-\theta t} \xi_{nh},$$

$0 \leq t \leq k$.

Observe that the adjoint ρ_θ^* of ρ_θ is given by

$$\rho_\theta^*(m)(x) = x(h) \int_0^h \mathrm{e}^{-\theta t} \mathrm{d}m(t), \ m \in B^*, \ x \in C([0,h]),$$

(recall that B^* is the space of bounded signed measure over $[0,h]$).

Then, the Dirac measure $\delta_{(h)}$ is the one and only eigenvector of ρ_θ^* and the corresponding eigenvalue is $\mathrm{e}^{-\theta h}$.

From property (c), it follows that $(X_n(h), \ n \in \mathbb{Z})$ is an AR(1) process such that

$$X_n(h) = \mathrm{e}^{-\theta h} X_{n-1}(h) + \int_{nh}^{(n+1)h} \mathrm{e}^{-\theta((n+1)h-s)} \mathrm{d}W(s), \ n \in \mathbb{Z}. \qquad \diamond$$

Example 10.10 *(CAR processes)*
Consider the following stochastic differential equation of order $k \geq 2$:

$$\sum_{j=0}^{k} a_j \mathrm{d}\xi^{(j)}(t) = \mathrm{d}W_t \tag{10.35}$$

where $a_0, \ldots, a_k$ are real ($a_k \neq 0$) and W is a bilateral Wiener process.

In (10.35), differentiation up to order $k-1$ is ordinary, whereas the order k derivative is taken in *Ito* sense.

Suppose that the roots $-r_1, \ldots, -r_k$ of the polynomial equation $\sum_{j=0}^{k} a_j r^j = 0$ are real and such that $-r_k < \cdots < -r_1 < 0$.

Then it can be established that the only stationary solution of (10.34) is the Gaussian process

$$\xi_t = \int_{-\infty}^{+\infty} g(t-s) \ \mathrm{d}W(s), \ t \in \mathbb{R},$$

where g is a Green function associated with (10.35), that is $g(t) = 0$ for $t < 0$ and, for $t \geq 0$, g is the unique solution of the problem

$$\sum_{j=0}^{k} a_j x^{(j)}(t) = 0, \ x(0) = \cdots = x^{(k-2)}(0) = 0, \ x^{(k-1)}(0) = a_k^{-1}.$$

By choosing a version of (ξ_t) such that every sample path possesses $k-1$ continuous derivatives, one may define a sequence (X_n) of random variables with values in $B = C_{k-1}([0,h])$, that is, the space of real functions on $[0,h]$ with $k-1$ continuous derivatives, equipped with the norm

$$\| x \|_{k-1} = \sum_{j=0}^{k-1} \sup_{0 \le t \le h} |x^{(j)}(t)|.$$

This space is complete, and each element y^* of B^* is uniquely representable under the form

$$y^*(x) = \sum_{j=0}^{k-2} \alpha_j x^{(j)}(0) + \int_0^h x^{(k-1)}(t)\, \mathrm{d}m(t),$$

when $\alpha_0, \ldots, \alpha_{k-2} \in \mathbb{R}$ and m is a bounded signed measure on $[0,h]$ (see, for example Dunford and Schwartz 1958).

Now, noting that

$$\mathrm{E}\left(\xi_{nh+t} | \xi_s,\ s \le nh\right) = \sum_{j=0}^{k-1} \xi^{(j)}(nh)\varphi_j(t),$$

$0 \le t \le h$, $n \in \mathbb{Z}$, where φ_j is the unique solution of $\sum_{j=0}^{k} a_j x^{(j)}(t) = 0$ satisfying the conditions $\varphi_j^{(\ell)}(0) = \delta_{j\ell}$; $\ell = 0, \ldots, k-1$ we are led to define ρ by

$$\rho(x)(t) = \sum_{j=0}^{k-1} x^{(j)}(h)\varphi_j(t),\ 0 \le t \le h,\ x \in B,$$

and $(X_n(\cdot) = \xi_{nh+\cdot},\ n \in \mathbb{Z})$ becomes an ARB(1). Since the only eigenelements of ρ are the functions $t \mapsto \mathrm{e}^{-r_i t}$ associated with the eigenvalues $\mathrm{e}^{-r_i h}$, $1 \le i \le k$, we have

$$\| \rho^j \|_{\mathcal{L}} = \mathcal{O}(\mathrm{e}^{-r_1 (j-1)h}),\ j \ge 1.$$

Concerning innovation, it is defined as

$$\varepsilon_n(t) = \int_{nh}^{nh+t} g(nh + t - u)\, \mathrm{d}W(u),\ 0 \le t \le h,\ n \in \mathbb{Z}.$$

Finally, it is noteworthy that (ξ_t) is not Markovian while (X_n) is a Markov process. ◇

Notes

The idea of representation (10.1) and the associated modelling by autoregressive processes appears in Bosq (1991).

Linearly closed subspaces and dominance come from Fortet (1995). Our definition of dominance is slightly different from the original one.

Content of Sections 10.2, 10.3 and 10.4 is taken from Bosq (2007). Properties of autoregressive processes in Banach and Hilbert spaces are studied in Bosq (2000).

The ARH(1) and MAH(1) models are rich enough to explain many time-dependent random experiments: we refer e.g. to works of Antoniadis and Sapatinas (2003); Bernard (1997); Besse and Cardot (1996); Besse *et al.* (2000); Cavallini *et al.* (1994); Marion and Pumo (2004) among others. Other results and applications appear in Ferraty and Vieu (2006). However more sophisticated models appear as useful in various situations. We now briefly indicate some of them.

Mourid (1995) has developed the theory of standard ARH(p) processes, and Merlevéde (1996) has studied standard linear processes.

Marion and Pumo (2004) have introduced the ARHD process for modelling processes with regular sample paths. Further study of this model appears in Mas and Pumo (2007).

Damon and Guillas (2005) have studied ARH with exogenous variables.

11

Estimation and prediction of functional linear processes

11.1 Introduction

Prediction for FLP contains various stages. First, in general, these processes are not centred, then it is necessary to estimate the mean of X_n. For this purpose we study asymptotic behaviour of the empirical mean. We shall see that classical limit theorems hold.

The second step is estimation of autocovariance operators by the corresponding empirical operators. Using representation of these empirical operators by empirical mean of FLP with values in spaces of operators, one derives asymptotic results.

Finally, the best linear predictor depends on the autocovariance operators. This dependance is rather delicate to handle due to the role of the inverse of the covariance operator which is not bounded in general. A possible solution consists of projecting the problem over a finite-dimensional space that varies with size of data. Then, under some regularity conditions, one may construct a statistical predictor which converges at a good rate. Applications show that this rate is not only a theoretical one. Proofs are not detailed, or are omitted, since most of them are rather technical. For complete proofs and further results we refer to Bosq (2000).

Inference and Prediction in Large Dimensions D. Bosq and D. Blanke

11.2 Estimation of the mean of a functional linear process

Consider a standard linear Hilbertian process (SLHP) defined as

$$X_n = m + \varepsilon_n + \sum_{j=1}^{\infty} \ell_j(\varepsilon_{n-j}), \quad n \in \mathbb{Z} \tag{11.1}$$

where $m \in H$, $\sum_{j\geq 1} \| \ell_j \|_{\mathcal{L}} < \infty$ and $\varepsilon = (\varepsilon_n)$ is a H-white noise. For convenience we suppose that ε is strong, but some results below remain valid for a weak white noise. Clearly the series in (11.1) converges in L_H^2 and almost surely.

One says that $X = (X_n, n \in \mathbb{Z})$ is *invertible* if there exists $(\rho_j, j \geq 1) \subset \mathcal{L}$ such that $\sum_{j\geq 1} \| \rho_j \|_{\mathcal{L}} < \infty$ and

$$X_n - m = \sum_{j\geq 1} \rho_j(X_{n-j} - m) + \varepsilon_n, \quad n \in \mathbb{Z}.$$

Now, given the data $X_1, \ldots, X_n$, a natural estimator of $m = \mathrm{E}X_n$ is the empirical mean

$$\bar{X}_n = \frac{S_n}{n} = \frac{X_1 + \cdots + X_n}{n}.$$

The next statement summarizes its principal asymptotic properties.

Theorem 11.1
Let $X = (X_n, n \in \mathbb{Z})$ be an invertible SLHP.

(1) One has

$$n\mathrm{E} \| \bar{X}_n - m \|^2 \to \sum_{k\in\mathbb{Z}} \mathrm{E}(\langle X_0 - m, X_k - m\rangle), \tag{11.2}$$

$$\bar{X}_n \to m \text{ a.s.},$$

and

$$\sqrt{n}\,(\bar{X}_n - m) \xrightarrow{\mathcal{D}} N \tag{11.3}$$

where N is a zero-mean Gaussian random variable and $\mathcal{D}$ denotes convergence in distribution in H.

(2) Moreover, if $\mathrm{E}(\exp \gamma \| X_0 \|) < \infty$ $(\gamma > 0)$, *then*

$$P(\| \bar{X}_n - m \| \geq \eta) \leq 4\exp\left(-\frac{n\eta^2}{\alpha + \beta\eta}\right), \quad \eta > 0 \tag{11.4}$$

where $\alpha > 0$ and $\beta > 0$ only depend on P_X, and

$$\| \overline{X}_n - m \| = \mathcal{O}\left(\left(\frac{\ln \ln n}{n}\right)^{1/2}\right) \quad a.s. \tag{11.5}$$

PROOF: (*Sketch*)
First, one may establish that the sequence $(X_n, X_{n-1}, \ldots)$, $n \in \mathbb{Z}$ of H^∞-valued random variables is an ARH$^\infty$ (1) provided H^∞ is equipped with a suitable Hilbertian norm. Moreover one may take $m = 0$ without loss of generality. Then, it suffices to prove Theorem 11.1 for a standard ARH(1):

$$X_n = \rho(X_{n-1}) + \varepsilon_n, \quad n \in \mathbb{Z} \tag{11.6}$$

where $\| \rho^{j_0} \|_{\mathcal{L}} < 1$ for some $j_0 \geq 1$, and (ε_n) is a strong white noise.

Now, (11.6) and an easy induction give

$$|\mathrm{E}\langle X_0, X_k\rangle| \leq \| \rho^k \|_{\mathcal{L}} \mathrm{E} \| X_0 \|^2, \quad k \geq 0,$$

hence (11.2) by a direct calculation of $\lim_{n\to\infty} n \cdot \mathrm{E} \| \overline{X}_n \|^2$.

Concerning pointwise convergence one may use a Hilbertian variant of a Doob lemma (see Bosq 2000, Corollary 2.3).

Next, the central limit theorem comes from the decomposition

$$\sqrt{n}\overline{X}_n = (I - \rho)^{-1}\sqrt{n}\overline{\varepsilon}_n + \sqrt{n}\Delta_n. \tag{11.7}$$

On one hand $\sqrt{n}\Delta_n \xrightarrow{p} 0$, on the other hand, the central limit theorem for H-valued i.i.d. random variables yields

$$\sqrt{n}\overline{\varepsilon}_n \xrightarrow{\mathcal{D}} N,$$

hence (11.3) by using a common property of ⋆weak convergence⋆.

Now the exponential type inequality (11.4) is a consequence of (11.7) and Pinelis–Sakhanenko inequality (see Lemma 4.2, p.106).

Finally, a classical method allows us to establish an exponential inequality for $\max_{1\leq i\leq n} \| S_i \|$ and (11.5) follows. ■

Notice that (11.5) is also a consequence of the law of the iterated logarithm for ARH(1) processes (cf. Bosq 2000).

11.3 Estimation of autocovariance operators

Let $X = (X_n, n \in \mathbb{Z})$ be a centred SLHP. The best linear predictor of X_{n+h}, given $X_n, X_{n-1}, \ldots$ depends on the sequence of autocovariance operators defined as

$$C_k = C_{X_0, X_k}, k \geq 0.$$

Estimation of these operators is then crucial for constructing statistical predictors.

Since the estimators of C_k are $\mathcal{S}$-valued, where $\mathcal{S}$ is the space of Hilbert–Schmidt operators on H, we begin with some facts concerning $\mathcal{S}$.

11.3.1 The space $\mathcal{S}$

An element s of $\mathcal{L}$ is said to be a *Hilbert–Schmidt operator* if it admits a decomposition of the form

$$s(x) = \sum_{j=1}^{\infty} \alpha_j \langle x, e_j \rangle f_j, \quad x \in H \tag{11.8}$$

where (e_j) and (f_j) are orthonormal bases of H and (α_j) is a real sequence such that $\sum_{j=1}^{\infty} \alpha_j^2 < \infty$. We may and do suppose that $|\alpha_1| \geq |\alpha_2| \geq \cdots$.

Now the space $\mathcal{S}$ of Hilbert–Schmidt operators becomes a Hilbert space if it is equipped with the scalar product

$$\langle s_1, s_2 \rangle_{\mathcal{S}} = \sum_{1 \leq i,j < \infty} \langle s_1(g_i), h_j \rangle \langle s_2(g_i), h_j \rangle; s_1, s_2 \in \mathcal{S}$$

where (g_i) and (h_j) are orthonormal bases of H. This scalar product does not depend on the choice of these orthonormal bases. Thus, we have

$$\| s \|_{\mathcal{S}} = \left(\sum_{j=1}^{\infty} \alpha_j^2 \right)^{1/2}$$

where (α_j) appears in (11.8), and $\| s \|_{\mathcal{S}} \geq \| s \|_{\mathcal{L}}$.

In particular, note that if $s \in \mathcal{S}$ is symmetric, (11.8) becomes

$$s(x) = \sum_{j=1}^{\infty} \alpha_j \langle x, e_j \rangle e_j, \quad x \in H,$$

therefore $s(e_j) = \alpha_j e_j, j \geq 1$; that is to say that $(\alpha_j, e_j), j \geq 1$ are the eigenelements of s.

Finally $\ell \in \mathcal{L}, s \in \mathcal{S} \Rightarrow \ell \circ s \in \mathcal{S}$ and $s \circ \ell \in \mathcal{S}$.

Cross-covariance operators are elements of $\mathcal{S}$ and covariance operators are symmetric elements of $\mathcal{S}$. Finally if a linear operator in H has a finite-dimensional range, it is in $\mathcal{S}$.

11.3.2 Estimation of C_0

A natural estimator of C_0 is *the empirical covariance operator*

$$C_{0,n}(x) = \frac{1}{n} \sum_{i=1}^{n} \langle X_i, x \rangle X_i, \quad x \in H$$

Note that $C_{0,n}$ is an $\mathcal{S}$-valued random variable.

In order to study $C_{0,n}$ we introduce some notation and assumptions: $a \otimes b$ will denote the application $x \mapsto \langle a, x\rangle b, x \in H$ $(a, b \in H)$, $\| \cdot \|_{\mathcal{L}}^{(S)}$ denotes the usual uniform norm on the bounded linear operators over $\mathcal{S}$. Finally we only consider the case where X is an ARH(1) by making the following assumption:

Assumptions 11.1 (A11.1)
$X = (X_n, n \in \mathbb{Z})$ is an ARH(1)

$$X_n = \rho(X_{n-1}) + \varepsilon_n, \quad n \in \mathbb{Z},$$

such that ρ satisfies (c_0) (see Chapter 10, p. 244), (ε_n) is a strong white noise and $\mathrm{E} \| X_0 \|^4 < \infty$.

Now we set $Z_n = X_n \otimes X_n - C_0, n \in \mathbb{Z}$. The next lemma shows that (Z_n) has an AR$\mathcal{S}$ autoregressive representation.

Lemma 11.1
If A11.1 holds, the $\mathcal{S}$-process (Z_n) is an AR$\mathcal{S}(1)$:

$$Z_n = R(Z_{n-1}) + E_n, \quad n \in \mathbb{Z} \tag{11.9}$$

where R is the linear bounded operator on $\mathcal{S}$ defined by

$$R(s) = \rho s \rho^*, \quad s \in \mathcal{S} \tag{11.10}$$

and (E_n) is an $\mathcal{S}$-white noise and is the innovation of (Z_n).

Moreover, R is such that

$$\| R^j \|_{\mathcal{L}}^{(S)} \leq \| \rho^j \|_{\mathcal{L}}^2, \ j \geq 1, \tag{11.11}$$

and $E_n \in L^2_{\mathcal{S}}(\Omega, \mathcal{B}_n, P)$, where $\mathcal{B}_n = \sigma(X_j, j \leq n)$, and

$$\mathrm{E}^{\mathcal{B}_{n-1}}(E_n) = 0, \quad n \in \mathbb{Z} \tag{11.12}$$

i.e. (E_n) is a martingale difference.

PROOF: (*Sketch*)
From the basic relation $X_n = \rho(X_{n-1}) + \varepsilon_n$, it follows that

$$Z_n = (\rho(X_{n-1}) + \varepsilon_n) \otimes (\rho(X_{n-1}) + \varepsilon_n) - C_0,$$

then, noting that $C_{\varepsilon_0} = C_0 - \rho C_0 \rho^*$, one obtains (11.9), with R given by (11.10)

and

$$E_n = (X_{n-1} \otimes \varepsilon_n)\rho^* + \rho(\varepsilon_n \otimes X_{n-1}) + (\varepsilon_n \otimes \varepsilon_n - C_\varepsilon).$$

Now, boundedness of R follows from

$$\| R(s) \|_{\mathcal{S}} = \| \rho s \rho^* \|_{\mathcal{S}} \leq \| \rho \|_{\mathcal{L}}^2 \| s \|_{\mathcal{S}}$$

and an easy induction entails $R^j(s) = \rho^j s \rho^{*j}$, $s \in \mathcal{S}$, $j \geq 1$, hence

$$\| R^j \|_{\mathcal{L}}^{(\mathcal{S})} \leq \| \rho^j \|_{\mathcal{L}}^2, \quad j \geq 1,$$

thus, R satisfies (c_0).

Finally, since $\sigma(\varepsilon_j, j \leq n) = \sigma(X_j, j \leq n)$ it is easy to verify that (E_n) is a martingale difference with respect to $(\mathcal{B}_n)$. Then, it is a white noise, and Proposition 10.4 shows that (E_n) is the innovation of (Z_n). ■

In the particular case where $H = \mathbb{R}$, Lemma 11.1 means that $(X_n^2 - \mathrm{E}X_n^2)$ is an AR(1) process.

Note that regularity properties of ρ induce similar properties of R. For example, if ρ is a symmetric compact operator with spectral decomposition

$$\rho = \sum_{j \geq 1} \alpha_j e_j \otimes e_j,$$

then R is symmetric compact on $\mathcal{S}$ with decomposition

$$R = \sum_{i,j} \alpha_i \alpha_j s_{ij} \otimes s_{ij}$$

where $s_{ij} = e_i \otimes e_j$.

Now, the representation Lemma 11.1 allows us to derive asymptotic results concerning

$$\overline{Z}_n = C_{0,n} - C_0;$$

we cannot directly apply Theorem 11.1 since (E_n) is not a strong white noise but almost similar properties can be obtained:

Theorem 11.2

(1) If A11.1 holds then

$$n\mathrm{E} \| C_{0,n} - C_0 \|_{\mathcal{S}}^2 \to \sum_{j \in \mathbb{Z}} \mathrm{E}\langle Z_0, Z_j \rangle_{\mathcal{S}}, \tag{11.13}$$

$$\| C_{0,n} - C_0 \|_{\mathcal{S}} \to 0 \ a.s., \tag{11.14}$$

and

$$\sqrt{n}\,(C_{0,n} - C_0) \xrightarrow{\mathcal{D}} N_1$$

where N_1 is an $\mathcal{S}$-valued centred Gaussian random operator and $\mathcal{D}$ is convergence in distribution on $\mathcal{S}$.

(2) *If in addition $\| X_0 \|$ is bounded, then*

$$P(\| C_{0,n} - C_0 \|_{\mathcal{S}} \geq \eta) \leq 4 \exp\left(-\frac{n\eta^2}{\alpha_1 + \beta_1 \eta} \right), \quad \eta > 0$$

where $\alpha_1 > 0$ and $\beta_1 > 0$ only depends on P_X, and

$$\| C_{0,n} - C_0 \|_{\mathcal{S}} = \mathcal{O}\left(\left(\frac{\ln \ln n}{n} \right)^{1/2} \right) \; a.s. \tag{11.15}$$

PROOF:
Similar to the proof of Theorem 11.1. In particular one applies the central limit theorem for H-martingales and the Pinelis–Sakhanenko exponential inequality (1985) for bounded *martingales*. ■

From Chapter 3, Example 3.4, it is known that the best 'linear' rate for estimating C_0 is n^{-1}; (11.13) shows that $C_{0,n}$ reaches this rate.

11.3.3 Estimation of the eigenelements of C_0

Since C_0 is a symmetric Hilbert–Schmidt operator it can be written under the form

$$C_0 = \sum_{j=1}^{\infty} \lambda_j v_j \otimes v_j, \quad (\lambda_1 \geq \lambda_2 \cdots \geq 0) \tag{11.16}$$

where (v_j) is an orthonormal basis of H and $\lambda_j = \mathrm{E}(\langle X_0, v_j \rangle^2)$, $j \geq 1$, therefore

$$\sum_{j=1}^{\infty} \lambda_j = \mathrm{E} \| X_0 \|^2 < \infty.$$

The empirical eigenelements are defined by the relations

$$C_{0,n}(v_{jn}) = \lambda_{jn} v_{jn}, \quad j \geq 1 \tag{11.17}$$

with $\lambda_{1n} \geq \lambda_{2n} \geq \cdots \geq \lambda_{nn} \geq 0 = \lambda_{n+1,n} = \lambda_{n+2,n} = \cdots$, and where (v_{jn}) is a random orthonormal basis of H.

Consistency and rates for λ_{jn} are directly derived from the following crucial lemma which is an application of a lemma concerning *symmetric operators*.

Lemma 11.2

$$\sup_{j\geq 1} |\lambda_{jn} - \lambda_j| \leq \| C_{0,n} - C \|_{\mathcal{L}} .$$

Now the asymptotic behaviour of $\sup_{j\geq 1} |\lambda_{jn} - \lambda_j|$ is an obvious consequence of (11.10)–(11.11) and (11.13)–(11.14); statements are left to the reader.

Concerning eigenvectors the situation is a little more intricate. First, identifiability of v_j is not ensured since, for example, $-v_j$ is also an eigenvector associated with λ_j. Actually the unknown parameter is the subspace V_j associated with λ_j.

If $\dim V_j = 1$, we set $v'_{jn} = \mathrm{sgn}\langle v_{jn}, v_j\rangle v_j, j \geq 1$ where $\mathrm{sgn}\, u = \mathbb{1}_{u\geq 0} - \mathbb{1}_{u<0}, u \in \mathbb{R}$.
Then, we have the following inequalities.

Lemma 11.3
If $\lambda_1 > \lambda_2 > \cdots > \lambda_j > \cdots > 0$, *then*

$$\| v_{jn} - v'_{jn} \| \leq a_j \| C_{0,n} - C_0 \|_{\mathcal{L}}, \quad j \geq 1,$$

where

$$\begin{cases} a_1 = 2\sqrt{2}(\lambda_1 - \lambda_2)^{-1} \\ a_j = 2\sqrt{2}\max[(\lambda_{j-1} - \lambda_j)^{-1}, (\lambda_j - \lambda_{j+1})^{-1}], \quad j \geq 2. \end{cases} \tag{11.18}$$

Again, the asymptotic behaviour of $\| v_{jn} - v'_{jn} \|$ is then an obvious consequence of (11.10)–(11.11), (11.12), (11.14) and statements are left to the reader.

Now, if $\dim V_j > 1$ the same rates may be obtained for $\| v_{jn} - \Pi^{V_j}(v_{jn}) \|$.

Finally Mas (2002) has established asymptotic normality for $\sqrt{n}(\lambda_{jn} - \lambda_j)$ and $\sqrt{n}(v_{jn} - v'_{jn})$.

11.3.4 Estimation of cross-autocovariance operators

The natural estimator of C_k $(k \geq 1)$ is

$$C_{k,n} = \frac{1}{n-k}\sum_{i=1}^{n-k} X_i \otimes X_{i+k}, \quad n > k.$$

The study of $C_{k,n}$ is based on the following extension of Lemma 11.1.

Lemma 11.4
If A11.1 holds, the $\mathcal{S}$-valued process $Z_n^{(k)} = X_{n-k} \otimes X_n - C_k$, $n \in \mathbb{Z}$, admits the pseudo-autoregressive representation

$$Z_n^{(k)} = R(Z_{n-1}^{(k)}) + E_n^{(k)}, \quad n \in \mathbb{Z}$$

where $R(s) = \rho s \rho^*$ and $E_n^{(k)} \in L^2_{\mathcal{S}}(\Omega, \mathcal{B}_n, P)$ is such that

$$\mathrm{E}^{\mathcal{B}_j}(E_i^{(k)}) = 0; \quad i, j \in \mathbb{Z}, \quad j \leq i - k - 1. \tag{11.19}$$

Relation (11.19) means that $(E_n^{(k)}, n \in \mathbb{Z})$ is 'almost' a martingale difference. Actually, the subsequences $(E_{j(k+1)+h}^{(k)}, j \in \mathbb{Z}), 0 \leq h \leq k$ are martingale differences.

Asymptotic behaviour of $C_{k,n} - C_k$ follows from Lemma 11.4:

Theorem 11.3
If A11.1 holds, we have

$$n\mathrm{E} \parallel C_{k,n} - C_k \parallel_{\mathcal{S}} \to \sum_{\ell=-\infty}^{+\infty} \mathrm{E}\langle Z_0^{(k)}, Z_\ell^{(k)} \rangle_{\mathcal{S}}$$

$$\parallel C_{k,n} - C_k \parallel_{\mathcal{S}} \to 0 \ a.s.$$

If, in addition, $\parallel X_0 \parallel$ is bounded, then

$$P(\parallel C_{k,n} - C_k \parallel_{\mathcal{S}} > \eta) \leq 8 \exp\left(-\frac{n\eta^2}{\gamma + \delta\eta}\right), \quad \eta > 0$$

where $\gamma > 0$ and $\delta > 0$ only depend on P_X, and

$$\parallel C_{k,n} - C_k \parallel_{\mathcal{S}} = \mathcal{O}\left(\left(\frac{\ln \ln n}{n}\right)^{1/2}\right) a.s. \tag{11.20}$$

11.4 Prediction of autoregressive Hilbertian processes

Suppose that the process X satisfies Assumptions A11.1. Then the best probabilistic predictor of X_{n+1} given $X_n, X_{n-1}, \ldots$ is

$$X_{n+1}^* = \mathrm{E}^{\mathcal{B}_n}(X_{n+1}) = \Pi^{\mathcal{M}_n}(X_{n+1}) = \rho(X_n),$$

thus prediction of X_{n+1} may be performed by estimating ρ. Now, from (11.6) it follows that $C_1 = \rho C_0$. *A priori* this relation suggests the following estimation method:
if C_0 is invertible one writes

$$\rho = C_1 C_0^{-1}$$

and, if $C_{0,n}$ is invertible, one sets

$$\rho_n = C_{1,n} C_{0,n}^{-1}.$$

This program may succeed in a finite-dimensional framework, but it completely fails in an infinite-dimensional context since, in that case, C_0^{-1} and $C_{0,n}^{-1}$ are not defined on the whole space H.

In fact, if $\lambda_j > 0$ for all j, one has

$$C_0(H) = \left\{ y : y \in H, \sum_{j=1}^{\infty} \langle y, v_j \rangle^2 \lambda_j^{-2} < \infty \right\},$$

thus $C_0(H)$ is strictly included in H; concerning $C_{0,n}(H)$, since it is included in $\mathrm{sp}(X_1, \ldots, X_n)$, we have $\dim C_{0,n}(H) \le n$.

A solution consists in projection on a suitable finite-dimensional subspace. Given a dimension, say k, it is well known that the subspace which furnishes the best information is $\mathrm{sp}(v_1, \ldots, v_k)$ where the v_j's are defined in (11.16) (see Grenander 1981).

This remark leads to the choice of the subspace

$$H_{k_n} = \mathrm{sp}(v_{1n}, \ldots, v_{k_n n})$$

where (k_n) is such that $k_n \to \infty$ and $k_n/n \to 0$, and where the v_{jn}'s are defined in (11.17).

Now, let Π^{k_n} be the orthogonal projector of the random subspace H_{k_n}. Set

$$\tilde{C}_{0,n} = \Pi^{k_n} C_{0,n} = \sum_{j=1}^{k_n} \lambda_{jn} v_{jn} \otimes v_{jn},$$

and, if $\lambda_{k_n,n} > 0$,

$$\tilde{C}_{0,n}^{-1} = \sum_{j=1}^{k_n} \lambda_{jn}^{-1} v_{jn} \otimes v_{jn}.$$

Note that $\tilde{C}_{0,n}^{-1}$ is the inverse of $\tilde{C}_{0,n}$ only on H_{k_n} but it is defined on the whole space H.

Then, the estimator of ρ is defined as $\tilde{\rho}_n = \Pi^{k_n} C_{1,n} \tilde{C}_{0,n}^{-1}$. Consistency of $\tilde{\rho}_n$ is described in the next statement.

Theorem 11.4

(1) Under A11.1 and the following assumptions

- $\lambda_1 > \lambda_2 > \cdots > \lambda_j > \cdots > 0$

- *$\lambda_{k_n,n} > 0$ a.s., $n \geq 1$*
- *ρ is a Hilbert–Schmidt operator*
-

$$\lambda_{k_n}^{-1} \sum_{j=1}^{k_n} a_j = \mathcal{O}(n^{1/4}(\ln n)^{-\beta}), \quad \beta > \frac{1}{2} \tag{11.21}$$

where (a_j) is defined in Lemma 11.3, we have

$$\| \tilde{\rho}_n - \rho \|_{\mathcal{L}} \to 0 \ a.s.$$

(2) *If, in addition, $\| X_0 \|$ is bounded, then*

$$P(\| \tilde{\rho}_n - \rho \|_{\mathcal{L}} \geq \eta) \leq c_1(\eta) \exp\left(-c_2(\eta) n \lambda_{k_n}^2 \left(\sum_1^{k_n} a_j \right)^{-2} \right) \tag{11.22}$$

$\eta > 0$, $n \geq \eta_1$, where $c_1(\eta)$ and $c_2(\eta)$ are positive constants.

Example 11.1
If $\lambda_j = ar^j, j \geq 1$ $(a > 0, 0 < r < 1)$, $k_n = o(\ln n)$ yields (11.21). If $k_n \simeq \ln \ln n$ the bound in (11.22) takes the form

$$P(\| \tilde{\rho}_n - \rho \|_{\mathcal{L}} \geq \eta) \leq c_1(\eta) \exp\left(-c_2'(\eta) \frac{n}{(\ln n)^4} \right). \qquad \diamond$$

Example 11.2
If $\lambda_j = aj^{-\gamma}, j \geq 1$ $(a > 0, \gamma > 1)$ and $k_n \simeq \ln n$, (11.22) becomes

$$P(\| \tilde{\rho}_n - \rho \|_{\mathcal{L}} \geq \eta) \leq c_1(\eta) \exp\left(-c_2''(\eta) \frac{n}{(\ln n)^{4(1+\gamma)}} \right). \qquad \diamond$$

Assumption of boundedness for $\| X_0 \|$ can be replaced by existence of some exponential moment provided X is geometrically strongly mixing (see Bosq 2000).

Prediction

To $\tilde{\rho}_n$ one associates the statistical predictor $\tilde{X}_n = \tilde{\rho}_n(X_n)$; the next corollary is an easy consequence of Theorem 11.4.

Corollary 11.1

(1) *Under the assumptions in Theorem 11.4 we have*

$$\| \tilde{X}_{n+1} - X^*_{n+1} \| \xrightarrow{p} 0.$$

(2) If, in addition, $\| X_0 \|$ is bounded, then

$$P\big(\| \tilde{X}_{n+1} - X^*_{n+1} \| \geq \eta\big) \leq c_1(M^{-1}\eta) \exp\left(-c_2(M^{-1}\eta) n \lambda_{k_n}^2 \left(\sum_1^{k_n} a_j \right)^{-2} \right)$$

where $M = \| X_0 \|_\infty$.

11.5 Estimation and prediction of ARC processes

In a general Banach space it is possible to obtain asymptotic results concerning the empirical mean and empirical autocovariance operators but sharp rates are not available.

In the important case of $C = C[0,1]$ special results appear provided the following regularity conditions hold: X is a (strong) ARC(1) process (see Definition 10.3, p. 255) such that

$$(L_\alpha)\ |X_n(t) - X_n(s)| \leq M_n |t-s|^\alpha;\ s,t \in [0,1]; \quad n \in \mathbb{Z}$$

where $\alpha \in]0,1]$ and $M_n \in L^2(\Omega, \mathcal{A}, P)$.

Then, one may evaluate the level of information associated with observations of X at discrete instants and obtain rates of convergence.

Set

$$M_n = \sup_{s \neq t} \frac{|X_n(t) - X_n(s)|}{|t-s|^\alpha}.$$

Owing to strict stationarity of X, $P_{M_n} = P_{M_0}$ and one puts $\mathrm{E}M_n^2 = V$, $n \in \mathbb{Z}$.

Theorem 11.5
Let $X = (X_n, n \in \mathbb{Z})$ be an ARC(1) with mean m and innovation $\varepsilon = (\varepsilon_n, n \in \mathbb{Z})$.

(1) If X satisfies (L_α), then

$$\mathrm{E}(\sup_{0 \leq t \leq 1} |\overline{X}_n(t) - m(t)|)^2 = \mathcal{O}\left(n^{-\frac{2\alpha}{2\alpha+1}}\right). \tag{11.23}$$

(2) if, in addition $\mathrm{E}(\exp \gamma \sup_{0 \leq t \leq 1} |X_0(t)|) < \infty$ for some $\gamma > 0$, and M_0 is bounded, then, for all $\eta > 0$ there exist $a(\eta) > 0$ and $b(\eta) > 0$ such that

$$P\left(\sup_{0 \leq t \leq 1} |\overline{X}_n(t) - m(t)| \geq \eta \right) \leq a(\eta) \exp\big(-nb(\eta)\big). \tag{11.24}$$

(3) If ε satisfies (L_1), then

$$\sqrt{n}(\overline{X}_n - m) \xrightarrow{\mathcal{D}} N$$

where N is a Gaussian zero-mean C-valued random variable, and $\mathcal{D}$ is convergence in distribution on C.

PROOF:
We may and do suppose that $m = 0$.

(1) For each $x^* \in C^*$ we have

$$x^*(X_k) = \sum_{j=0}^{k-1} x^* \rho^j(\varepsilon_{k-j}) + x^* \rho^k(X_0), \quad k \geq 1,$$

hence

$$\mathrm{E}(x^*(X_0)x^*(X_k)) = \mathrm{E}(x^*(X_0)x^* \rho^k(X_0)),$$

then

$$|\mathrm{E}(x^*(X_0)x^*(X_k))| \leq \| \rho^k \|_{\mathcal{L}} \| x^* \|_*^2 \, \mathrm{E} \| X_0 \|^2 \tag{11.25}$$

where $\| \cdot \|_*$ is the usual norm in C^*. Then, (11.25) and stationarity of X yield

$$\begin{aligned}\mathrm{E}|x^*(X_1) + \cdots + x^*(X_n)|^2 &\leq 2n \| x^* \|_*^2 \sum_{j=0}^{n-1} \| \rho^j \|_{\mathcal{L}} \, \mathrm{E} \| X_0 \|^2 \\ &\leq n \cdot r_{x^*}\end{aligned}$$

where $r_{x^*} = 2 \| x^* \|^2 \sum_{j=0}^{\infty} \| \rho^j \|_{\mathcal{L}} \cdot \mathrm{E} \| X_0 \|^2$.
Choosing $x^* = \delta_{(t)}$ $(0 \leq t \leq 1)$ one obtains

$$\mathrm{E}|\overline{X}_n(t)|^2 \leq \frac{r_{\delta(t)}}{n} \leq \frac{2 \sum_{j=0}^{\infty} \| \rho^j \|_{\mathcal{L}} \, \mathrm{E} \| X_0 \|^2}{n} = \frac{r}{n}. \tag{11.26}$$

Now, set $t_j = j/\nu_n$, $1 \leq j \leq \nu_n$; from (L_α) we get

$$|\overline{X}_n(t) - \overline{X}_n(t_j)| \leq \left(\frac{1}{n} \sum_{i=1}^{n} M_i \right) \nu_n^{-\alpha},$$

$(j-1)/\nu_n \leq t \leq j/\nu_n$; $i = 1, \ldots, \nu_n$. Therefore

$$\left| \sup_{0 \leq t \leq 1} |\overline{X}_n(t)| - \sup_{1 \leq j \leq \nu_n} |\overline{X}_n(t_j)| \right| \leq \overline{M}_n \nu_n^{-\alpha}$$

which gives

$$\begin{aligned}
\mathrm{E} \parallel \overline{X}_n \parallel^2 &\leq \mathrm{E}\left(\sup_{1\leq j\leq \nu_n} |\overline{X}_n(t_j)| \right)^2 + 2\mathrm{E}(\bar{M}_n^2)\nu_n^{-2\alpha} \\
&\leq 2\sum_{j=1}^{\nu_n} \mathrm{E}|\overline{X}_n(t_j)|^2 + 2\mathrm{E}\left(\frac{1}{n}\sum_{1}^{n} M_i^2\right)\nu_n^{-2\alpha};
\end{aligned}$$

from (11.26) it follows that

$$\mathrm{E} \parallel \overline{X}_n \parallel^2 \leq \frac{2r\nu_n}{n} + 2V\nu_n^{-2\alpha},$$

and the choice $\nu_n \simeq n^{1/(2\alpha+1)}$ gives (11.23).

(2) Consider the decomposition

$$\begin{aligned}
\overline{X}_n = \frac{1}{n}[&\varepsilon_n + (I+\rho)(\varepsilon_{n-1}) + \cdots + (I+\rho+\cdots+\rho^{n-1})(\varepsilon_1) \\
&+ (\rho+\cdots+\rho^n)(X_0)],
\end{aligned}$$

$n \geq 1$, which shows that $n\overline{X}_n$ is the sum of $(n+1)$ independent B-valued random variables: $Y_n, Y_{n-1}, \ldots, Y_0$.

Then, if $\mathrm{E}(\exp\gamma \parallel \gamma_0 \parallel) < \infty$, since $\parallel \varepsilon_1 \parallel \leq \parallel X_1 \parallel + \parallel \rho \parallel_{\mathcal{L}} \parallel X_0 \parallel$ it follows that $\mathrm{E}(\exp\gamma' \parallel \varepsilon_1 \parallel) < \infty$ where $\gamma' = (\gamma/2)(\max(1, \parallel \rho \parallel_{\mathcal{L}}))^{-1}$, hence $\mathrm{E}(\exp\gamma'' \parallel Y_j \parallel) < \infty$ where $\gamma'' = \gamma'/r'$ with $r' = \sum_{j=0}^{\infty} \parallel \rho^j \parallel_{\mathcal{L}}$.

Now, it can be shown that $\mathrm{E}(\exp\gamma'' \parallel Y_j \parallel) < \infty$ is equivalent to

$$\sum_{j=0}^{n} \mathrm{E} \parallel Y_j \parallel^k \leq \frac{k!}{2} n\ell_0^2 b^{k-2}, k \geq 2, j = 0, \ldots, n, (\ell_0 > 0, b > 0),$$

thus, one may apply the Pinelis–Sakhanenko inequality (see Lemma 4.2, p. 106) to obtain for $\eta > 0$

$$P(\parallel \overline{X}_n \parallel - \mathrm{E} \parallel \overline{X}_n \parallel \geq \eta) \leq 2\exp\left(-\frac{n\eta^2}{2\ell_0 + 2b\eta}\right).$$

From (11.23) it follows that, for n large enough,

$$\mathrm{E} \parallel \overline{X}_n \parallel \leq \left(\mathrm{E} \parallel \bar{X}_n \parallel^2\right)^{1/2} \leq \frac{\eta}{2}$$

then

$$\parallel \overline{X}_n \parallel \geq \eta \Rightarrow \parallel \overline{X}_n \parallel - \mathrm{E} \parallel \overline{X}_n \parallel \geq \frac{\eta}{2},$$

hence (11.24).

(3) From the decomposition $\bar{X}_n = (I - \rho)^{-1}\bar{\varepsilon}_n + \Delta_n$, and the fact that $\Delta_n \xrightarrow{p} 0$, it follows that X satisfies the CLT if and only if ε satisfies it. Since ε satisfies (L_1) the Jain–Marcus (1975) theorem gives the CLT for ε and hence for X. ■

11.5.1 Estimation of autocovariance

We now turn to estimation of (C_k). For convenience we suppose that X is centred. In $C([0,1])$, C_0 is completely determined by the covariance function

$$c_0(s,t) = \mathrm{E}(X_0(s)X_0(t)); s,t \in [0,1]. \tag{11.27}$$

This property follows from $c_0(s,t) = \delta_{(t)}[C_0(\delta_{(s)})]$ and the fact that discrete distributions are dense in the dual space of C, i.e. the space of bounded signed measures on $[0,1]$.

Similarly C_k is determined by

$$c_k(s,t) = \mathrm{E}(X_0(s)X_k(t)); s,t \in [0,1].$$

The next statement summarizes some properties of the *empirical covariance function*

$$c_{0,n}(s,t) = \frac{1}{n}\sum_{i=1}^{n} X_i(s)X_i(t); s,t \in [0,1].$$

Theorem 11.6
Let X be an ARC(1) that satisfies L_α.

(1) If $E(\| X_0 \|^2 M_0^2) < \infty$, one has

$$\mathrm{E}\int_{[0,1]^2} [c_{0,n}(s,t) - c(s,t)]^2 \mathrm{d}s\,\mathrm{d}t = \mathcal{O}\left(n^{-\frac{\alpha}{1+\alpha}}\right).$$

(2) If X_0 and M_0 are bounded, then

$$P\left(\int_{[0,1]^2} [c_{0,n}(s,t) - c(s,t)]^2 \mathrm{d}s\mathrm{d}t \geq \eta^2\right) \leq d\eta^{-2/\alpha} \exp\left(-\frac{n\eta^2}{4\gamma + 2\delta\eta}\right),$$

$\eta > 0.$

Similar results may be obtained for $c_{k,n} - c_k$ where

$$c_{k,n}(s,t) = \frac{1}{n-k}\sum_{i=1}^{n-k} X_i(s)X_{i+k}(t).$$

11.5.2 Sampled data

A common situation in practice is to dispose of observations of X at discrete instants. Suppose that data have the form $X_i(j\nu_n^{-1})$, $1 \leq j \leq \nu_n$, $1 \leq i \leq n$ and that (L_α) holds. It is then possible to construct good approximations of $X_1, \ldots, X_n$ by interpolating these data. Here we only deal with *linear interpolation* defined for each i as

$$Y_i(t) = X_i(\nu_n^{-1}), 0 \leq t \leq \nu_n^{-1}$$
$$Y_i(t) = X_i(j\nu_n^{-1}) + (\nu_n t - j)[X_i((j+1)\nu_n^{-1}) - X_i(j\nu_n^{-1})],$$

$j\nu_n^{-1} \leq t \leq (j+1)\nu_n^{-1}$, $1 \leq j \leq \nu_n - 1$, then, (L_α) yields

$$\| X_i - Y_i \|_\infty \leq 2M_i\nu_n^{-\alpha}, \quad 1 \leq i \leq n$$

hence

$$\| \overline{X}_n - \overline{Y}_n \|_\infty \leq 2\overline{M}_n\nu_n^{-\alpha}; \tag{11.28}$$

thus from Theorem 11.5, if $\nu_n = n^{1/(2\alpha+1)}$, (11.23) and (11.28) we get

$$\mathrm{E} \| \overline{Y}_n - m \|_\infty^2 = \mathcal{O}\left(n^{-\frac{2\alpha}{2\alpha+1}}\right).$$

Similarly (11.24) and (11.27) give

$$P(\| \overline{Y}_n - m \|_\infty \geq \eta) \leq a\left(\frac{\eta}{2}\right) \exp\left(-nb\left(\frac{\eta}{2}\right)\right)$$

provided ν_n is large enough.

Finally if (ε) satisfies (L_1) and $\sqrt{n}\nu_n^{-\alpha} \to 0$, $\sqrt{n}(\overline{Y}_n - m)$ and $\sqrt{n}\,(\overline{X}_n - m)$ have the same limit in distribution.

Concerning the empirical covariance function associated with $Y_1, \ldots, Y_n$ we have, in the centred case,

$$c_{0,n}^{(Y)}(s,t) = \frac{1}{n}\sum_{i=1}^{n} Y_i(s)Y_i(t)$$
$$\| c_{0,n}^{(X)} - c_{0,n}^{(Y)} \|_2 \leq \frac{2}{n}\sum_{1}^{n} \| X_i \| M_i\nu_n^{-\alpha}.$$

If $\mathrm{E}(\| X_0 \|^2 M_0^2) < \infty$ one obtains

$$\mathrm{E} \| c_{0,n}^{(X)} - c_{0,n}^{(Y)} \|_2^2 \leq 4\mathrm{E}\left(\frac{1}{n}\sum_{1}^{n} \| X_i \| M_i\right)^2 \nu_n^{-2\alpha}$$

hence $\mathrm{E} \parallel c_{0,n}^{(Y)} - c_0 \parallel_2^2 = \mathcal{O}(n^{-\alpha/(1+\alpha)})$ provided $\nu_n \simeq n^{2/(1+\alpha)}$.

Finally, if X_0 and M_0 are bounded,

$$P(\parallel c_n^{(Y)} - c_0 \parallel_2 \geq \eta) \leq a\left(\frac{\eta}{2}\right) \exp\left(-nb\left(\frac{\eta}{2}\right)\right), \eta > 0$$

for large enough n and provided $\nu_n \to \infty$.

11.5.3 Estimation of ρ and prediction

In order to simplify estimation of ρ we suppose here that it is associated with a continuous bounded *kernel* $r(\cdot, \cdot)$, that is

$$\rho(x)(t) = \int_0^1 r(s,t)x(s)\mathrm{d}s, \quad x \in C,$$

with $\parallel \rho \parallel_{\mathcal{L}} < 1$.

Clearly, ρ induces a bounded linear operator ρ' on $L^2 := L^2([0,1], \mathcal{B}_{[0,1]}, \lambda)$ where λ is Lebesgue measure, defined by an analogous formula, and such that $\parallel \rho' \parallel_{\mathcal{L}(L^2)} < 1$ where $\parallel \cdot \parallel_{\mathcal{L}(L^2)}$ denotes norm of linear operators on L^2.

We also suppose that X_n is *bounded* (hence ε_n is bounded) and that (L_α) is satisfied with M_n bounded.

Now, to X we may associate a standard ARL^2(1) process $X' = (X_n', n \in \mathbb{Z})$, where

$$X_n' = \sum_{j=1}^{\infty} \left(\int_0^1 X_n(s)e_j(s)\mathrm{d}s \right) e_j, \quad n \in \mathbb{Z}$$

with $(e_j, j \geq 1)$ an orthonormal basis of L^2. Then

$$X_n' = \rho'(X_{n-1}') + \varepsilon_n', \quad n \in \mathbb{Z}$$

where $\varepsilon_n' = \sum_{j=1}^{\infty} \left(\int_0^1 \varepsilon_n(s)e_j(s)\mathrm{d}s \right) e_j$, $n \in \mathbb{Z}$. Since (ε_n) is a strong white noise on C, (ε_n') is a strong white noise on L^2.

This construction will allow us to use some tools concerning estimation of ρ' to estimate ρ.

Finally, from the Mercer lemma (see Riesz and Nagy 1955) and the above considerations it follows that

$$c(s,t) = \mathrm{E}(X_s X_t) = \sum_{j=1}^{\infty} \lambda_j v_j(s) v_j(t), \quad 0 \leq s,t \leq 1$$

where $(\lambda_j, v_j, j \geq 1)$ are the eigenelements of the covariance operator C of X'. Note that v_j is continuous and therefore bounded.

Secondly, we may define empirical operators associated with X by setting

$$C_{0,n} = \frac{1}{n}\sum_{i=1}^{n} X_i' \otimes X_i', \quad \text{and} \quad C_{1,n} = \frac{1}{n-1}\sum_{i=1}^{n-1} X_i' \otimes X_{i+1}',$$

observe that $C_{0,n}$ has the continuous kernel

$$c_{0,n}(s,t) = \frac{1}{n}\sum_{i=1}^{n} X_i(s)X_i(t),$$

which allows it to operate in C as well as in L^2. As usual we will denote by (λ_{jn}, v_{jn}), $n \geq 1$ its eigenelements.

We are now in a position to define the estimator ρ_n of ρ; we suppose that $\lambda_1 > \lambda_2 > \cdots > 0$ and $\lambda_{1,n} > \lambda_{2,n} > \cdots > \lambda_{n,n} > 0$ a.s. and put

$$\rho_n(x) = \tilde{\Pi}^{k_n} C_{1,n} C_{0,n}^{-1} \tilde{\Pi}^{k_n}(x), \quad x \in C$$

where $\tilde{\Pi}^{k_n}(x) = \sum_{j=1}^{k_n} \langle x, v_{jn} \rangle v_{jn}$, $x \in C$, $\langle \cdot, \cdot \rangle$ denoting the scalar product on L^2, thus $\tilde{\Pi}^{k_n}$ operates on C but also on L^2.

If the above conditions hold, and, in addition, $v = \sup_{j \geq 1} \| v_j \|_\infty < \infty$, then

$$P(\| \rho_n - \rho \|_{\mathcal{L}(C)} \geq \eta) \leq d_1(\eta) \exp\left(-d_2(\eta) \frac{n\lambda^2 k_n}{\left(\sum_1^{k_n} a_j\right)^2} \right) \tag{11.29}$$

where $d_1(\eta) > 0$ and $d_2(\eta) > 0$ are constants, and (a_j) is defined by (11.18).

Now one may define the statistical predictor $\tilde{X}_{n+1} = \rho_n(X_n)$ and, if the series in (11.29) converges for all $\eta > 0$ it follows that

$$\tilde{X}_{n+1} - X_{n+1}^* \to 0 \text{ a.s.}$$

where $X_{n+1}^* = \mathrm{E}^{\mathcal{B}_n}(X_{n+1}) = \rho(X_n)$ and $\mathcal{B}_n = \sigma(X_j, j \leq n)$.

Notes

Most of the results in this chapter come from Bosq (2000). Results (11.5), (11.15) and (11.20) appear in Bosq (2002a). Various results in $C[0,1]$ have been obtained by Pumo (1999).

A more complete study is developed in Bosq (2000). Note that the Berry–Esséen bound indicated in this monograph is incorrect since an additional technical assumption is needed for it (see Bosq 2003, 2004a). A simple but less sharp Berry–Esséen bound appears in Dehling and Sharipov (2005).

Estimation of moving average

Estimation of a standard MAH is somewhat difficult: consider the MAH(1)

$$X_n = \varepsilon_n + \ell(\varepsilon_{n-1}), \quad n \in \mathbb{Z}$$

where $\ell \in \mathcal{L}$ and (ε_n) is a strong H-white noise. Then ℓ is solution of

$$\ell^2 C_1^* - \ell C_0 + C_1 = 0, \tag{11.30}$$

which suggests estimation of ℓ from the equation

$$\ell^2 C_{1,n}^* - \ell C_{0,n} + C_{1,n} = 0.$$

However such an equation has no explicit solution.

In the special case where ℓ and C_{ε_0} commute one may replace (11.30) with a system of equations and solve it component by component. The obtained estimator has good asymptotic properties (cf. Bosq 2004b).

Other methods for predicting an ARH(1)

In this chapter we have studied a prediction method based on a preliminary estimation of ρ. Antoniadis and Sapatinas (2003) propose a direct construction of the predictor. They show how this approach is related to linear ill-posed inverse problems and use wavelet methods to solve the question. Karguine and Onatski (2005) claim that projection over $\overline{\text{sp}}(v_{1n}, \ldots, v_{k_n n})$ is optimal for estimation but not for prediction. So they work on decomposition of ρ instead of decomposition of C_0.

Implementation

Prediction by functional autoregressive processes has been used in various practical situations, with some variants. We give a list below:

- simulations: Besse and Cardot (1996); Pumo (1992)
- prediction of traffic: Besse and Cardot (1996)
- forecasting climate variations: Antoniadis and Sapatinas (2003); Besse *et al.* (2000); Pumo (1992); Valderrama *et al.* (2002)
- prediction of electrocardiograms: Bernard (1997)
- prediction of eurodollar rates: Karguine and Onatski (2005)
- ozone forecasting: Damon and Guillas (2002).

Appendix

A.1 Measure and probability

A *σ-algebra* (or σ-field) over a nonvoid set Ω is a family $\mathcal{A}$ of subsets of Ω such that $\Omega \in \mathcal{A}$, $A \in \mathcal{A}$ implies $A^c \in \mathcal{A}$, and if $(A_n,\ n \geq 1)$ is a sequence of elements of $\mathcal{A}$, then $\bigcup_{n=1}^{\infty} A_n \in \mathcal{A}$.

A *measure* on the *measurable space* $(\Omega, \mathcal{A})$ is a set function $\mu\colon \mathcal{A} \to \bar{\mathbb{R}}_+ := [0, \infty]$ such that if $(A_n,\ n \geq 1)$ is a sequence of disjoint elements in $\mathcal{A}$, then

$$\mu\left(\bigcup_{n=1}^{\infty} A_n\right) = \sum_{n=1}^{\infty} \mu(A_n), \qquad \textit{(countable additivity)}.$$

A *probability* is a measure P such that $P(\Omega) = 1$. If P is a probability on $(\Omega, \mathcal{A})$, $(\Omega, \mathcal{A}, P)$ is said to be a *probability space*.

Let Ω be a metric space, the σ-field $\mathcal{B}_\Omega = \sigma(\mathcal{O})$ generated by the open sets is the *Borel σ-field* of Ω. If Ω is *separable* (i.e. if it contains a countable dense set) $\mathcal{B}_\Omega$ is also the σ-field generated by the open balls. In particular $\mathcal{B}_\mathbb{R}$ is generated by the open intervals.

The Lebesgue measure λ on $(\mathbb{R}, \mathcal{B}_\mathbb{R})$ is characterized by $\lambda(]a, b]) = b - a$; $a, b \in \mathbb{R}$, $a < b$. It is translation invariant.

A measure is *σ-finite* if there exists a sequence (A_n) in $\mathcal{A}$ such that $A_n \subset A_{n+1}$, $n \geq 1$, $\bigcup_{n \geq 1} A_n = \Omega$ and $\mu(A_n) < \infty$, $n \geq 1$.

Let μ and ν be two measures on $(\Omega, \mathcal{A})$. μ is *absolutely continuous* with respect to ν $(\mu \prec \nu)$ if $\nu(A) = 0$ implies $\mu(A) = 0$, $A \in \mathcal{A}$. If $\mu \prec \nu$ and $\nu \prec \mu$, μ and ν are said to be equivalent. μ and ν are said to be orthogonal $(\mu \perp \nu)$ if there exists $A \in \mathcal{A}$ such that $\mu(A) = 0$ and $\nu(A^c) = 0$. If μ and ν are σ-finite and $\mu \prec \nu$ then there exists $f\colon (\Omega, \mathcal{A}) \to (\mathbb{R}, \mathcal{B}_\mathbb{R})$ positive and measurable such that

$$\mu(A) = \int_A f \mathrm{d}\nu, \quad A \in \mathcal{A} \quad \text{(Radon Nikodym theorem)}$$

Inference and Prediction in Large Dimensions D. Bosq and D. Blanke

f is ν-almost everywhere unique. More generally, if μ and ν are σ-finite, there exist two uniquely determined σ-finite measures μ_1 and μ_2 such that

$$\mu = \mu_1 + \mu_2, \quad \mu_1 \prec \nu, \quad \mu_2 \perp \nu \qquad \text{(Lebesgue Nikodym theorem).}$$

Let $(\Omega_i, \mathcal{A}_i, \mu_i)$, $1 \leq i \leq k$ be spaces equipped with σ-finite measures. Their *product* is defined as $(\Omega_1 \times \cdots \times \Omega_k, \mathcal{A}_1 \otimes \cdots \otimes \mathcal{A}_k, \mu_1 \otimes \cdots \otimes \mu_k)$ where $\mathcal{A}_1 \otimes \cdots \otimes \mathcal{A}_k$ is the σ-field generated by the sets $A_1 \times \cdots \times A_k$; $A_1 \in \mathcal{A}_1, \cdots, A_k \in \mathcal{A}_k$ and $\mu_1 \otimes \cdots \otimes \mu_k$ is the unique measure such that

$$(\mu_1 \otimes \cdots \otimes \mu_k)(A_1 \times \cdots \times A_k) = \prod_{i=1}^{k} \mu_i(A_i).$$

A family $(\mathcal{B}_i,\, i \in I)$ of sub-σ-fields of $\mathcal{A}$ is (*stochastically*) *independent* (noted $\perp\!\!\!\perp$) if $(\forall J \text{ finite} \subseteq I)$,

$$P\left(\bigcap_{j\in J} B_j\right) = \prod_{j\in J} P(B_j); \quad B_j \in \mathcal{B}_j, \quad j \in J.$$

A.2 Random variables

An E-valued *random variable* (r.v.) defined on $(\Omega, \mathcal{A}, P)$ is a *measurable* application from $(\Omega, \mathcal{A})$ to $(E, \mathcal{B})$ where $\mathcal{B}$ is a σ-algebra over E, that is $X\colon \Omega \to E$ is such that $X^{-1}(B) \in \mathcal{A}$ for all $B \in \mathcal{B}$.

The σ-algebra generated by X is

$$\sigma(X) = X^{-1}(\mathcal{B}) = \{X^{-1}(B),\ B \in \mathcal{B}\}.$$

Two r.v.'s X and Y are said to be independent if $\sigma(X) \perp\!\!\!\perp \sigma(Y)$.

The *distribution* P_X of X is defined by

$$P_X(B) = P(X^{-1}(B)), \quad B \in \mathcal{B},$$

it is a probability on $(E, \mathcal{B})$.

An r.v. X is said to be *real* (respectively *positive*) if $(E, \mathcal{B}) = (\mathbb{R}, \mathcal{B}_{\mathbb{R}})$ (respectively $(\mathbb{R}_+, \mathcal{B}_{\mathbb{R}_+})$). A *simple* real random variable has the form

$$X = \sum_{j=1}^{k} x_j \mathbb{1}_{A_j} \tag{A.1}$$

where $x_j \in \mathbb{R}$, $A_j \in \mathcal{A}$, $1 \leq j \leq k$ and $\mathbb{1}_{A_j}(\omega) = 1$ if $\omega \in A_j$, $= 0$ if $\omega \in A_j^c$.

The decomposition in equation (A.1) is said to be canonical if $x_j \neq x_{j'}, j \neq j'$ and $A_j = \{\omega: X(\omega) = x_j\}$, $1 \leq j \leq k$.

If X is simple and (A.1) is canonical, the *expectation* (or *mean*) of X is defined as

$$\mathrm{E}X = \sum_{j=1}^{k} x_j P(A_j).$$

Now, let X be a positive random variable, one sets

$$\mathrm{E}X = \sup\{\mathrm{E}Y,\ 0 \leq Y \leq X,\ Y \in \mathcal{E}\}$$

where $\mathcal{E}$ denotes the family of simple random variables.

An r.v. such that $\mathrm{E}|X| < \infty$ is said to be integrable and its expectation is defined by

$$\mathrm{E}X = \mathrm{E}X^+ - \mathrm{E}X^- = \int X\mathrm{d}P = \int_\Omega X\mathrm{d}P = \int_\Omega X(\omega)dp\,(\omega)$$

where $X^+ = \max(0, X)$ and $X^- = \max(0, -X)$.

$X \mapsto \mathrm{E}X$ is a positive linear functional on the space $\mathcal{L}^1(P)$ of integrable real random variables. Integral with respect to a measure may be defined similarly (see Billingsley 1995).

If $\mathrm{E}X^2 < \infty$ the *variance* of X is defined as

$$\mathrm{Var}X = \mathrm{E}(X - \mathrm{E}X)^2 = \mathrm{E}X^2 - (\mathrm{E}X)^2.$$

A random d-dimensional vector $X = (X_1, \cdots, X_d)'$ is a measurable application from $(\Omega, \mathcal{A})$ to $(\mathbb{R}^d, \mathcal{B}_{\mathbb{R}^d})$. Its expectation is defined coordinatewise: $\mathrm{E}X = (\mathrm{E}X_1, \cdots, \mathrm{E}X_d)'$. The variance is replaced by the *covariance matrix*:

$$C_X = [\mathrm{E}(X_i - \mathrm{E}X_i)(X_j - \mathrm{E}X_j)]_{1 \leq i,j \leq d} := [\mathrm{Cov}(X_i, X_j)]_{1 \leq i,j \leq d}.$$

The *correlation matrix* is given by $R_X = (\rho_{i,j})_{1 \leq i,j \leq d}$ where ρ_{ij} is the *correlation coefficient* of X_i and X_j:

$$\mathrm{corr}(X_i, X_j) := \rho_{ij} = \frac{\mathrm{Cov}(X_i, X_j)}{\sqrt{\mathrm{Var}X_i}\sqrt{\mathrm{Var}X_j}}$$

provided X_i and X_j are not degenerate.

Finally if X and Y are two d-dimensional random vectors such that $\mathrm{E}X_i^2 < \infty$ and $\mathrm{E}Y_i^2 < \infty$, $1 \leq i \leq d$, their *cross-covariance matrices* are defined as

$$C_{X,Y} = [\mathrm{Cov}(X_i, Y_j)]_{1 \leq i,j \leq d}, \quad C_{Y,X} = [\mathrm{Cov}(Y_i, X_j)]_{1 \leq i,j \leq d}.$$

A.3 Function spaces

A *Banach space* B is a linear space on $\mathbb{R}$ with a norm $\| \cdot \|$ for which it is complete. Its (topological) *dual* B^* is the space of continuous linear functionals on B. It is equipped with the norm (cf. Dunford and Schwartz 1958)

$$\| x^* \|^* = \sup_{\substack{x \in B \\ \| x \| \leq 1}} |x^*(x)|, \quad x^* \in B^*.$$

A real *Hilbert space* H is a Banach space, the norm of which is associated with a scalar product $\langle \cdot, \cdot \rangle$. Thus $\| x \| = (\langle x, x \rangle)^{1/2}$ and

$$|\langle x, y \rangle| \leq \| x \| \| y \|; \quad x, y \in H \qquad \text{(Cauchy–Schwarz inequality)}.$$

H^* may be identified to H in the sense that, for each $x^* \in H^*$, there exists a unique $y \in H$, such that

$$x^*(x) = \langle y, x \rangle, \quad x \in H \qquad \text{(Riesz theorem)}.$$

Let G be a closed linear subspace of H. Its *orthogonal projector* is the application $\Pi^G : H \to G$ characterized by

$$\langle x - \Pi^G(x), z \rangle = 0, \quad z \in G; \quad x \in H,$$

or by $\| x - \Pi^G(x) \| = \inf_{z \in G} \| x - z \|$, $x \in H$.

A family $(e_j,\ j \in J)$ of elements in H is said to be *orthonormal* if

$$\begin{cases} \langle e_i, e_j \rangle & = 0 \quad i \neq j, \\ & = 1 \quad i = j. \end{cases}$$

Let H be an infinite-dimensional separable real Hilbert pace. Then, there exists an orthonormal sequence $(e_j,\ j \geq 1)$ such that

$$x = \sum_{j=1}^{\infty} \langle x, e_j \rangle e_j, \quad x \in H$$

where the series converges in H norm, thus $\| x \|^2 = \sum_{j=1}^{\infty} \langle x, e_j \rangle^2$; such a sequence is called an *orthonormal basis* in H.

Let $(H_i,\ i \geq 1)$ be a finite or infinite sequence of closed subspaces of H. H is the direct sum of the H_i's ($H = \bigoplus_{i \geq 1} H_i$) if

$$\langle x, y \rangle = \sum_{i \geq 1} \langle \Pi^{H_i}(x), \Pi^{H_i}(y) \rangle; \quad x, y \in H.$$

A.4 Common function spaces

Let $p \in [1, +\infty]$ and let ℓ^p be the space of real sequences $x = (x_i,\ i \geq 0)$ such that $\| x \|_p = (\sum_i |x_i|^p)^{1/p} < \infty$ (if $1 \leq p < \infty$); $\| x \|_\infty = \sup_i |x_i| < \infty$ (if $p = \infty$), then ℓ^p is a separable Banach space. In particular ℓ^2 is a Hilbert space with scalar product

$$\langle x, y \rangle_2 = \sum_{i=1}^{\infty} x_i y_i; \quad x, y \in \ell^2.$$

Let μ be a measure on $(\Omega, \mathcal{A})$, f and g two measurable real functions defined on Ω. One identifies f with g if $\mu(f \neq g) = 0$. Then, one defines the spaces $L^p(\mu) = L^p(\Omega, \mathcal{A}, \mu)$, $1 \leq p \leq \infty$ of functions $f\colon \Omega \to \mathbb{R}$ such that

$$\| f \|_p = \left(\int |f|^p \mathrm{d}\mu \right)^{1/p} < \infty \quad (\text{if} \quad 1 \leq p < \infty)$$

and $\| f \|_\infty = \inf\{c\colon \mu\{|f| > c\} = 0\} < \infty$ (if $p = \infty$).

For each p, $L^p(\mu)$ is a Banach space. If $1 < p < \infty$, $[L^p(\mu)]^*$ is identified with $L^q(\mu)$ where $1/p + 1/q = 1$. In particular $L^2(\mu)$, equipped with the scalar product

$$\langle f, g \rangle = \int fg \, \mathrm{d}\mu; \quad f, g \in L^2(\mu)$$

is a Hilbert space. If $f \in L^p(\mu)$ and $g \in L^q(\mu)$ with $1/p + 1/q = 1$, then $fg \in L^1(\mu)$ and

$$\int |fg| \mathrm{d}\mu \leq \| f \|_p \| g \|_q \qquad \text{(Hölder inequality)}$$

if $p = q = 2$ one obtains the Cauchy–Schwarz inequality.

Note that $\ell^p = L^p(\mathbb{N}, \mathcal{P}(\mathbb{N}), \mu)$ where $\mathcal{P}(N)$ is the σ-field of all subsets of $\mathbb{N}$ and μ is the counting measure on $\mathbb{N}$: $\mu(\{n\}) = 1$, $n \in \mathbb{N}$.

If μ is Lebesgue measure on $\mathbb{R}$, the associated L^p-spaces are separable.

Let $k \geq 0$ be an integer and $[a, b]$ a compact interval in $\mathbb{R}$. The space $C_k([a, b])$ of functions $x\colon [a, b] \to \mathbb{R}$ with k continuous derivatives is a separable Banach space for the norm

$$\| x \|_k = \sum_{\ell=0}^{k} \sup_{a \leq t \leq b} |x^{(\ell)}(t)|.$$

The dual of $C_0([a, b])$ is the space of bounded signed measures on $[a, b]$.

A.5 Operators on Hilbert spaces

Let H be a separable Hilbert space and $\mathcal{L}$ be the space of linear continuous operators from H to H. $\mathcal{L}$, equipped with the norm

$$\| \ell \|_{\mathcal{L}} = \sup_{\|x\|\leq 1} \| \ell(x) \|,$$

is a Banach space.

$\ell \in \mathcal{L}$ is said to be *compact* if there exist two orthonormal bases (e_i) and (f_j) of H and a sequence (λ_j) of scalars tending to zero such that

$$\ell(x) = \sum_{j=1}^{\infty} \lambda_j \langle x, e_j \rangle f_j, \quad x \in H.$$

Now the space $\mathcal{S}$ of *Hilbert–Schmidt operators* is the family of compact operators such that $\sum \lambda_j^2 < \infty$. The space $\mathcal{N}$ of *nuclear operators* is the subspace of $\mathcal{S}$ for which $\sum |\lambda_j| < \infty$. Thus $\mathcal{N} \subset \mathcal{S} \subset \mathcal{L}$.

On $\mathcal{S}$ one defines the Hilbert–Schmidt norm

$$\| s \|_{\mathcal{S}} = \left(\sum_j \lambda_j^2 \right)^{1/2}, \quad s \in \mathcal{S},$$

then $\mathcal{S}$ becomes a Hilbert space, with scalar product

$$\langle s_1, s_2 \rangle = \sum_{1\leq i,j<\infty} \langle s_1(g_i), h_j \rangle \langle s_2(g_i), h_j \rangle,$$

$s_1, s_2 \in \mathcal{S}$; where (g_i) and (h_j) are arbitrary orthonormal bases of H.

Concerning $\mathcal{N}$, it is a Banach space with respect to the norm

$$\| \nu \|_{\mathcal{N}} = \sum_j |\lambda_j|, \quad \nu \in \mathcal{N}.$$

Clearly,

$$\| \nu \|_{\mathcal{N}} \geq \| \nu \|_{\mathcal{S}} \geq \| \nu \|_{\mathcal{L}}, \quad \nu \in \mathcal{N}.$$

Finally, let $\ell \in \mathcal{L}$; the *adjoint* ℓ^* of ℓ is defined by

$$\langle \ell^*(x), y \rangle = \langle x, \ell(y) \rangle; \quad x, y \in H.$$

ℓ is said to be a *symmetric operator* if $\ell^* = \ell$; it is said to be *positive* if

$$\langle \ell(x), x \rangle \geq 0, \quad x \in H.$$

On the class of symmetric operators one defines an order relation $\preceq$ by

$$\ell_1 \preceq \ell_2 \Leftrightarrow \ell_2 - \ell_1 \quad \text{is positive.}$$

A.6 Functional random variables

Let H be a separable Hilbert space, an H-random variable defined on the probability space $(\Omega, \mathcal{A}, P)$ is a measurable application from $(\Omega, \mathcal{A})$ to $(H, \mathcal{B}_H)$.

An H-random variable X is said to be *integrable* (in Bochner sense) if $\mathrm{E} \parallel X \parallel < \infty$, and one characterizes $\mathrm{E}X$ by

$$\langle \mathrm{E}X, x \rangle = \mathrm{E}(\langle X, x \rangle), \quad x \in H.$$

If $\mathrm{E} \parallel X \parallel^2 < \infty$, the *covariance operator* of X is defined by setting

$$C_X(x) = \mathrm{E}(\langle X - \mathrm{E}X, x \rangle (X - \mathrm{E}X)), \quad x \in H.$$

C_X is symmetric, positive and nuclear with $\parallel C_X \parallel_{\mathcal{N}} = \mathrm{E} \parallel X \parallel^2$.

If $\mathrm{E} \parallel X \parallel^2 < \infty$ and $\mathrm{E} \parallel Y \parallel^2 < \infty$ the cross-covariance operators of X and Y are defined as

$$C_{X,Y}(x) = \mathrm{E}(\langle X - \mathrm{E}X, x \rangle (Y - \mathrm{E}Y)), \quad x \in H$$

and

$$C_{Y,X} = C_{X,Y}^*.$$

These operators are nuclear.

A.7 Conditional expectation

Let $(\Omega, \mathcal{A}, P)$ be a probability space and $\mathcal{A}_0$ a sub-σ-field of $\mathcal{A}$. The *conditional expectation* given $\mathcal{A}_0$ is the application $\mathrm{E}^{\mathcal{A}_0} : L^1(\Omega, \mathcal{A}, P) \to L^1(\Omega, \mathcal{A}_0, P)$ characterized by

$$\int_B \mathrm{E}^{\mathcal{A}_0}(X) \, \mathrm{d}P = \int_B X \mathrm{d}P, \quad B \in \mathcal{A}_0, \quad X \in L^1(\Omega, \mathcal{A}, P).$$

The restricted application $\mathrm{E}^{\mathcal{A}_0} : L^2(\Omega, \mathcal{A}, P) \to L^2(\Omega, \mathcal{A}_0, P)$ is nothing but the orthogonal projector of $L^2(\Omega, \mathcal{A}_0, P)$. Thus $\mathrm{E}^{\mathcal{A}_0} X$ is the best approximation of $X \in L^2(\Omega, \mathcal{A}, P)$ by an $\mathcal{A}_0$-measurable random variable.

If $\mathcal{A}_0$ is generated by the family of random variables $(X_i, \ i \in I)$, $\mathrm{E}^{\mathcal{A}_0} X$ is said to be the conditional expectation given $(X_i, \ i \in I)$ or the *best predictor* given $(X_i, \ i \in I)$.

If $X = \mathbb{1}_A$, one sets

$$P^{\mathcal{A}_0}(A) = \mathrm{E}^{\mathcal{A}_0}(\mathbb{1}_A), \quad A \in \mathcal{A}.$$

$P^{\mathcal{A}_0}$ is the *conditional probability* given $\mathcal{A}_0$. Under some conditions there exists a *regular* version of $P^{\mathcal{A}_0}$, i.e. a version such that $A \mapsto P^{\mathcal{A}_0}(A)$ is a probability measure on some σ-field (see Rao 1984).

A.8 Conditional expectation in function spaces

B being a separable real Banach space, $L^1_B(\Omega, \mathcal{A}, P)$ denotes the Banach space of (classes of) integrable random variables X, defined on $(\Omega, \mathcal{A}, P)$ and with values in $(B, \mathcal{B}_B)$, with norm $\| X \|_1 = \mathrm{E} \| X \|$.

The conditional expectation given $\mathcal{A}_0$ is defined as in A.6, replacing $L^1(\Omega, \mathcal{A}_0, P)$ by $L^1_B(\Omega, \mathcal{A}_0, P)$. $\mathrm{E}^{\mathcal{A}_0}$ possesses the following properties:

- $\mathrm{E}^{\mathcal{A}_0}$is linear and $\| \mathrm{E}^{\mathcal{A}_0} X \| \leq \mathrm{E}^{\mathcal{A}_0} \| X \|$ (a.s.)
- for all ℓ, linear continuous operators from B to B, we have

$$\mathrm{E}^{\mathcal{A}_0}(\ell(X)) = \ell(\mathrm{E}^{\mathcal{A}_0} X)$$

- if X and $\mathcal{A}_0$ are independent (i.e. $\sigma(X) \perp\!\!\!\perp \mathcal{A}_0$) then $\mathrm{E}^{\mathcal{A}_0} X = \mathrm{E}X$, in particular $\mathrm{E}^{\{\varnothing, \Omega\}} X = \mathrm{E}X$
- if $X \in L^1_B(\Omega, \mathcal{A}_0, P)$ then $\mathrm{E}^{\mathcal{A}_0} X = X$
- finally, if $X = \sum_{j=1}^{k} \mathbb{1}_{A_j} x_j$, $A_j \in \mathcal{A}$, $x_j \in B$, $j = 1, \cdots, k$, then

$$\mathrm{E}^{\mathcal{A}_0} X = \sum_{j=1}^{k} P^{\mathcal{A}_0}(A_j) x_j.$$

Now, if H is a real separable Hilbert space, $L^2_H(\Omega, \mathcal{A}, P)$ denotes the Hilbert space of (classes of) random variables X with values in $(H, \mathcal{B}_H)$ and such that $\mathrm{E} \| X \|^2 < \infty$, equipped with the scalar product

$$[X, Y] = \mathrm{E}\langle X, Y \rangle; \quad X, Y \in L^2_H(\Omega, \mathcal{A}, P).$$

In that situation, the restriction of $\mathrm{E}^{\mathcal{A}_0}$ to $L^2_H(\Omega, \mathcal{A}, P)$ is the orthogonal projector of $L^2_H(\Omega, \mathcal{A}_0, P)$, thus $\mathrm{E}^{\mathcal{A}_0}$ is characterized by

(i) $\mathrm{E}^{\mathcal{A}_0}(X) \in L^2_H(\Omega, \mathcal{A}_0, P)$, $X \in L^2_H(\Omega, \mathcal{A}, P)$

(ii) $\mathrm{E}\langle X - \mathrm{E}^{\mathcal{A}_0}(X), Z \rangle = 0$, $X \in L^2_H(\Omega, \mathcal{A}, P)$, $Z \in L^2_H(\Omega, \mathcal{A}_0, P)$.

A.9 Stochastic processes

A family $X = (X_t,\ t \in I)$ of random variables, defined on $(\Omega, \mathcal{A}, P)$ and with values in some measurable space $(E, \mathcal{B})$ is a *stochastic process* with time range I. If I is countable ($I = \mathbb{Z}, \mathbb{N}, \cdots$) X is a *discrete time* process, if I is an interval in $\mathbb{R}$, X is said to be a *continuous time* process. The random variable $\omega \mapsto X_t(\omega)$ is the *state* of X at time t, the function $t \mapsto X_t(\omega)$ is the *sample path* of ω.

If the cartesian product E^I is equipped with the σ-algebra $\mathcal{S} = \sigma(\Pi_s,\ s \in I)$ where $\Pi_s(x_t,\ t \in I) = x_s,\ s \in I$, then X defines an E^I-valued random variable. The distribution P_X of that variable is called the *distribution* of the stochastic process. If E is a Polish space (i.e. metric, separable, complete), P_X is entirely determined by the finite-dimensional distributions $P_{(X_{t_1}, \cdots, X_{t_k})}$, $k \geq 1,\ t_1, \cdots, t_k \in I$.

A *white noise* is a sequence $\varepsilon = (\varepsilon_n,\ n \in \mathbb{Z})$ of real random variables such that $0 < \sigma^2 = \mathrm{E}\varepsilon_n^2 < \infty$, $\mathrm{E}\varepsilon_n = 0$, $\mathrm{E}(\varepsilon_n \varepsilon_m) = \sigma^2 \delta_{n,m}$; $n, m \in \mathbb{Z}$. If, in addition, the ε_n's are i.i.d. (i.e. independent and identically distributed) one says that ε is a *strong white noise*.

A *linear process* is defined as

$$X_n = \sum_{j=0}^{\infty} a_j \varepsilon_{n-j}, \quad n \in \mathbb{Z} \tag{A.2}$$

where (ε_n) is a white noise, and (a_j) is a sequence of scalars such that $a_0 = 1$ and $\sum_j a_j^2 < \infty$. The series converges in mean square.

If $a_j = \rho^j$, $j \geq 0$, where $|\rho| < 1$, then

$$X_n = \rho X_{n-1} + \varepsilon_n, \quad n \in \mathbb{Z},$$

and one says that (X_n) is an *(AR(1))*.

If $a_j = 0, j > q$ and $a_q \neq 0$, one says that (X_n) is a *moving average of order q* (MA(q)).

A real process $(X_t,\ t \in I)$ is said to be Gaussian if all the random variables $\sum_{j=1}^{k} \alpha_j X_{t_j}$, $k \geq 1$; $\alpha_1, \cdots, \alpha_k \in \mathbb{R}$; $t_1, \cdots, t_k \in I$, are Gaussian (possibly degenerate). The *mean* $t \mapsto \mathrm{E}X_t$ and the *covariance* $(s, t) \mapsto \mathrm{Cov}(X_s, X_t)$ of a Gaussian process X determine P_X.

A *Wiener process* $W = (W_t,\ t \geq 0)$ is a zero-mean Gaussian process with covariance $(s, t) \mapsto \sigma^2 \min(s, t)$ where $\sigma^2 > 0$ is a constant. If $\sigma^2 = 1$, W is said to be *standard*. W models the trajectory of a particle subject to Brownian motion.

A Wiener process has *stationary independent increments*:

$$P_{(W_{t_1+h}, W_{t_2+h} - W_{t_1+h}, \cdots, W_{t_k+h} - W_{t_{k-1}+h})} = P_{W_{t_1}} \otimes P_{(W_{t_2} - W_{t_1})} \otimes \cdots \otimes P_{(W_{t_k} - W_{t_{k-1}})},$$

$k \geq 2$, $0 \leq t_1 < \cdots < t_k$, $h > 0$.

Let $(T_n, n \geq 1)$ be a discrete time real process such that $0 < T_1 < T_2 < \cdots < T_n \uparrow \infty$ (a.s.); it may be interpreted as a sequence of random arrival times of events and is called a *point process*.

The *counting process* associated with (T_n) is defined as

$$N_t = \sum_{n=1}^{\infty} \mathbb{1}_{[0,t]}(T_n), t \geq 0.$$

If (N_t) has independent stationary increments, then there exists $\lambda > 0$ such that

$$P(N_t - N_s = k) = \mathrm{e}^{-\lambda(t-s)} \frac{[\lambda(t-s)]^k}{k!}, \quad k \geq 0; \quad 0 \leq s < t.$$

In this case (N_t) is called a (homogeneous) *Poisson process* with intensity λ.

Let $(\mathcal{F}_n, n \geq 0)$ be a sequence of sub-σ-fields of $\mathcal{A}$, such that $\mathcal{F}_{n-1} \subseteq \mathcal{F}_n$, $n \geq 1$. A sequence $(X_n, n \geq 0)$ of integrable B-random variables is a *martingale* adapted to $(\mathcal{F}_n)$ if

$$\mathrm{E}^{\mathcal{F}_n}(X_{n+1}) = X_n, \quad n \geq 0.$$

Then, $(X_{n+1} - X_n, n \geq 0)$ is called a *martingale difference*. In continuous time the definition is similar:

$$\mathrm{E}^{\mathcal{F}_s}(X_t) = X_s, \quad 0 \leq s \leq t, \quad s, t \in \mathbb{R}_+.$$

The Wiener process and the centred Poisson process $(N_t - \mathrm{E}N_t)$ are martingales.

A.10 Stationary processes and Wold decomposition

Let $X = (X_t, t \in I)$ be a real stochastic process. X is said to be (weakly) *stationary* if $\mathrm{E}X_t^2 < \infty$, $t \in I$, $\mathrm{E}X_t$ does not depend on t, and

$$\mathrm{Cov}(X_{s+h}, X_{t+h}) = \mathrm{Cov}(X_s, X_t); \quad s, t, s+h, t+h \in I.$$

Then, if $0 \in I$, the autocovariance is defined as

$$\gamma(t) = \mathrm{Cov}(X_0, X_t); \quad t \in I.$$

Now, let $X = (X_t, t \in I)$ be a stochastic process with values in some measurable space $(E, \mathcal{B})$, and such that I is an additive subgroup of $\mathbb{R}$. Let us set

$$X^{(h)} = (X_{t+h}, t \in I), \quad h \in I,$$

then, X is said to be *strictly stationary* if

$$P_X = P_{X^{(h)}}, \quad h \in I.$$

A Gaussian stationary process is strictly stationary.

Let $X = (X_t,\ t \in \mathbb{Z})$ be a zero-mean stationary discrete time process. Let $\mathcal{M}_n = \overline{\text{sp}}(X_s,\ s \leq n)$ be the closed linear subspace of $L^2(\Omega, \mathcal{A}, P)$ generated by $(X_s,\ s \leq n)$, $n \in \mathbb{Z}$. Set

$$\varepsilon_n = X_n - \Pi^{\mathcal{M}_{n-1}}(X_n), \quad n \in \mathbb{Z};$$

then X is said to be *regular* if $\mathrm{E}\varepsilon_n^2 = \sigma^2 > 0$, $n \in \mathbb{Z}$. If X is regular, (ε_n) is a white noise called the *innovation (process)* of X and the following *Wold decomposition* holds:

$$X_n = \sum_{j=0}^{\infty} a_j \varepsilon_{n-j} + Y_n, \ \ n \in \mathbb{Z}$$

with $a_0 = 1$, $\sum_j a_j^2 < \infty$, the *singular* part (Y_n) of X being such that

$$Y_n \in \bigcap_{j=0}^{\infty} \mathcal{M}_{n-j}, \quad n \in \mathbb{Z} \quad \text{and} \quad \mathrm{E}(Y_n \varepsilon_m) = 0; \quad n, m \in \mathbb{Z}.$$

If $Y_n = 0$, (X_n) is said to be *(purely) nondeterministic.*

A linear process, defined by Equation (A.2), is said to be *invertible* if $\varepsilon_n \in \mathcal{M}_n$, $n \in \mathbb{Z}$. In that situation (A.2) is its *Wold decomposition.*

An autoregressive process of order p (AR(p)) satisfies the equation

$$X_n = \sum_{j=1}^{p} \pi_j X_{n-j} + \varepsilon_n, \quad n \in \mathbb{Z} \quad (\pi_p \neq 0),$$

it is linear and invertible if $1 - \sum_{j=1}^{p} \pi_j z^j \neq 0$ for $|z| \leq 1$, $z \in \mathbb{C}$.

A.11 Stochastic integral and diffusion processes

A real continuous time stochastic process $X = (X_t,\ a \leq t \leq b)$ $(-\infty \leq a < b \leq \infty)$ is said to be *measurable* if $(t, \omega) \mapsto X_t(\omega)$ is measurable with respect to the σ-algebras $\mathcal{B}_{[a,b]} \otimes \mathcal{A}$ and $\mathcal{B}_{\mathbb{R}}$. In the current section all the processes are supposed to be measurable.

Now, given a standard Wiener process W, one defines the *filtration*

$$\mathcal{F}_t = \sigma(W_s,\ 0 \leq s \leq t), \quad t \geq 0.$$

A process $Y = (Y_t,\ t \geq 0)$ belongs to the class $\mathcal{C}$ if

(i) Y_t is $\mathcal{F}_t$-measurable for all $t \geq 0$,

(ii) $(t,\omega) \mapsto Y_t(\omega)$ is in $L^2([0,a] \otimes P, \mathcal{B}_{[0,a]} \otimes \mathcal{A}, \lambda \otimes P)$ where λ is Lebesgue measure on $[0,a]$.

Let $\mathcal{E}$ be the subfamily of $\mathcal{C}$ of (simple) stochastic processes $Y = (Y_t)$ satisfying

$$Y_t(\omega) = \sum_{j=1}^{k-1} Y_{t_j}(\omega) \mathbb{1}_{[t_j, t_{j+1}[}(t) + Y_{t_k}(\omega) \mathbb{1}_{[t_k, t_{k+1}]}(t),$$

$0 \leq t \leq a,\ \ \omega \in \Omega;\ \ 0 = t_1 < \cdots < t_{k+1} = a\ \ (k \geq 2)$ where $Y_{t_j} \in L^2(\Omega, \mathcal{F}_{t_j}, P)$, $1 \leq j \leq k$.

Now, one puts

$$I(Y) = \int_0^a Y_t \, \mathrm{d}W_t = \sum_{j=1}^{k} Y_{t_j}(W_{t_j+1} - W_{t_j}),$$

then $I(Y) \in L^2(P)$, $Y \mapsto I(Y)$ is linear and such that

$$\mathrm{E}I(Y_1) = 0, \quad \mathrm{E}(I(Y_1)I(Y_2)) = \int_0^a \mathrm{E}(Y_{1,t}\ Y_{2,t}); \quad Y_1, Y_2 \in \mathcal{E}. \tag{A.3}$$

It can be shown that $\mathcal{E}$ is dense in $\mathcal{C}$, this property allows us to extend the linear isometry $Y \mapsto I(Y)$ to $\mathcal{C}$. This extension is called *Ito integral* or *stochastic integral.* Properties given by relation (A.3) remain valid on $\mathcal{C}$.

The definition can be extended to intervals of infinite length.

Now a *diffusion process* is solution of the stochastic differential equation

$$\mathrm{d}X_t = S(X_t, t)\,\mathrm{d}t + \sigma(X_t, t)\mathrm{d}W_t,$$

that is $X_t - X_0 = \int_0^t S(X_s, s)\,\mathrm{d}s + \int_0^t \sigma(X_s, s)\,\mathrm{d}W(s)$ where W is a standard Wiener process, S and σ satisfy some regularity conditions, and the second integral is an Ito one. Details may be found in various works, for example Ash and Gardner (1975); Kutoyants (2004); Sobczyk (1991).

In particular the *Ornstein–Uhlenbeck process* is solution of the stochastic differential equation

$$\mathrm{d}X_t = -\theta X_t\,\mathrm{d}t + \sigma\,\mathrm{d}W_t \quad (\theta > 0,\ \sigma > 0).$$

A stationary solution of this equation is

$$X_t = \sigma \int_{-\infty}^{t} \mathrm{e}^{-\theta(t-s)} \mathrm{d}W(s), \quad t \in \mathbb{R}$$

where W_t is a standard *bilateral* Wiener process: $W_t = \mathbb{1}_{t\geq 0}W_{1,t} + \mathbb{1}_{t<0}W_{2,-t}$, $t \in \mathbb{R}$, where W_1 and W_2 are two independent standard Wiener processes.

A.12 Markov process

Let $X = (X_t,\ t \in I)$ be an $(E, \mathcal{B})$-valued stochastic process, with $I \subset \mathbb{R}$. Set $\mathcal{F}_t = \sigma(X_s, s \leq t,\ s \in I)$, $\mathcal{F}^t = \sigma(X_s,\ s \geq t,\ s \in I)$, and $\mathcal{B}_t = \sigma(X_t)$, $t \in I$. X is said to be a *Markov process* if

$$P^{\mathcal{B}_t}(A \cap B) = P^{\mathcal{B}_t}(A)P^{\mathcal{B}_t}(B), \quad A \in \mathcal{F}_t, \quad B \in \mathcal{F}^t, \tag{A.4}$$

that is, $\mathcal{F}_t$ and $\mathcal{F}^t$ are *conditionally independent* given $\mathcal{B}_t$.

The following condition is equivalent to (A.4):

$$P^{\mathcal{F}_s}(X_t \in B) = P^{\mathcal{B}_s}(X_t \in B), \quad B \in \mathcal{B}, \quad s < t; \quad s, t \in I. \tag{A.5}$$

Note that the symmetry of (A.4), and (A.5) yield

$$P^{\mathcal{F}^s}(X_t \in B) = P^{\mathcal{B}_s}(X_t \in B), \quad B \in \mathcal{B}, \quad s > t; \quad s, t \in I. \tag{A.6}$$

Let $X = (X_t, t \in I)$ be a family of square integrable real random variables. Set

$$\mathcal{M}_{(t_1,\cdots,t_h)} = \text{sp}\{X_{t_1}, \cdots, X_{t_h}\}, \quad t_1, \cdots, \quad t_h \in I, \quad h \geq 1,$$

then X is called a *Markov process in the wide sense* if

$$\Pi^{\mathcal{M}_{(t_1,\cdots,t_{n-1})}}(X_{t_n}) = \Pi^{\mathcal{M}_{t_{n-1}}}(X_{t_n}),$$

$t_1 < \cdots < t_{n-1} < t_n$, $t_1, \cdots, t_n \in I$, $n \geq 2$.

If $0 < \mathrm{E}X_t^2 < \infty$, $t \in I$, one may set

$$\rho(s, t) = \text{corr}(X_s, X_t); \quad s, t \in I,$$

then, X is a Markov process in the wide sense iff

$$\rho(r, t) = \rho(r, s)\rho(s, t), \quad r < s < t; \quad r, s, t \in I. \tag{A.7}$$

By using relation (A.7) it is possible to prove the following: let $X = (X_t,\ t \in \mathbb{R})$ be a real Gaussian stationary Markov process such that $\rho(t) = \text{corr}(X_0, X_t)$, $t \in \mathbb{R}$ is continuous and such that $|\rho(t)| < 1$ for $t \neq 0$; then X is an Ornstein–Uhlenbeck process.

A.13 Stochastic convergences and limit theorems

Let $(Y_n, n \geq 0)$ be a sequence of B-random variables, where B is a separable Banach space.

$Y_n \to Y_0$ almost surely (a.s.) if, as $n \to \infty$,

$$P\{\omega : \| Y_n(\omega) - Y_0(\omega) \| \to 0\} = 1,$$

$Y_n \to Y_0$ in probability (p) if

$$\forall \varepsilon > 0, \quad P(\| Y_n - Y_0 \| > \varepsilon) \to 0,$$

$Y_n \to Y_0$ in $L_B^2(\Omega, \mathcal{A}, P)$ (or in *quadratic mean*) if

$$\mathrm{E} \| Y_n - Y_0 \|^2 \to 0,$$

finally $Y_n \to Y_0$ in distribution ($\mathcal{D}$) (or weakly) if

$$\int f \mathrm{d}P_{Y_n} \to \int f \mathrm{d}P_Y$$

for all continuous bounded functions $f: B \to \mathbb{R}$.

The following implications hold:

$$Y_n \xrightarrow{\text{a.s.}} Y_0 \Rightarrow Y_n \xrightarrow{p} Y_0 \Rightarrow Y_n \xrightarrow{\mathcal{D}} Y_0$$

and

$$Y_n \xrightarrow{L_B^2(P)} Y_0 \Rightarrow Y_n \xrightarrow{p} Y_0.$$

Strong law of large numbers

Let $(X_n, n \geq 1)$ be a sequence of i.i.d. B-valued integrable random variables, then

$$\bar{X}_n := \frac{S_n}{n} = \frac{X_1 + \cdots + X_n}{n} \xrightarrow{\text{a.s.}} \mathrm{E}X_1.$$

Central limit theorem

Let $(X_n, n \geq 1)$ be a sequence of i.i.d. H-valued random variables, where H is a separable Hilbert space. Then, if $\mathrm{E}X_n^2 < \infty$,

$$\sqrt{n}(\bar{X}_n - \mathrm{E}X_1) \xrightarrow{\mathcal{D}} N$$

where $N \approx \mathcal{N}(0, C_{X_1})$.

A.14 Strongly mixing processes

Let $(\Omega, \mathcal{A}, P)$ be a probability space and $\mathcal{B}$, $\mathcal{C}$ two sub-σ-algebras of $\mathcal{A}$. Their *strong mixing coefficient* is defined as

$$\alpha := \alpha(\mathcal{B}, \mathcal{C}) = \sup_{B\in\mathcal{B}, C\in\mathcal{C}} |P(B \cap C) - P(B)P(C)|,$$

then $\mathcal{B} \perp\!\!\!\perp \mathcal{C} \Leftrightarrow \alpha = 0$, and $0 \leq \alpha \leq 1/4$.

A process $X = (X_t,\ t \in I)$ where $I = \mathbb{Z}$ or $\mathbb{R}$, is said to be *α-mixing* or *strongly mixing* if

$$\alpha(u) = \sup_{t\in I} \alpha(\mathcal{F}_t, \mathcal{F}^{t+u}) \underset{u\to+\infty}{\longrightarrow} 0$$

where $\mathcal{F}_t = \sigma(X_s,\ s \leq t,\ s \in I)$, and $\mathcal{F}^{t+u} = \sigma(X_s,\ s \geq t+u,\ s \in I)$. It is said to be 2-$\alpha$-mixing if

$$\alpha^{(2)}(u) = \sup_{t\in I} \alpha(\sigma(X_t), \sigma(X_{t+u})) \underset{u\to+\infty}{\longrightarrow} 0.$$

Consider the real linear process

$$X_t = \sum_{j=0}^{\infty} a_j \varepsilon_{t-j}, \quad t \in \mathbb{Z},$$

where $|a_j| \leq a\,\rho^j$, $j \geq 0$ ($a > 0$, $0 < \rho < 1$), and (ε_t) is a strong white noise such that ε_0 has a density with respect to Lebesgue measure. Then (X_t) is strongly mixing. It is even *geometrically strongly mixing* (GSM), that is

$$\alpha(u) \leq b\,r^u, \quad u \in \mathbb{N}, \quad (b > 0, \quad 0 < r < 1).$$

Diffusion processes regular enough, in particular the Ornstein–Uhlenbeck process, are GSM in continuous time.

Classical limit theorems hold for stationary strongly mixing, regular enough, processes (see Doukhan 1994; Rio 2000).

Finally one has the following inequalities.

Billingsley inequality

If X and Y are bounded real random variables, then

$$|\mathrm{Cov}(X, Y)| \leq 4 \parallel X \parallel_\infty \parallel Y \parallel_\infty \alpha(\sigma(X), \sigma(Y)).$$

Davydov inequality
If $X \in L^q(P)$, $Y \in L^r(P)$ with $q > 1$, $r > 1$ and $1/q + 1/r < 1$, then

$$|\mathrm{Cov}(X, Y)| \leq 2p[2\alpha(\sigma(X), \sigma(Y))]^{1/p} \parallel X \parallel_{L^q(P)} \parallel Y \parallel_{L^r(P)}$$

where $1/p + 1/q + 1/r = 1$.

Rio inequality
Let X and Y be two integrable real random variables, and let $Q_X(u) = \inf\{t: P(|X| > t) \leq u\}$ be the quantile function of X. Then, if $Q_X Q_Y$ is integrable over $[0, 1]$ we have

$$|\mathrm{Cov}(X, Y)| \leq 2 \int_0^{2\alpha(\sigma(X),\sigma(Y))} Q_X(u) Q_Y(u) \mathrm{d}u.$$

A.15 Some other mixing coefficients

The *β-mixing coefficient* between the σ-algebras $\mathcal{B}$ and $\mathcal{C}$ is given by

$$\beta = \beta(\mathcal{B}, \mathcal{C}) = \mathrm{E} \sup_{C \in \mathcal{C}} |P^{\mathcal{B}}(C) - P(C)|$$

where $P^{\mathcal{B}}$ is a version of the conditional probability given $\mathcal{B}$.

The *φ-mixing coefficient* is defined as

$$\varphi = \varphi(\mathcal{B}, \mathcal{C}) = \sup_{\substack{B \in \mathcal{B}, P(B) > 0 \\ C \in \mathcal{C}}} |P(C) - P^B(C)|,$$

finally, the *ρ-mixing coefficient* is

$$\rho = \rho(\mathcal{B}, \mathcal{C}) = \sup_{\substack{X \in L^2(\mathcal{B}) \\ Y \in L^2(\mathcal{C})}} |\mathrm{corr}(X, Y)|.$$

The following inequalities hold:

$$2\alpha \leq \beta \leq \varphi, \quad \text{and} \quad 4\alpha \leq \rho \leq 2\varphi^{1/2}.$$

For a continuous or discrete time process, φ-mixing, β-mixing and ρ-mixing are defined similarly to α-mixing (cf. Section A.14); ρ-mixing and α-mixing are equivalent for a stationary Gaussian process. Moreover, a Gaussian stationary φ-mixing process is *m-dependent* (i.e. for some m, $\sigma(X_s,\ s \leq t)$ and $\sigma(X_s,\ s > t + m)$ are independent). A comprehensive study of mixing conditions appears in Bradley (2005).

A.16 Inequalities of exponential type

Let $X_1, \cdots, X_n$ be independent zero-mean real random variables; the three following inequalities hold.

Hoeffding inequality
If $a_i \leq X_i \leq b_i$, $1 \leq i \leq n$ (a.s.), where $a_1, b_1, \cdots, a_n, b_n$ are constants, then

$$P\left(\left|\sum_{i=1}^{n} X_i\right| \geq t\right) \leq 2\exp\left(-\frac{2t^2}{\sum_{i=1}^{n}(b_i - a_i)^2}\right), \quad t > 0.$$

Bernstein inequality

(1) If $|X_i| \leq M$ (a.s.), $i = 1, , n$, M constant, then

$$P\left(\left|\sum_{i=1}^{n} X_i\right| \geq t\right) \leq 2\exp\left(-\frac{t^2}{2\sum_{i=1}^{n} \mathrm{E}X_i^2 + \dfrac{2Mt}{3}}\right), \quad t > 0.$$

(2) If there exists $c > 0$ such that $\mathrm{E}|X_i|^k \leq c^{k-2} k!\, \mathrm{E}X_i^2 < +\infty$, $i = 1, \cdots, n$; $k = 3, 4, \cdots$ (Cramér's conditions) then

$$P\left(\left|\sum_{i=1}^{n} X_i\right| \geq t\right) \leq 2\exp\left(-\frac{t^2}{4\sum_{i=1}^{n} \mathrm{E}X_i^2 + 2ct}\right), \quad t > 0.$$

The dependent case (*Rio coupling lemma*)
Let X be a real-valued r.v. taking a.s. its values in $[a, b]$ and defined on a probability space $(\Omega, \mathcal{A}, P)$. Suppose that $\mathcal{A}_0$ is a sub-σ-algebra of $\mathcal{A}$ and that U is an r.v. with uniform distribution over $[0, 1]$, independent on the σ-algebra generated by $\mathcal{A}_0 \cup \sigma(X)$. Then, there is a random variable X^*, which is measurable with respect to the σ-algebra generated by $\mathcal{A}_0 \cup \sigma(X) \cup \sigma(U)$ with the following properties:

(a) X^* is independent of $\mathcal{A}_0$,

(b) X^* has the same distribution as the variable X

(c) $\mathrm{E}|X - X^*| \leq 2(b - a)\,\alpha(\sigma(X), \mathcal{A}_0)$

The Rio coupling lemma (or Bradley coupling lemma, Bradley 1983) allows us to extend Hoeffding and Bernstein inequalities, by approximating dependent variables with independent ones.

Bibliography

Adke SR and Ramanathan TV 1997 On optimal prediction for stochastic processes. *J. Statist. Plann. Inference* **63**(1), 1–7.

Ango Nze P, Bühlmann P and Doukhan P 2002 Weak dependence beyond mixing and asymptotics for nonparametric regression. *Ann. Statist.* **30**(2), 397–430.

Antoniadis A and Sapatinas T 2003 Wavelet methods for continuous-time prediction using Hilbert-valued autoregressive processes. *J. Multivariate Anal.* **87**(1), 133–58.

Armstrong JS 2001 *Principle of forecasting: a handbook for researchers and practitioners.* Kluwer Academic Publishers, Nowell MA.

Ash R and Gardner M 1975 *Topics in stochastic processes.* Academic Press, New York.

Aubin JB 2005 *Estimation fonctionnelle par projection adaptative et applications.* PhD thesis, Université Paris 6.

Bahadur RR 1954 Sufficiency and statistical decision functions. *Ann. Math. Statist.* **25**, 423–62.

Baker CR 1973 Joint measures and cross-covariance operators. *Trans. Amer. Math. Soc.* **186**, 273–89.

Banon G 1978 Nonparametric identification for diffusion processes. *SIAM J. Control Optim.* **16**(3), 380–95.

Banon G and Nguyen HT 1978 Sur l'estimation récurrente de la densité et de sa dérivée pour un processus de Markov. *C. R. Acad. Sci. Paris Sér. A Math.* **286**(16), 691–4.

Banon G and Nguyen HT 1981 Recursive estimation in diffusion model. *SIAM J. Control Optim.* **19**(5), 676–85.

Berlinet A and Thomas-Agnan C 2004 *Reproducing kernel Hilbert spaces in probability and statistics.* Kluwer Academic Publishers, Dordrecht.

Bernard P 1997 *Analyse de signaux physiologiques.* Mémoire, Univ. Cathol. Angers.

Bertrand-Retali M 1974 Convergence uniforme de certains estimateurs de la densité par la méthode du noyau. *C R. Acad. Sci. Paris Sér. I Math.* **279**, 527–9.

Besse P and Cardot H 1996 Approximation spline de la prévision d'un processus fonctionnel autorégressif d'ordre 1. *Canad. J. Statist.* **24**(4), 467–87.

Inference and Prediction in Large Dimensions D. Bosq and D. Blanke

Besse P, Cardot H and Stephenson D 2000 Autoregressive forecasting of some climatic variation. *Scand. J. Statist.* **27**(4), 673–87.

Bianchi A 2007 *Problems of statistical inference for multidimensional diffusions*. PhD thesis, Milano and Paris 6 University.

Billingsley P 1968 *Convergence of probability measures.* John Wiley & Sons, Inc., New York.

Billingsley P 1995 *Probability and measure.* Wiley series in probability and mathematical statistics 3rd edn. John Wiley & Sons, Inc., New York.

Blackwell D 1956 An analogy of the minimax theorem for vector payoffs. *Pacific J. Math.* **6**, 1–8.

Blanke D 2004 Sample paths adaptive density estimator. *Math. Methods Statist.* **13**(2), 123–52.

Blanke D 2006 Adaptive sampling schemes for density estimation. *J. Statist. Plann. Inference* **136**(9), 2898–917.

Blanke D and Bosq D 1997 Accurate rates of density estimators for continuous time processes. *Statist. Probab. Letters* **33**(2), 185–91.

Blanke D and Bosq D 2000 A family of minimax rates for density estimators in continuous time. *Stochastic Anal. Appl.* **18**(6), 871–900.

Blanke D and Merlevède F 2000 Estimation of the asymptotic variance of local time density estimators for continuous time processes. *Math. Methods Statist.* **9**(3), 270–96.

Blanke D and Pumo B 2003 Optimal sampling for density estimation in continuous time. *J. Time Ser. Anal.* **24**(1), 1–24.

Bosq D 1978 Tests hilbertiens et test du χ^2. *C. R. Acad Sci. Paris Sér. A-B* **286**(20), 945–9.

Bosq D 1980 Sur une classe de tests qui contient le test du χ^2. *Publ. Inst. Statist. Univ. Paris* **25**(1-2), 1–16.

Bosq D 1983a Lois limites et efficacité asymptotique des tests hilbertiens de dimension finie sous des hypothèses adjacentes. *Statist. Anal. Données* **8**(1), 1–40.

Bosq D 1983b Sur la prédiction non paramétrique de variables aléatoires et de mesures aléatoires. *Z. Warschein Vern. Gebiet.* **64**(4), 541–53.

Bosq D 1989 Tests du χ^2 généralisés. Comparaison avec le χ^2 classique. *Rev. Statist. Appl.* **37**(1), 43–52.

Bosq D 1991 Modelization, nonparametric estimation and prediction for continuous time processes. In *Nonparametric functional estimation and related topics* (ed. Roussas) vol. 335 of *NATO Adv. Sci. Ser. C Math. Phys. Sci.* Kluwer Academic Publishers, Dordrecht pp. 509–29.

Bosq D 1995 Sur le comportement exotique de l'estimateur à noyau de la densité marginale d'un processus à temps continu. *C.R. Acad. Sci. Paris Sér. I Math.* **320**(3), 369–72.

Bosq D 1997 Parametric rates of nonparametric estimators and predictors for continuous time processes. *Ann. Statist.* **25**(3), 982–1000.

Bosq D 1998 *Nonparametric statistics for stochastic processes. Estimation and prediction.* Lecture notes in statistics, 110 2nd edn. Springer-Verlag, New York.

Bosq D 2000 *Linear processes in function spaces.* Lecture notes in statistics, 149. Springer-Verlag, New York.

Bosq D 2002a Estimation of mean and covariance operator of autoregressive processes in Banach spaces. *Stat. Inference Stoch. Process.* **5**(3), 287–306.

Bosq D 2002b Functional tests of fit. In *Goodness-of-fit tests and model validity.* Statistics for industry and technology Birkhäuser, (ed. Muker-Carol). Boston, MA pp. 341–56.

Bosq D 2003 Berry-Esséen inequality for linear processes in Hilbert spaces. *Statist. Probab. Letters* **63**(3), 243–7.

Bosq D 2004a Erratum and complements to: 'Berry–Esséen inequality for linear processes in Hilbert spaces' [Statist. Probab. Letters **63** (2003), no. 3, 243–7]. *Statist. Probab. Letters* **70**(2), 171–4.

Bosq D 2004b Moyennes mobiles hilbertiennes standard. *Ann. ISUP.* **48**(3), 17–28.

Bosq D 2005a Estimation suroptimale de la densité par projection. *Canad. J. Statist.* **33**(1), 1–18.

Bosq D 2005b *Inférence et prévision en grandes dimensions.* Economica, Paris.

Bosq D 2007 General linear processes in Hilbert spaces and prediction. *J. Statist. Plann. Inference* **137**(3), 879–94.

Bosq D and Blanke D 2004 Local superefficiency of data-driven projection density estimators in continuous time. *Statist. Oper. Research Trans.* **28**(1), 37–54.

Bosq D and Cheze-Payaud N 1999 Optimal asymptotic quadratic error of nonparametric regression function estimates for a continuous-time process from sampled-data. *Statistics* **32**(3), 229–47.

Bosq D and Davydov Y 1999 Local time and density estimation in continuous time. *Math. Methods Statist.* **8**(1), 22–45.

Bosq D and Delecroix M 1985 Prediction of a Hilbert valued random variable. *Stochastic Process. Appl.* **19**, 271–80.

Bosq D, Merlevède F and Peligrad M 1999 Asymptotic normality for density kernel estimators in discrete and continuous time. *J. Multivariate Anal.* **68**(1), 78–95.

Bradley RC 1983 Approximation theorems for strongly mixing random variables. *Michigan Math. J.* **30**(1), 69–81.

Bradley RC 2005 Basic properties of strong mixing conditions. A survey and some open questions. *Probab. Surv.* **2**, 107–44 (electronic). Update of, and a supplement to, the 1986 original.

Brockwell PJ 1993 Threshold ARMA processes in continuous time. In *Dimension estimation and models* (ed. Tong) World Scientific Publishing, River Edge pp. 170–90.

Brockwell PJ and Davis RA 1991 *Time series: theory and methods* 2nd edn. Springer-Verlag, New York.

Brockwell PJ and Stramer O 1995 On the approximation of continuous time threshold ARMA processes. *Ann. Inst. Statist. Math.* **47**, 1–20.

Bucklew JA 1985 A note on the prediction error for small time lags into the future. *IEEE Trans. Inform. Theory* **31**(5), 677–9.

Carbon M 1982 *Sur l'estimation asymptotique d'une classe de paramètres fonctionnels pour un processus stationnaire.* PhD thesis Thèse 3ème cycle, Université Lille 1.

Carbon M and Delecroix M 1993 Nonparametric forecasting in time series: a computational point of view. *Comput. Statist. Data Anal.* **9**(8), 215–29.

Castellana JV and Leadbetter MR 1986 On smoothed probability density estimation for stationary processes. *Stochastic Process. Appl.* **21**(2), 179–93.

Cavallini A, Montanari G, Loggini M, Lessi O and Cacciari M 1994 Nonparametric prediction of harmonic levels in electrical networks *IEEE ICHPS VI,* Bologna pp. 165–71.

Chatfield C 2000 *Time-series forecasting.* Chapman and Hall, Boca Raton.

Cheze-Payaud N 1994 Nonparametric regression and prediction for continuous time processes. *Publ. Inst. Stat. Univ. Paris* **38**, 37–58.

Chung KL and Williams RJ 1990 *Introduction to stochastic integration.* Probability and its applications 2nd edn. Birkhäuser, Boston, MA.

Comte F and Merlevède F 2002 Adaptive estimation of the stationary density of discrete and continuous time mixing processes. *ESAIM Prob. and Stat.* **6**, 211–38.

Comte F and Merlevède F 2005 Super optimal rates for nonparametric density estimation via projection estimators. *Stochastic Process. Appl.* **115**(5), 797–826.

Cramér H and Leadbetter MR 1967 *Stationary and related stochastic processes. Sample function properties and their applications.* John Wiley & Sons, Inc., New York.

Dabo-Niang S and Rhomari N 2002 *Nonparametric regression estimation when the regressor takes its values in a metric space.* Technical Report 9, Prépublications L.S.T.A.

Damon J and Guillas S 2002 The inclusion of exogenous variables in functional autoregressive ozone forecasting. *Environmetrics* **13**(7), 759–74.

Damon J and Guillas S 2005 Estimation and simulation of autoregressive hilbertian processes with exogenous variables. *Stat. Inf. Stoch. Proc.* **8**(2), 185–204.

Davydov Y 2001 Remarks on estimation problem for stationary processes in continuous time. *Stat. Inf. Stoch. Proc.* **4**(1), 1–15.

Deheuvels P 2000 Uniform limit laws for kernel density estimators on possibly unbounded intervals. In *Recent advances in reliability theory.* Statistics for industry and technology Birkhäuser (ed. Linnios). Boston, MA pp. 477–92.

Dehling H and Sharipov OS 2005 Estimation of mean and covariance operator for Banach space valued autoregressive processes with dependent innovations. *Stat. Inf. Stoch. Proc.* **8**(2), 137–49.

Delecroix M 1980 Sur l'estimation des densités d'un processus stationnaire a temps continu. *Publ. Inst. Stat. Univ. Paris* **25**(1-2), 17–39.

Devroye L 1987 *A course in density estimation.* Progress in probability and statistics, 14. Birkhäuser, Boston, MA.

Donoho DL, Johnstone I, Kerkyacharian G and Picard D 1996 Density estimation by wavelet thresholding. *Ann. Statist.* **24**(2), 508–39.

Doob J 1953 *Stochastic processes.* John Wiley & Sons, Inc., New York.

Doukhan P 1994 *Mixing: properties and examples.* Lecture notes in statistics, 85. Springer-Verlag.

Doukhan P and Léon J 1994 Asymptotics for the local time of a strongly dependent vector-valued Gaussian random field. *Acta Math. Hung.* **70** (4), 329–51.

Doukhan P and Louhichi S 2001 Functional estimation of a density under a new weak dependence condition. *Scand. J. Statist.* **28**(2), 325–41.

Dunford N and Schwartz JT 1958 *Linear operators, part I: general theory.* Interscience Publishers, Inc., New York.

Efromovich S 1999 *Nonparametric curve estimation.* Springer series in statistics. Springer-Verlag, New York.

Ferraty F and Vieu P 2006 *Nonparametric functional data analysis.* Springer series in statistics. Springer, New York.

Fortet R 1995 *Vecteurs, fonctions et distributions aléatoires dans les espaces de Hilbert.* Hermes, Paris.

Frenay A 2001 Sur l'estimation de la densité marginale d'un processus à temps continu par projection orthogonale. *Ann. ISUP.* **45**(1), 55–92.

Gadiaga D 2003 *Sur une classe de tests qui contient le test du khi-deux: le cas d'un processus stationnaire.* PhD thesis Doctorat d'état, Université de Ouagadougou, Burkina Faso.

Gadiaga D and Ignaccolo R 2005 Tests du χ^2 généralisés. Application au processus autorégressif. *Rev. Stat. Appl.* **53**(2), 67–84.

Geman D and Horowitz J 1973 Occupation times for smooth stationary processes. *Ann. Probab.* **1**(1), 131–7.

Geman D and Horowitz J 1980 Occupation densities. *Ann. Probab.* **8**(1), 1–67.

Giné E and Guillou A 2002 Rates of strong uniform consistency for multivariate kernel density estimators. *Ann. Inst. H. Poincaré B.* **38**(6), 907–21. Special issue dedicated to J. Bretagnolle, D. Dacunha-Castelle, I. Ibragimov.

Gourieroux C and Monfort A 1983 *Cours de série temporelle.* Economica, Paris.

Grenander U 1981 *Abstract inference.* John Wiley & Sons, Inc., New York.

Guilbart C 1979 Produits scalaires sur l'espace des mesures. *Ann. Inst. H. Poincaré B.* **15**(4), 333–54.

Györfi L, Kohler M, Krzyzak A and Walk H 2002 *A distribution-free theory of nonparametric regression.* Springer series in statistics. Springer-Verlag, New York.

Hall P, Lahiri SN and Truong YK 1995 On bandwidth choice for density estimation with dependent data. *Ann. Statist.* **23**(6), 2241–63.

Härdle W 1990 *Applied nonparametric regression.* Cambridge University Press, Cambridge, UK.

Hart JD 1997 *Nonparametric smoothing and lack-of-fit tests.* Springer series in statistics. Springer-Verlag, New York.

Hart JD and Vieu P 1990 Data-driven bandwidth choice for density estimation based on dependent data. *Ann. Statist.* **18**(2), 873–90.

Ibragimov IA and Hasminskii RZ 1981 *Statistical estimation, asymptotic theory.* Springer, New York.

Ignaccolo R 2002 *Tests d'ajustement fonctionnels pour des observations correlées.* PhD thesis, Université Paris 6.

Jain NC and Marcus MB 1975 Central limit theorem for C(S)-valued random variables. *J. Funct. Anal.* **19**,216–31.

Johansson B 1990 Unbiased prediction in Poisson and Yule processes. *Scand. J. Statist.* **17**(2), 135–45.

Karguine V and Onatski A 2005 Curve forecasting by functional autoregression. *Comp. Econ. Finance* **59** (electronic), 33 pages.

Kerstin C 2003 From forecasting to foresight processes. New participative foresight activities in Germany. *J. Forecasting* **22**(2-3), 93–111.

Kim TY 1997 Asymptotically optimal bandwidth selection rules for the kernel density estimator with dependent observations. *J. Statist. Plann. Inference* **59**(2), 321–36.

Klokov SA and Veretennikov AY 2005 On subexponential mixing rate for markov processes. *Theory Prob. App.* **49**(1), 110–22.

Kutoyants YA 1997a On unbiased density estimation for ergodic diffusion. *Statist. Probab. Letters* **34**(2), 133–40.

Kutoyants YA 1997b Some problems of nonparametric estimation by observations of ergodic diffusion processes. *Statist. Probab. Letters* **32**(3), 311–20.

Kutoyants YA 1998 Efficiency density estimation for ergodic diffusion process. *Stat. Inference Stoch. Process.* **1**(2), 131–55.

Kutoyants YA 2004 *Statistical inference for ergodic diffusion processes.* Springer series in statistics. Springer-Verlag, New York.

Labrador B 2006 Almost sure convergence of the k_T-occupation time density estimator. *C. R. Math. Acad. Sci. Paris* **343**(10), 665–9.

Leadbetter MR, Lindgren G and Rootzén H 1983 *Extremes and related properties of random sequences and processes.* Springer series in statistics. Springer-Verlag, New York.

Leblanc F 1995 Discretized wavelet density estimators for continuous time stochastic processes. In *Wavelets and statistics* Lecture notes in statistics, 103 (ed. Antoniadis). Springer-Verlag, New York pp. 209–224.

Leblanc F 1997 Density estimation for a class of continuous time processes. *Math. Methods Statist.* **6**(2), 171–99.

Ledoux M and Talagrand M 1991 *Probability in Banach spaces.* Springer-Verlag, Berlin.

Lehmann EL 1991 *Theory of point estimation.* Chapman and Hall, New York.

Lehmann EL and Casella G 1998 *Theory of point estimation* 2nd edn. Springer, New York.

Lejeune FX 2006 Propriétés des estimateurs par histogrammes et polygones de fréquences de la densité marginale d'un processus à temps continu. *Ann. I.S.U.P.* **50** (1–2), 47–77.

Lerche HR and Sarkar J 1993 The Blackwell prediction algorithm for infinite 0-1 sequences, and a generalization. In *Statistical decision theory and related topics V* (ed. Gupta). Springer-Verlag pp. 503–11.

Liebscher E 2001 Estimation of the density and the regression function under mixing conditions. *Statist. Decisions* **19**(1), 9–26.

Liptser RS and Shiraev AN 2001 *Statistics of random processes I, II*, 2nd edn. Springer, New York.

Marion JM and Pumo B 2004 Comparaison des modéles ARH(1) et ARHD(l) sur des données physiologiques. *Ann. ISUP.* **48**(3), 29–38.

Mas A 2002 Weak convergence for the covariance operators of a Hilbertian linear process. *Stochastic Process. Appl.* **99**(1), 117–35.

Mas A and Pumo B 2007 The ARHD model. *J. Statist. Plann. Inference* **137**(2), 538–53.

Masry E 1983 Probability density estimation from sampled data. *IEEE Trans. Inform. Theory* **29**(5), 696–709.

Merlevède F 1996 *Processus linéaires Hilbertiens: inversibilité, théorèmes limites.* PhD thesis, Université Paris 6.

Merlevède F, Peligrad M and Utev S 1997 Sharp conditions for the CLT of linear processes in a Hilbert space. *J. Theor. Probab.* **10**(3), 681–93.

Mourid T 1995 *Contribution à la statistique des processus autorégressifs à temps continu.* PhD thesis, Université Paris 6.

Nadaraja ÈA 1964 On a regression estimate. *Teor. Verojatnost. i Primenen.* **9**, 157–9.

Neveu J 1972 *Martingales à temps discret.* Masson, Paris.

Neyman J 1937 'Smooth test' for goodness of fit. *Skand. Aktuarie Tidskv.* **20**, 149–99.

Nguyen HT 1979 Density estimation in a continuous-time stationary Markov process. *Ann. Statist.* **7**(2), 341–8.

Nguyen HT and Pham TD 1980 Sur l'utilisation du temps local en statistique des processus. *C. R. Acad. Sci. Paris Sér. A Math.* **290**(3), 165–8.

Nikitin Y 1995 *Asymptotic efficiency of nonparametric tests.* Cambridge University Press, Cambridge.

Parzen E 1962 On estimation of probability density function and mode. *Annals Math. Statist.* **33**, 1065–76.

Pearson K 1900 On the criterion that a given system of deviations from the probable in the case of a correlated system of variables is such that it can be reasonably supposed to have arisen from random sampling. *Philos. Mag. Ser.* **50**, 157–175.

Pinelis IF and Sakhanenko I 1985 Remarks on inequalities for probabilities of large deviations. *Theor. Probab. Appl.* **30**(1), 143–8.

Prakasa Rao BLS 1990 Nonparametric density estimation for stochastic process from sampled data. *Publ. Inst. Stat. Univ. Paris* **35**(3), 51–83.

Pumo B 1992 *Estimation et prévision de processus autorégressifs fonctionnels. Application aux processus à temps continu.* PhD thesis, Université Paris 6.

Pumo B 1999 Prediction of continuous time processes by $C[0, 1]$-valued autoregressive processes. *Stat. Inference Stoch. Process.* **1**(3), 1–13.

Qian Z and Zheng W 2004 A representation formula for transition probability densities of diffusions and applications. *Stochastic Process. Appl.* **111**(1), 57–76.

Qian Z, Russo F and Zheng W 2003 Comparison theorem and estimates for transition probability densities of diffusion processes. *Probab. Theory Rel. Fields* **127**(3), 388–406.

R Development Core Team 2006 *R: a language and environment for statistical computing* R Foundation for Statistical Computing, Vienna, Austria. ISBN 3-900051-07-0.

Rao MM 1984 *Probability theory with applications.* Academic Press, Orlando.

Rhomari N 1994 *Filtrage non paramétrique pour les processus non Marhoviens. Applications.* PhD thesis, Université Paris 6.

Riesz F and Nagy B 1955 *Leçons d'analyse fonctionnelle.* Gauthier-Villars, Paris.

Rio E 1995 The functional law of the iterated logarithm for stationary strongly mixing sequences. *Ann. Probab.* **23**(3), 1188–1209.

Rio E 2000 *Théorie asymptotique des processus aléatoires faiblements dépendants.* Mathématiques et applications, 31. Springer-Verlag.

Robbins E and Siegmund D 1971 *A convergence theorem for nonnegative almost supermartingales and some applications.* Academic Press, New York pp. 233–57.

Rosenblatt M 1956 Remarks on some nonparametric estimates of a density function. *Ann. Math. Statist.* **27**, 832–7.

Rosenblatt M 2000 *Gaussian and non-Gaussian linear time series and random fields.* Springer series in statistics. Springer-Verlag, New York.

Roussas GG 1972 *Contiguity of probability measures: some applications in statistics.* Cambridge tracts in mathematics and mathematical physics, 63. Cambridge University Press, London.

Sazonov VV 1968a On ω^2 criterion. *Sankhyā Ser. A* **30**, 205–10.

Sazonov VV 1968b On the multi-dimensional central limit theorem. *Sankhyā Ser. A* **30**, 181–204.

Serot I 2002 Temps local et densités d'occupation: panorama. *Ann. ISUP* **46**(3), 21–41.

Silverman BW 1986 *Density estimation for statistics and data analysis.* Monographs on statistics and applied probability. Chapman and Hall, London.

Silverman BW and Jones MC 1989 E. Fix and J. L. Hodges (1951): an important contribution to nonparametric discriminant analysis and density estimation. *Int. Stat. Rev.* **57**(3) 233–47.

Sköld M 2001 The asymptotic variance of the continuous-time kernel estimator with applications to bandwidth selection. *Stat. Inference Stoch. Process.* **4**(1), 99–117.

Sköld M and Hössjer O 1999 On the asymptotic variance of the continuous-time kernel density estimator. *Statist. Probab. Letters* **44**(1), 97–106.

Sobczyk K 1991 *Stochastic differential equations, with applications to physics and engineering.* Kluwer Academic Publishers, Dordrecht.

Stein ML 1988 An application of the theory of equivalence gaussian measures to a prediction problem. *IEEE Trans. Inform. Theory* **34**(3), 580–2.

Stute W 1984 The oscillation behavior of empirical processes: the multivariate case. *Ann. Probab.* **12**(2), 361–79.

Takeuchi K and Akahira M 1975 Characterization of prediction sufficiency (adequacy) in terms of risk functions. *Ann. Statist.* **3**(4), 1018–24.

Torgersen EN 1977 Prediction sufficiency when the loss function does not depend on the unknown parameter. *Ann. Statist.* **5**(1), 155–63.

Tsai H and Chan KS 2000 A note on the covariance structure of a continuous-time process. *Statist. Sinica* **10**, 989–98.

Tusnády G 1973 An asymptotically efficient test for fitting a distribution. *Period. Math. Hungar.* **3**, 157–65. Collection of articles dedicated to the memory of Alfréd Rényi, II.

Vakhania NN, Tarieladze VI and Chobanyan SA 1987 *Probability distributions on Banach spaces.* Reidel, Dordrecht.

Valderrama M, Ocaña FA and Aguilera AM 2002 Forecasting PC-ARIMA models for functional data. In *COMPSTAT 2002 (Berlin).* Physica-Verlag, Heidelberg pp. 25–36.

van Zanten JH 2000 Uniform convergence of curve estimators for ergodic diffusion processes. Technical report PNA-R0006, Centrum voor Wiskunde en Informatica.

Veretennikov AY 1987 Bounds for the mixing rate in the theory of stochastic equations. *Theory Probab. Appl.* **32**(2), 273–81.

Veretennikov AY 1999 On Castellana-Leadbetter's condition for diffusion density estimation. *Stat. Inference Stoch. Process.* **2**(1), 1–9.

Viennet G 1997 Inequalities for absolutely regular sequences: application to density estimation. *Probab. Theory Rel. Fields* **107**(4), 467–92.

Watson GS 1964 Smooth regression analysis. *Sankhyā Ser. A.* **26**, 359–72.

Wentzel AD 1975 *Stochastic processes* in russian edn. Nauka, Moscow.

Wu B 1997 Kernel density estimation under weak dependence with sampled data. *J. Statist. Plann. Inference.* **61**(1), 141–54.

Yatracos YG 1992 On prediction and mean-square error. *Canad. J. Statist.* **20**(2), 187–200.

Index

Inference and Prediction in Large Dimensions D. Bosq and D. Blanke

WILEY SERIES IN PROBABILITY AND STATISTICS

established by WALTER A. SHEWHART and SAMUEL S. WILKS

Editors
David J. Balding, Peter Bloomfield, Noel A. C. Cressie, Nicholas I. Fisher, Iain M. Johnstone, J.B. Kadane, Geert Molenberghs, Louise M. Ryan, David W. Scott, Adrian F.M. Smith Editors Emeriti Vic Barnett, J. Stuart Hunter, David G. Kendall, Jozef L. Teugels

The Wiley Series in Probability and Statistics is well established and authoritative. It covers many topics of current research interest in both pure and applied statistics and probability theory. Written by leading statisticians and institutions, the titles span both state-of-the-art developments in the field and classical methods.

Reflecting the wide range of current research in statistics, the series encompasses applied, methodological and theoretical statistics, ranging from applications and new techniques made possible by advances in computerized practice to rigorous treatment of theoretical approaches.

This series provides essential and invaluable reading for all statisticians, whether in academia, industry, government, or research.

ABRAHAM and LEDOLTER Statistical Methods for Forecasting
AGRESTI Analysis of Ordinal Categorical Data
AGRESTI An Introduction to Categorical Data Analysis
AGRESTI Categorical Data Analysis, Second Edition
ALTMAN, GILL, and McDONALD Numerical Issues in Statistical Computing for the Social Scientist
AMARATUNGA and CABRERA Exploration and Analysis of DNA Microarray and Protein Array Data
ANDEL Mathematics of Chance
ANDERSON An Introduction to Multivariate Statistical Analysis, Third Edition
ANDERSON The Statistical Analysis of Time Series
ANDERSON, AUQUIER, HAUCK, OAKES, VANDAELE, and WEISBERG Statistical Methods for Comparative Studies
ANDERSON and LOYNES The Teaching of Practical Statistics
ARMITAGE and DAVID (EDITORS) Advances in Biometry
ARNOLD, BALAKRISHNAN, and NAGARAJA Records
ARTHANARI and DODGE Mathematical Programming in Statistics

BAILEY The Elements of Stochastic Processes with Applications to the Natural Sciences*
BALAKRISHNAN and KOUTRAS Runs and Scans with Applications
BALAKRISHNAN AND NG Precedence-Type Tests and Applications
BARNETT Comparative Statistical Inference, Third Edition
BARNETT Environmental Statistics: Methods & Applications
BARNETT and LEWIS Outliers in Statistical Data, Third Edition

*Now available in a lower priced paperback edition in the Wiley Classics Library.

BARTOSZYNSKI and NIEWIADOMSKA-BUGAJ Probability and Statistical Inference
BASILEVSKY Statistical Factor Analysis and Related Methods: Theory and Applications
BASU and RIGDON Statistical Methods for the Reliability of Repairable Systems
BATES and WATTS Nonlinear Regression Analysis and Its Applications
BECHHOFER, SANTNER, and GOLDSMAN Design and Analysis of Experiments for Statistical Selection, Screening, and Multiple Comparisons
BELSLEY Conditioning Diagnostics: Collinearity and Weak Data in Regression
BELSLEY, KUH, and WELSCH Regression Diagnostics: Identifying Influential Data and Sources of Collinearity
BENDAT and PIERSOL Random Data: Analysis and Measurement Procedures, Third Edition
BERNARDO and SMITH Bayesian Theory
BERRY, CHALONER, and GEWEKE Bayesian Analysis in Statistics and Econometrics: Essays in Honor of Arnold Zellner
BHAT and MILLER Elements of Applied Stochastic Processes, Third Edition
BHATTACHARYA and JOHNSON Statistical Concepts and Methods
BHATTACHARYA and WAYMIRE Stochastic Processes with Applications
BIEMER, GROVES, LYBERG, MATHIOWETZ, and SUDMAN Measurement Errors in Surveys
BILLINGSLEY Convergence of Probability Measures, Second Edition
BILLINGSLEY Probability and Measure, Third Edition
BIRKES and DODGE Alternative Methods of Regression
BLISCHKE and MURTHY (editors) Case Studies in Reliability and Maintenance
BLISCHKE and MURTHY Reliability: Modeling, Prediction, and Optimization
BLOOMFIELD Fourier Analysis of Time Series: An Introduction, Second Edition
BOLLEN Structural Equations with Latent Variables
BOLLEN and CURRAN Latent Curve Models: A Structural Equation Perspective
BOROVKOV Ergodicity and Stability of Stochastic Processes
BOSQ and BLANKE Inference and Prediction in Large Dimensions
BOULEAU Numerical Methods for Stochastic Processes
BOX Bayesian Inference in Statistical Analysis
BOX R. A. Fisher, the Life of a Scientist
BOX and DRAPER Empirical Model-Building and Response Surfaces
BOX and DRAPER Evolutionary Operation: A Statistical Method for Process Improvement
BOX, HUNTER, and HUNTER Statistics for Experimenters: An Introduction to Design, Data Analysis, and Model Building
BOX, HUNTER, and HUNTER Statistics for Experimenters: Design, Innovation and Discovery, Second Edition
BOX and LUCE~NO Statistical Control by Monitoring and Feedback Adjustment
BRANDIMARTE Numerical Methods in Finance: A MATLAB-Based Introduction
BROWN and HOLLANDER Statistics: A Biomedical Introduction
BRUNNER, DOMHOF, and LANGER Nonparametric Analysis of Longitudinal Data in Factorial Experiments
BUCKLEW Large Deviation Techniques in Decision, Simulation, and Estimation

CAIROLI and DALANG Sequential Stochastic Optimization
CASTILLO, HADI, BALAKRISHNAN and SARABIA Extreme Value and Related Models with Applications in Engineering and Science*

*Now available in a lower priced paperback edition in the Wiley Classics Library.

CHAN Time Series: Applications to Finance
CHATTERJEE AND HADI Regression Analysis by Example, Fourth Edition
CHATTERJEE and HADI Sensitivity Analysis in Linear Regression
CHATTERJEE and PRICE Regression Analysis by Example, Third Edition
CHERNICK Bootstrap Methods: A Practitioner's Guide
CHERNICK and FRIIS Introductory Biostatistics for the Health Sciences
CHIL'ES and DELFINER Geostatistics: Modeling Spatial Uncertainty
CHOW and LIU Design and Analysis of Clinical Trials: Concepts and Methodologies, Second Edition
CLARKE and DISNEY Probability and Random Processes: A First Course with Applications, Second Edition
COCHRAN and COX Experimental Designs, Second Edition
CONGDON Applied Bayesian Modelling
CONGDON Bayesian Models for Categorical Data
CONGDON Bayesian Statistical Modelling
CONGDON Bayesian Statistical Modelling, Second Edition
CONOVER Practical Nonparametric Statistics, Second Edition
COOK Regression Graphics
COOK and WEISBERG An Introduction to Regression Graphics
COOK and WEISBERG Applied Regression Including Computing and Graphics
CORNELL Experiments with Mixtures, Designs, Models, and the Analysis of Mixture Data, Third Edition
COVER and THOMAS Elements of Information Theory
COX A Handbook of Introductory Statistical Methods
COX Planning of Experiments
CRESSIE Statistics for Spatial Data, Revised Edition footnote*
CSORGO and HORV'ATH Limit Theorems in Change Point Analysis

DANIEL Applications of Statistics to Industrial Experimentation
DANIEL Biostatistics: A Foundation for Analysis in the Health Sciences, Sixth Edition
DANIEL Fitting Equations to Data: Computer Analysis of Multifactor Data, Second Edition
DASU and JOHNSON Exploratory Data Mining and Data Cleaning
DAVID and NAGARAJA Order Statistics, Third Edition
DEGROOT, FIENBERG, and KADANE Statistics and the Law
DEL CASTILLO Statistical Process Adjustment for Quality Control
DEMARIS Regression with Social Data: Modeling Continuous and Limited Response Variables
DEMIDENKO Mixed Models: Theory and Applications
DENISON, HOLMES, MALLICK, and SMITH Bayesian Methods for Nonlinear Classification and Regression
DETTE and STUDDEN The Theory of Canonical Moments with Applications in Statistics, Probability, and Analysis
DEY and MUKERJEE Fractional Factorial Plans
DILLON and GOLDSTEIN Multivariate Analysis: Methods and Applications
DODGE Alternative Methods of Regression
DODGE and ROMIG Sampling Inspection Tables, Second Edition

*Now available in a lower priced paperback edition in the Wiley Classics Library.

DOOB Stochastic Processes
DOWDY, WEARDEN, and CHILKO Statistics for Research, Third Edition
DRAPER and SMITH Applied Regression Analysis, Third Edition
DRYDEN and MARDIA Statistical Shape Analysis
DUDEWICZ and MISHRA Modern Mathematical Statistics
DUNN and CLARK Applied Statistics: Analysis of Variance and Regression, Second Edition
DUNN and CLARK Basic Statistics: A Primer for the Biomedical Sciences, Third Edition
DUPUIS and ELLIS A Weak Convergence Approach to the Theory of Large Deviations

EDLER and KITSOS (editors) Recent Advances in Quantitative Methods in Cancer and Human Health Risk Assessment
ELANDT-JOHNSON and JOHNSON Survival Models and Data Analysis
ENDERS Applied Econometric Time Series
ETHIER and KURTZ Markov Processes: Characterization and Convergence
EVANS, HASTINGS, and PEACOCK Statistical Distribution, Third Edition
FELLER An Introduction to Probability Theory and Its Applications, Volume I, Third Edition, Revised; Volume II, Second Edition

FISHER and VAN BELLE Biostatistics: A Methodology for the Health Sciences
FITZMAURICE, LAIRD, and WARE Applied Longitudinal Analysis
FLEISS The Design and Analysis of Clinical Experiments
FLEISS Statistical Methods for Rates and Proportions, Second Edition
FLEMING and HARRINGTON Counting Processes and Survival Analysis
FULLER Introduction to Statistical Time Series, Second Edition
FULLER Measurement Error Models

GALLANT Nonlinear Statistical Models.
GEISSER Modes of Parametric Statistical Inference
GELMAN and MENG (editors) Applied Bayesian Modeling and Casual Inference from Incomplete-data Perspectives
GEWEKE Contemporary Bayesian Econometrics and Statistics
GHOSH, MUKHOPADHYAY, and SEN Sequential Estimation
GIESBRECHT and GUMPERTZ Planning, Construction, and Statistical Analysis of Comparative Experiments
GIFI Nonlinear Multivariate Analysis
GIVENS and HOETING Computational Statistics
GLASSERMAN and YAO Monotone Structure in Discrete-Event Systems
GNANADESIKAN Methods for Statistical Data Analysis of Multivariate Observations, Second Edition
GOLDSTEIN and LEWIS Assessment: Problems, Development, and Statistical Issues
GREENWOOD and NIKULIN A Guide to Chi-Squared Testing
GROSS and HARRIS Fundamentals of Queueing Theory, Third Edition

HAHN and SHAPIRO Statistical Models in Engineering
HAHN and MEEKER Statistical Intervals: A Guide for Practitioners
HALD A History of Probability and Statistics and their Applications Before 1750
HALD A History of Mathematical Statistics from 1750 to 1930
HAMPEL Robust Statistics: The Approach Based on Influence Functions

HANNAN and DEISTLER The Statistical Theory of Linear Systems
HEIBERGER Computation for the Analysis of Designed Experiments
HEDAYAT and SINHA Design and Inference in Finite Population Sampling
HEDEKER AND GIBBONS Longitudinal Data Analysis
HELLER MACSYMA for Statisticians
HINKELMANN and KEMPTHORNE Design and Analysis of Experiments, Volume 1: Introduction to Experimental Design
HINKELMANN and KEMPTHORNE Design and analysis of experiments, Volume 2: Advanced Experimental Design
HOAGLIN, MOSTELLER, and TUKEY Exploratory Approach to Analysis of Variance
HOAGLIN, MOSTELLER, and TUKEY Exploring Data Tables, Trends and Shapes
HOAGLIN, MOSTELLER, and TUKEY Understanding Robust and Exploratory Data Analysis
HOCHBERG and TAMHANE Multiple Comparison Procedures
HOCKING Methods and Applications of Linear Models: Regression and the Analysis of Variance, Second Edition
HOEL Introduction to Mathematical Statistics, Fifth Edition
HOGG and KLUGMAN Loss Distributions
HOLLANDER and WOLFE Nonparametric Statistical Methods, Second Edition
HOSMER and LEMESHOW Applied Logistic Regression, Second Edition
HOSMER and LEMESHOW Applied Survival Analysis: Regression Modeling of Time to Event Data
HUBER Robust Statistics
HUBERTY Applied Discriminant Analysis
HUNT and KENNEDY Financial Derivatives in Theory and Practice, Revised Edition
HUSKOVA, BERAN, and DUPAC Collected Works of Jaroslav Hajek—with Commentary
HUZURBAZAR Flowgraph Models for Multistate Time-to-Event Data

IMAN and CONOVER A Modern Approach to Statistics

JACKSON A User's Guide to Principle Components
JOHN Statistical Methods in Engineering and Quality Assurance
JOHNSON Multivariate Statistical Simulation
JOHNSON and BALAKRISHNAN Advances in the Theory and Practice of Statistics: A Volume in Honor of Samuel Kotz
JOHNSON and BHATTACHARYYA Statistics: Principles and Methods, Fifth Edition
JUDGE, GRIFFITHS, HILL, LU TKEPOHL, and LEE The Theory and Practice of Econometrics, Second Edition
JOHNSON and KOTZ Distributions in Statistics
JOHNSON and KOTZ (editors) Leading Personalities in Statistical Sciences: From the Seventeenth Century to the Present*
JOHNSON, KOTZ, and BALAKRISHNAN Continuous Univariate Distributions, Volume 1, Second Edition
JOHNSON, KOTZ, and BALAKRISHNAN Continuous Univariate Distributions, Volume 2, Second Edition
JOHNSON, KOTZ, and BALAKRISHNAN Discrete Multivariate Distributions

*Now available in a lower priced paperback edition in the Wiley Classics Library.

JOHNSON, KOTZ, and KEMP Univariate Discrete Distributions, Second Edition
JURECKOVA and SEN Robust Statistical Procedures: Asymptotics and Interrelations
JUREK and MASON Operator-Limit Distributions in Probability Theory

KADANE Bayesian Methods and Ethics in a Clinical Trial Design
KADANE and SCHUM A Probabilistic Analysis of the Sacco and Vanzetti Evidence
KALBFLEISCH and PRENTICE The Statistical Analysis of Failure Time Data, Second Edition
KARIYA and KURATA Generalized Least Squares
KASS and VOS Geometrical Foundations of Asymptotic Inference
KAUFMAN and ROUSSEEUW Finding Groups in Data: An Introduction to Cluster Analysis
KEDEM and FOKIANOS Regression Models for Time Series Analysis
KENDALL, BARDEN, CARNE, and LE Shape and Shape Theory
KHURI Advanced Calculus with Applications in Statistics, Second Edition
KHURI, MATHEW, and SINHA Statistical Tests for Mixed Linear Models
KISH Statistical Design for Research
KLEIBER and KOTZ Statistical Size Distributions in Economics and Actuarial Sciences
KLUGMAN, PANJER, and WILLMOT Loss Models: From Data to Decisions
KLUGMAN, PANJER, and WILLMOT Solutions Manual to Accompany Loss Models: From Data to Decisions
KOTZ, BALAKRISHNAN, and JOHNSON Continuous Multivariate Distributions, Volume 1, Second Edition
KOTZ and JOHNSON (editors) Encyclopedia of Statistical Sciences: Volumes 1 to 9 with Index
KOTZ and JOHNSON (editors) Encyclopedia of Statistical Sciences: Supplement Volume
KOTZ, READ, and BANKS (editors) Encyclopedia of Statistical Sciences: Update Volume 1
KOTZ, READ, and BANKS (editors) Encyclopedia of Statistical Sciences: Update Volume 2
KOVALENKO, KUZNETZOV, and PEGG Mathematical Theory of Reliability of Time-Dependent Systems with Practical Applications*
KUROWICKA and COOKE Uncertainty Analysis with High Dimensional Dependence Modelling

LACHIN Biostatistical Methods: The Assessment of Relative Risks
LAD Operational Subjective Statistical Methods: A Mathematical, Philosophical, and Historical Introduction
LAMPERTI Probability: A Survey of the Mathematical Theory, Second Edition
LANGE, RYAN, BILLARD, BRILLINGER, CONQUEST, and GREENHOUSE Case Studies in Biometry
LARSON Introduction to Probability Theory and Statistical Inference, Third Edition
LAWLESS Statistical Models and Methods for Lifetime Data, Second Edition
LAWSON Statistical Methods in Spatial Epidemiology, Second Edition
LE Applied Categorical Data Analysis LE Applied Survival Analysis
LEE and WANG Statistical Methods for Survival Data Analysis, Third Edition
LEPAGE and BILLARD Exploring the Limits of Bootstrap
LEYLAND and GOLDSTEIN (editors) Multilevel Modelling of Health Statistics

*Now available in a lower priced paperback edition in the Wiley - Interscience Paperback Series.

LIAO Statistical Group Comparison
LINDVALL Lectures on the Coupling Method
LINHART and ZUCCHINI Model Selection
LITTLE and RUBIN Statistical Analysis with Missing Data, Second Edition
LLOYD The Statistical Analysis of Categorical Data
LOWEN and TEICH Fractal-Based Point Processes

MAGNUS and Neudecker Matrix Differential Calculus with Applications in Statistics and Econometrics,
Revised Edition
MALLER and ZHOU Survival Analysis with Long Term Survivors
MALLOWS Design, Data, and Analysis by Some Friends of Cuthbert Daniel
MANN, SCHAFER, and SINGPURWALLA Methods for Statistical Analysis of Reliability and Life Data
MANTON, WOODBURY, and TOLLEY Statistical Applications Using Fuzzy Sets
MARCHETTE Random Graphs for Statistical Pattern Recognition
MARDIA and JUPP Directional Statistics
MARONNA, MARTIN, and YOHAI Robust Statistics: Theory and Methods
MASON, GUNST, and HESS Statistical Design and Analysis of Experiments with Applications to Engineering and Science, Second Edition
MCCULLOCH and SERLE Generalized, Linear, and Mixed Models
MCFADDEN Management of Data in Clinical Trials
MCLACHLAN Discriminant Analysis and Statistical Pattern Recognition
MCLACHLAN, DO, and AMBROISE Analyzing Microarray Gene Expression Data
MCLACHLAN and KRISHNAN The EM Algorithm and Extensions
MCLACHLAN and PEEL Finite Mixture Models
MCNEIL Epidemiological Research Methods
MEEKER and ESCOBAR Statistical Methods for Reliability Data
MEERSCHAERT and SCHEFFLER Limit Distributions for Sums of Independent Random Vectors: Heavy Tails in Theory and Practice
MICKEY, DUNN, AND CLARK Applied Statistics: Analysis of Variance and Regression, Third Edition
MILLER Survival Analysis, Second Edition
MONTGOMERY, PECK, and VINING Introduction to Linear Regression Analysis, Fourth Edition
MORGENTHALER and TUKEY Configural Polysampling: A Route to Practical Robustness
MUIRHEAD Aspects of Multivariate Statistical Theory
MULLER AND STEWART Linear Model Theory: Univariate, Multivariate, and Mixed Models
MURRAY X-STAT 2.0 Statistical Experimentation, Design Data Analysis, and Nonlinear Optimization
MURTHY, XIE, and JIANG Weibull Models
MYERS and Montgomery Response Surface Methodology: Process and Product Optimization Using Designed Experiments, Second Edition
MYERS, MONTGOMERY, and VINING Generalized Linear Models. With Applications in Engineering and the Sciences

NELSON Accelerated Testing, Statistical Models, Test Plans, and Data Analysis
NELSON Applied Life Data Analysis
NEWMAN Biostatistical Methods in Epidemiology

OCHI Applied Probability and Stochastic Processes in Engineering and Physical Sciences
OKABE, BOOTS, SUGIHARA, and CHIU Spatial Tesselations: Concepts and Applications of Voronoi Diagrams, Second Edition
OLIVER and SMITH Influence Diagrams, Belief Nets and Decision Analysis
PALTA Quantitative Methods in Population Health: Extentions of Ordinary Regression
PANJER Operational Risks: Modeling Analytics
PANKRATZ Forecasting with Dynamic Regression Models
PANKRATZ Forecasting with Univariate Box-Jenkins Models: Concepts and Cases
PARZEN Modern Probability Theory and Its Applications*
Now available in a lower priced paperback edition in the Wiley - Interscience Paperback Series.

PENA, TIAO, and TSAY A Course in Time Series Analysis
PIANTADOSI Clinical Trials: A Methodologic Perspective
PORT Theoretical Probability for Applications
POURAHMADI Foundations of Time Series Analysis and Prediction Theory
PRESS Bayesian Statistics: Principles, Models, and Applications
PRESS Subjective and Objective Bayesian Statistics, Second Edition
PRESS and TANUR The Subjectivity of Scientists and the Bayesian Approach
PUKELSHEIM Optimal Experimental Design
PURI, VILAPLANA, and WERTZ New Perspectives in Theoretical and Applied Statistics
PUTERMAN Markov Decision Processes: Discrete Stochastic Dynamic Programming

QIU Image Processing and Jump Regression Analysis

RAO Linear Statistical Inference and its Applications, Second Edition
RAUSAND and HOYLAND System Reliability Theory: Models, Statistical Methods and Applications, Second Edition
RENCHER Linear Models in Statistics
RENCHER Methods of Multivariate Analysis, Second Edition
RENCHER Multivariate Statistical Inference with Applications
RIPLEY Spatial Statistics
RIPLEY Stochastic Simulation
ROBINSON Practical Strategies for Experimenting
ROHATGI and SALEH An Introduction to Probability and Statistics, Second Edition
ROLSKI, SCHMIDLI, SCHMIDT, and TEUGELS Stochastic Processes for Insurance and Finance
ROSENBERGER and LACHIN Randomization in Clinical Trials: Theory and Practice
ROSS Introduction to Probability and Statistics for Engineers and Scientists
ROSSI, ALLENBY, and MCCULLOCH Bayesian Statistics and Marketing
ROUSSEEUW and LEROY Robust Regression and Outline Detection
RUBIN Multiple Imputation for Nonresponse in Surveys

*Now available in a lower priced paperback edition in the Wiley Classics Library

RUBINSTEIN Simulation and the Monte Carlo Method
RUBINSTEIN and MELAMED Modern Simulation and Modeling
RYAN Modern Regression Methods
RYAN Statistical Methods for Quality Improvement, Second Edition

SALEH Theory of Preliminary Test and Stein-Type Estimation with Applications
SALTELLI, CHAN, and SCOTT (editors) Sensitivity Analysis
SCHEFFE The Analysis of Variance
SCHIMEK Smoothing and Regression: Approaches, Computation, and Application
SCHOTT Matrix Analysis for Statistics
SCHOUTENS Levy Processes in Finance: Pricing Financial Derivatives
SCHUSS Theory and Applications of Stochastic Differential Equations
SCOTT Multivariate Density Estimation: Theory, Practice, and Visualization*
SEARLE Linear Models SEARLE Linear Models for Unbalanced Data
SEARLE Matrix Algebra Useful for Statistics
SEARLE and WILLETT Matrix Algebra for Applied Economics
SEBER Multivariate Observations
SEBER and LEE Linear Regression Analysis, Second Edition
SEBER and WILD Nonlinear Regression
SENNOTT Stochastic Dynamic Programming and the Control of Queueing Systems
SERFLING Approximation Theorems of Mathematical Statistics
SHAFER and VOVK Probability and Finance: Its Only a Game!
SILVAPULLE and SEN Constrained Statistical Inference: Inequality, Order, and Shape Restrictions
SINGPURWALLA Reliability and Risk: A Bayesian Perspective
SMALL and MCLEISH Hilbert Space Methods in Probability and Statistical Inference
SRIVASTAVA Methods of Multivariate Statistics
STAPLETON Linear Statistical Models
STAUDTE and SHEATHER Robust Estimation and Testing
STOYAN, KENDALL, and MECKE Stochastic Geometry and Its Applications, Second Edition
STOYAN and STOYAN Fractals, Random and Point Fields: Methods of Geometrical Statistics
STYAN The Collected Papers of T. W. Anderson: 1943–1985
SUTTON, ABRAMS, JONES, SHELDON, and SONG Methods for Meta-Analysis in Medical Research

TANAKA Time Series Analysis: Nonstationary and Noninvertible Distribution Theory
THOMPSON Empirical Model Building
THOMPSON Sampling, Second Edition
THOMPSON Simulation: A Modeler's Approach
THOMPSON and SEBER Adaptive Sampling
THOMPSON, WILLIAMS, and FINDLAY Models for Investors in Real World Markets
TIAO, BISGAARD, HILL, PENA, and STIGLER (editors) Box on Quality and Discovery: with Design, Control, and Robustness

*Now available in a lower priced paperback edition in the Wiley Classics Library.

TIERNEY LISP-STAT: An Object-Oriented Environment for Statistical Computing and Dynamic Graphics
TSAY Analysis of Financial Time Series

UPTON and FINGLETON Spatial Data Analysis by Example, Volume II: Categorical and Directional Data

VAN BELLE Statistical Rules of Thumb
VAN BELLE, FISHER, HEAGERTY, and LUMLEY Biostatistics: A Methodology for the Health Sciences, Second Edition
VESTRUP The Theory of Measures and Integration
VIDAKOVIC Statistical Modeling by Wavelets
VINOD and REAGLE Preparing for the Worst: Incorporating Downside Risk in Stock Market Investments

WALLER and GOTWAY Applied Spatial Statistics for Public Health Data
WEERAHANDI Generalized Inference in Repeated Measures: Exact Methods in MANOVA and Mixed Models
WEISBERG Applied Linear Regression, Second Edition
WELISH Aspects of Statistical Inference*
WESTFALL and YOUNG Resampling-Based Multiple Testing: Examples and Methods for p-Value Adjustment
WHITTAKER Graphical Models in Applied Multivariate Statistics
WINKER Optimization Heuristics in Economics: Applications of Threshold Accepting
WONNACOTT and WONNACOTT Econometrics, Second Edition
WOODING Planning Pharmaceutical Clinical Trials: Basic Statistical Principles
WOOLSON and CLARKE Statistical Methods for the Analysis of Biomedical Data, Second Edition
WU and HAMADA Experiments: Planning, Analysis, and Parameter Design Optimization
WU AND ZHANG Nonparametric Regression Methods for Longitudinal Data Analysis: Mixed-Effects Modeling Approaches

YANG The Construction Theory of Denumerable Markov Processes
YOUNG, VALERO-MORA, AND FRIENDLY Visual Statistics: Seeing Data with Dynamic Interactive Graphics

ZELLNER An Introduction to Bayesian Inference in Econometrics
ZELTERMAN Discrete Distributions: Applications in the Health Sciences
ZHOU, OBUCHOWSKI, and McCLISH Statistical Methods in Diagnostic Medicine

*Now available in a lower priced paperback edition in the Wiley Classics Library.